Modern Optimization Methods for Decision Making Under Risk and Uncertainty

Editors:

Alexei A. Gaivoronski
Professor, Department of Industrial
Economy and Technology Management
Norwegian University of Science and Technology
Trondheim, Norway

Pavel S. Knopov
Professor, Department of Mathematical Methods of Operations
V.M. Glushkov Institute of Cybernetics
National Academy of Sciences of Ukraine
Kyiv, Ukraine

Volodymyr A. Zaslavskyi
Professor, Department of Mathematical Informatics
Taras Shevchenko National University of Kyiv
Kyiv, Ukraine

CRC Press
Taylor & Francis Group
Boca Raton London New York

CRC Press is an imprint of the
Taylor & Francis Group, an **informa** business
A SCIENCE PUBLISHERS BOOK

First edition published 2024
by CRC Press
6000 Broken Sound Parkway NW, Suite 300, Boca Raton, FL 33487-2742

and by CRC Press
4 Park Square, Milton Park, Abingdon, Oxon, OX14 4RN

CRC Press is an imprint of Taylor & Francis Group, LLC

Library of Congress Cataloging-in-Publication Data (applied for)

ISBN: 978-1-032-19641-1 (hbk)
ISBN: 978-1-032-19643-5 (pbk)
ISBN: 978-1-003-26019-6 (ebk)

DOI: 10.1201/9781003260196

Typeset in Times New Roman
by Radiant Productions

Preface

The book represents the original articles on important issues of risk theory, decision making under uncertainty, statistical decisions, problems of cryptography and control of stochastic systems and their applications to energy systems, among others. They share the common approach of applied quantitative risk management. The articles are outcomes of the series of international projects involving the leading scholars of different research institutions from Austria, Georgia, Moldova, Norway, Ukraine, USA and other countries in the field of modern stochastic optimization and decision making. These projects were directed towards development of study program in the risk management of master and PhD level. The articles included in this book contain novel research results and at the same time they were written keeping in mind the requirements of research based education on these topics.

The book consists of three parts. The first part presents the results concerning problems in the field of risk theory, optimization and information security. The second part focuses on modern mathematical methods of risk theory, stochastic and non-smooth optimization. statistical evaluation. In the third part we consider several applied problems of financial and insurance mathematics, energy and economics solved with the methods described in the first two parts. The structure of stochastic optimization solvers which are central for analysis of applied problems presented in the book is described in the first part. These solvers implement stochastic quasi-gradient methods for optimization and identification of complex nonlinear models including simulation models. These models constitute an important methodology for finding optimal decisions under risk and uncertainty. While a large part of current approaches towards optimization under uncertainty stems from linear programming (LP) and often results in large LPS of special structure, stochastic quasi-gradient methods confront nonlinearities directly without need of linearization. This makes them an appropriate tool for solving complex nonlinear problems, concurrent optimization and simulation models, and equilibrium situations of different types, for instance, Nash or Stackelberg equilibrium situations. The solver finds the equilibrium solution when the optimization model describes the system with several actors. The solver is parallelizable, performing several simulation threads in parallel. It is capable of solving stochastic optimization problems, finding stochastic Nash equilibria, and of composite stochastic bilevel problems where each level may require the solution of stochastic optimization problem or finding Nash equilibrium. Several complex examples with applications to water resources management, energy markets, pricing of services on social networks are provided.

For example, in the case of power system, regulator makes decision on the final expansion plan, taking into account the strategic behavior of regulated companies and coordinating the interests of different economic entities. Such a plan can be an equilibrium – a planned decision where a company cannot increase its expected gain unilaterally. The solver mentioned above can compute this equilibrium.

We dedicate this book to the memory of

Yuri M. Ermoliev (1936–2022) was a D.Sc., Professor, Academician of the Ukrainian National Academy of Sciences, and researcher at the International Institute for Applied Systems Analysis in Laxenburg, Austria. He was also a member of the V.M. Glushkov Institute of Cybernetics. Yuri M. Ermoliev, as a "Pioneer of Stochastic Programming" (named award, 2005), made a fundamental contribution to stochastic optimization, a crucial tool for quantitative risk management. Project CPEA-LT-2016/10003 and this book largely adopt his perspectives on risk management and its applications. Yuri M. Ermoliev served as a mentor and instructor to many of the book's authors, as well as an inspiration.

Dimitri I. Solomon (1951–2022) was a D.Sc., Professor, and Rector of the Academy of Transport, Informatics, and Communications of the Republic of Moldova, the Laureate of the State Prize in Science and Technology of Moldova. He made a significant contribution to the development of mathematical modeling of economic processes in transport, solving problems of non-differential optimization, fractional linear programming and discrete optimal control in dynamic networks, while working as the head of the project working group at the Academy. Dmitry I. Solomon was an attentive, kind friend, like-minded and responsible colleague for all of us.

Evgeny A. Lebedev (1953–2021) was a D.Sc., Professor, Head of the Department of Applied Statistics of the Faculty of Computer Science and Cybernetics of Taras Shevchenko National University of Kyiv. He is a well-known scientist in the field of applied statistics, random processes, risk theory and queuing systems, a project participant, a teacher of risk management schools for Ph.D., and graduate students, an attentive teacher and leader.

Sergei G. Cataranciuc (1957–2019) was a D.Sc., Professor, Head of the Department of Informatics and Discrete Optimization, Head of the project working group at the State University of Moldova. Sergey G. Cataranciuc is a well-known scientist in the field of combinatorial topology, convex sets, graph theory and decision-making. He was a talented scientific leader for the next generation of researchers, and a winner of the National Prize in Science and Technology, Member of the Parliament of the Republic of Moldova.

We want to note the significant contribution of our dear colleagues to the scientific results obtained in the process of joint work on the project, which are reflected in this monograph.

Acknowledgements

The work was supported the grant CPEA-LT-2016/10003 "Advanced Collaborative Program for Research Based Education on Risk Management in Industry and Services under Global Economic, Technological and Environmental Changes: Enhanced Edition" of Norwegian Directorate for Higher Education and Skills (HKDIR), managed by Norwegian University of Science and Technology (NTNU), Trondheim, Norway.

Alexei A. Gaivoronski
Professor, Norwegian University
of Science and Technology
Trondheim, Norway

Pavel S. Knopov
Professor, V.M. Glushkov Institute
of Cybernetics National Academy
of Sciences of Ukraine, Kyiv, Ukraine

Volodymyr A. Zaslavskyi
Professor, Taras Shevchenko
National University
of Kyiv, Ukraine

Contents

CHAPTER 1

Optimization of Simulation Models and other Complex Problems with Stochastic Gradient Methods

Alexei A Gaivoronski[1,*] and *Yuri M Ermoliev*[2]

1. Introduction

This chapter describes the solver SQG, which implements stochastic (quasi) gradient methods for the solution of stochastic optimization problems. Its particular strength lies in its capability to solve problems with substantial nonlinearities and optimized functions, defined by complex stochastic simulation models. It can also solve the stochastic equilibrium problems of different types when the studied system includes constellations of independent actors, choosing their own decisions. Finding the Nash equilibrium and solving bilevel stochastic optimization problems and stochastic Stackelberg games are among the capabilities of this solver. Also, in the case of equilibrium problems the payoff functions of individual actors can be obtained by simulation models.

The simplest problem addressed by SQG is

$$\min_{x \in X} \mathbb{E}\, f(x, \omega) \tag{1}$$

which finds the values of the decision variables x of a single actor on the feasible set $X \subseteq \mathbb{R}^n$. The expectation in (1) is taken with respect to the vector

[1] Norwegian University of Science and Technology, Trondheim, Norway.
[2] International Institute for Applied Systems Analysis, Laxenburg, Austria.
* Corresponding author: Alexei.Gaivoronski@ntnu.no

of random parameters ω, which models the uncertainty and is defined on appropriate probability space. The SQG solves this problem by generating the sequence of points x^s starting from some initial point x^0 and applying the recursive rule

$$x^{s+1} = \Pi_X\left(x^s - \rho_s \xi^s\right) \tag{2}$$

where $\prod_X(\cdot)$ is the projection operator on set X, ξ^s is a statistical estimate of the gradient of function $F(x) = \mathbb{E}f(x, \omega)$ at point x^s, meaning that it satisfies the property

$$\mathbb{E}\left(\xi^s \mid \mathbb{B}_s\right) = F_x(x^s) + b_s \tag{3}$$

where the conditional expectation in (3) is taken with respect to the σ – field $\mathbb{B}_s$ describing the history of the process and b_s is some diminishing term. The step size ρ_s satisfies the property

$$\rho_s \geq 0, \sum_{s=0}^{\infty} \rho_s = \infty, \tag{4}$$

which is weaker than what is normally required in stochastic approximation (Pflug 1996) because for optimization purposes we need a weaker notion of convergence than what is normally expected in statistics. Under additional technical assumptions, the sequence x^s converges to the solution of (1) (Gaivoronski 1978, Ermoliev 1983, Gaivoronski 2005).

The basic problem in (1) is used in SQG as a building block for the construction of considerably more complex problems, including the equilibrium and bilevel problems mentioned above. Besides, SQG has the capability to process the problems where the expectation operators are present not only in the objective (1), but also in constraints. Such problems occur, for example, in portfolio optimization with risk constraints (Zenios 2007). This and other capabilities, like integration with simulation models, required substantial additional conceptual and algorithmic development, described in the rest of the chapter.

In Section 2 we define the three-level problem hierarchy implemented in SQG together with the basic algorithmic tools utilized there. This is followed by a discussion of the challenges of concurrent optimization and simulation, showing the approach taken by SQG on this issue in Section 3. Section 4 describes examples of applied problems solved by SQG, concentrating on concurrent simulation and optimization.

We do not present in this paper the mathematical results of the convergence of algorithms, which underlie the operation of SQG. For such results, we refer a reader to Gaivoronski (1978), Ermoliev (1983), Pflug (1996), Becker and Gaivoronski (2014).

2. Problem hierarchy in SQG

SQG is developed for solutions of stochastic optimization and equilibrium problems involving decision models of $I \geq 1$ actors. We refer to an instance of such problems as *stochastic decision problems* (StDP). Such an instance is defined by a collection of functions Φ, which is the basic structure in SQG:

$$\Phi = \left\{ f_{ij}\left(x^i, x^{-i}, y^i, \omega\right), i = 1:I, j = 0:J_i \right\} \tag{5}$$

Here $x^i \in \mathbb{R}^{n_i}$ is the vector of *decision variables* of actor i, while x^{-i} is the vector of decision variables of all other actors, $x^{-i} = \{x^l, l = 1: I, l \neq i\}$. We denote by x the vector of all decision variables: $x = (x^i, x^{-i})$ for any i. The vectors x^i take values from feasibility sets X_i and collection Ψ of these sets constitutes another basic structure in SQG:

$$\Psi = \{X_i, i = 1:I\} \tag{6}$$

The vector ω is the vector of all uncertain parameters in the decision models of all actors. We consider ω to be a random vector defined on appropriate probability space $(\Omega, \mathbb{B}, \mathbb{P})$ with Ω being the event set equipped with σ-field $\mathbb{B}$, on which the probability measure $\mathbb{P}$ is defined.

The vector y^i is the vector of state variables of the decision model of the actor i. It is assumed that the state y^i evolves in a discrete-time $t = 1: T$, where the time horizon T can be finite or infinite. Then the state y^i takes the value y^{it} at time period t. The subsequent values of state variables are connected with the state equation

$$y^{i,t+1} = \Theta^i\left(x^i, x^{-i}, y^{it}, \omega\right) \tag{7}$$

The functions Θ^i can be quite complex and be defined by a simulation model. We define by $\Theta = \Theta^i$, $i = 1{:}I$ the collection of all state equations for all actors. The SQG provides the meta language and conventions for defining the structure (Φ, Ψ, Θ) described above. Some of the elements of this structure can be empty. In particular, the state variables y^i can be absent, and then the state equations (7) and the notion of time disappear. In this case, the decision problems become static. Besides, the set of actors can contain a single actor, $I = 1$. In this case there will be no stochastic equilibrium problem and StDP will become a stochastic optimization problem. We shall simplify our notations appropriately in such cases. For example, the functions from (5) will be denoted $f_j(x, \omega)$, $j = 1{:}J$ in the case of the single actor without the state variables.

Starting from some initial point x^{i0}, $i = 1{:}I$ the SQG generates iteratively the sequence of points x^{is} possibly with additional auxiliary sequences, which converge in a certain probabilistic sense to the solution of collection of problems defined on the structure (Φ, Ψ, Θ) while the number of iterations s tends to infinity. Naturally, the SQG solver does not perform the infinite number of iterations, instead, it stops the iteration process when a certain stopping criterion is satisfied. The solved problems and corresponding iterative processes are organized in the hierarchy shown in Figure 1.

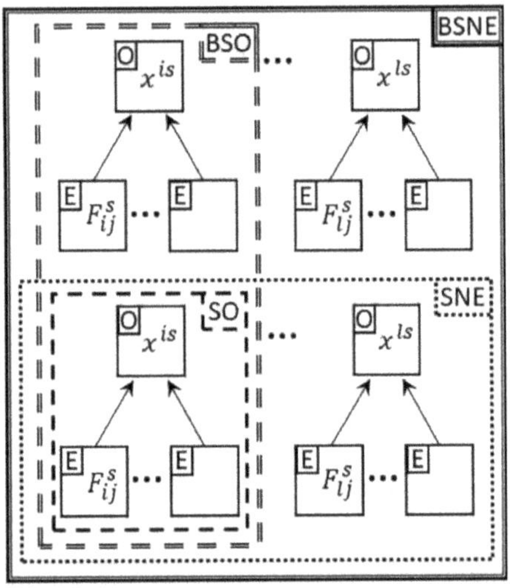

Figure 1. Problem and process hierarchy in SQG.

2.1 Lower level: estimation (E level)

Obtain statistical estimates of the values and gradients of functions $F_{ij}(x^i, x^{-i})$, which are the expected values of functions from (5). These estimates serve as input to the upper levels of the problem hierarchy as specified in what follows. More precisely, SQG computes these estimates for one of the following functions

$$F_{ij}\left(x^i, x^{-i}\right) = \mathbb{E} f_{ij}\left(x^i, x^{-i}, \omega\right) \tag{8}$$

$$F_{ij}\left(x^i, x^{-i}\right) = \mathbb{E}\frac{1}{T}\sum_{t=1}^{T} f_{ij}\left(x^i, x^{-i}, y^{it}, \omega\right) \tag{9}$$

$$F_{ij}\left(x^i, x^{-i}\right) = \mathbb{E}\lim_{T\to\infty}\frac{1}{T}\sum_{t=1}^{T} f_{ij}\left(x^i, x^{-i}, y^{it}, \omega\right) \tag{10}$$

where $\mathbb{E}$ is the expectation operator with respect to random variables ω. The case (8) occurs in the absence of the state variables y^i. The case (9) refers to the dynamic problem with the finite time horizon, while the case (10) deals with the estimation of the steady state function values on the infinite time horizon.

1a. Estimation of function values: These estimates are obtained by the moving average iterative process

$$F_{ij}^{s+1} = \left(1 - \alpha_i^s F_{ij}^s\right) + \alpha_i^s \varphi_{ij}^s \tag{11}$$

where φ_{ij}^s is an observation of the quantities under the expectation sign in (8)–(10) satisfying the property

$$\mathbb{E}\left(\varphi_{ij}^s \mid \mathbb{B}_s\right) = F_{ij}\left(x^{is}, x^{-is}\right) + a_{ij}^s \tag{12}$$

where a_{ij}^s tends to zero with $s \to \infty$ and $\mathbb{B}_s$ is the σ-field describing the history of the iterative process, which generates the sequences (x^{is}, F_{ij}^s) prior to iteration s. Some examples of observations φ_{ij}^s:

$$\varphi_{ij}^s = f_{ij}\left(x^{is}, x^{-is}, \omega^s\right) \tag{13}$$

$$\varphi_{ij}^s = \frac{1}{T}\sum_{t=1}^{T} f_{ij}\left(x^{is}, x^{-is}, y^{is}, \omega^{st}\right) \tag{14}$$

$$\varphi_{ij}^s = f_{ij}\left(x^{is}, x^{-is}, y^{is}, \omega^s\right) \tag{15}$$

where ω^s and ω^{st}, $s = 1, \ldots$, $t = 1, \ldots$ are independent observations of random parameters ω. In expressions above (13) corresponds to (8), (14) corresponds to (9) and (15) corresponds to (10). In the case (15), (10) the time step t corresponds to a single iteration step s. That is, in order to obtain the new update of the estimate F_{ij}^s through the process (11) it is enough to simulate the dynamics of the system for just one-time step obtaining the observation ω^s of random parameters, computing y^{is} as in (7):

$$y^{is} = \Theta^i\left(x^{is}, x^{-is}, y^{i,s-1}, \omega^s\right)$$

and obtaining φ_{ij}^s from (15). The SQG obtains the observations φ_{ij}^s by calling the external function `fun` written by the SQG user following the interface rules, which make it callable from SQG. This function takes as the input the values (x^{is}, x^{-is}) and additionally the value of $y^{i,s-1}$ in case (15). Its output comprises the value φ_{ij}^s and the new state y^{is} in the case (15). During the call, this function generates the new observation ω^s of the random parameters, if it is required by the estimation process. The value of the additional input parameter signals this necessity to the `fun`. Alternatively, in the case of several concurrent estimation processes with different inputs of decision parameters, this function provides the observations of φ_{ij}^s with the same values of random parameters. See the section on the estimation of the gradient for the explanation of such a necessity.

The estimates F_{ij}^s are fed to the upper levels of the problem hierarchy, where they are used in the iterative solution process of the upper-level problems. In particular, they are used in the stopping criterion of the whole iterative process, for the output of the optimal values of the objective functions, for estimation of gradients and for processing of constraints. They should satisfy the following fundamental convergence property

$$\left|F_{ij}^s - F_{ij}\left(x^{is}, x^{-is}\right)\right| \to 0 \tag{16}$$

in a certain probabilistic sense as $s \to \infty$. In other words, we do not need here the precise estimates of the function values from (8)–(10) for any fixed value of decision variables (x^i, x^{-i}). Instead, the estimates should trace with gradually increasing precision the changing value of these functions while (x^{is}, x^{-is}) changes in the course of iterations. This is a very mild requirement, allowing it to make only a single observation of random parameters and a single estimation step per one update of decision parameters (x^{is}, x^{-is}). However, it requires a specific coordination between the estimation process (11) and the iterative updating process of decision variables (x^i, x^{-i}). Namely, the steps α_i^s from (11) should be larger than the updating steps for the decision variables and their ratio should asymptotically approach infinity.

Besides tracing the values of functions (8)–(10) at changing (x^{is}, x^{-is}), other concurrent estimation processes of type (11) may compute the estimates F_{ij}^{ls}, which traces the values of these functions for other sequences of points (x^{ils}, x^{-ils}), which are connected with the original sequence. In particular, the output of such estimation processes may be used on the upper levels of the problem hierarchy for the estimation of the gradients of functions (8)–(10) using finite differences. In this case, one can take

$$x^{ils} = x^{is} + \delta_i^s e_l^i, l = 1 : n_i \tag{17}$$

where e_l^i is a unit vector of space $\mathbb{R}^{n_i}$.

1b. Estimation of gradient values: The SQG computes the estimates ξ_{ij}^s of the gradient of function $F_{ij}(x^i, x^{-i})$ at point (x^{is}, x^{-is}) from (8)–(10) with respect to variables x^i. These estimates are sent to the higher levels of the problem hierarchy and should satisfy the following condition

$$\mathbb{E}\left(\xi_{ij}^s \mid \mathbb{B}_s\right) = \nabla_{x^i} F_{ij}\left(x^{is}, x^{-is}\right) + b_{ij}^s \tag{18}$$

where b_{ij}^s tends to zero with $s \to \infty$. There are two possibilities

- The user includes the computation of ξ^s_{ij} in the body of the external function `fun`. Then this function returns the value of ξ^s_{ij} with each call of `fun` performed by SQG. This option is preferable when it is relatively easy to compute one of the following entities

$$\xi^s_{ij} = \nabla_{x^i} f_{ij}\left(x^{is}, x^{-is}, \omega^s\right) \tag{19}$$

$$\xi^s_{ij} = \frac{1}{T}\sum_{t=1}^{T} \nabla_{x^i} f_{ij}\left(x^{is}, x^{-is}, y^{is}, \omega^{st}\right) \tag{20}$$

$$\xi^s_{ij} = \nabla_{x^i} f_{ij}\left(x^{is}, x^{-is}, y^{is}, \omega^s\right), \tag{21}$$

which similarly to (13)–(15) corresponds to the cases (8)–(10).

- The complexity of functions (8)–(10) typical of simulation models makes it difficult to compute the values (19)–(21) directly. Then the estimates ξ^s_{ij} are computed by the estimation level of SQG using only the function observations (12) and function estimates (11). In order to do this $n_i + 1$ tracing moving average processes (11) are performed in parallel for the same sequence of observations of random parameters ω^s or ω^{st} and $n_i + 1$ sequences of decision parameters x^{is}, x^{ils} from (17). These processes produce the estimates F^s_{ij}, F^{ls}_{ij} and the SQG computes

$$\xi^s_{ij} = \frac{1}{\delta^s_i}\sum_{l=1}^{n_i}\left(F^{ls}_{ij} - F^s_{ij}\right) \tag{22}$$

2.2 *Intermediate level: optimization (O level)*

This level takes the values of the estimates F^s_{ij}, ξ^s_{ij} supplied by the estimation level and generates the sequences x^{is}, $i = 1: I$, $s = 1,...$ which solves the following problem

$$\underset{x^i}{\text{"min"}}(\text{"max"})\, F_{i0}\left(x^i, x^{-i}\right) \tag{23}$$

subject to

$$F_{ij}\left(x^i, x^{-i}\right) \le 0,\ j = 1:J \tag{24}$$

$$x^i \in X_i \tag{25}$$

That is, the generated sequence x^{is} attempts to minimize or maximize the objective $F_{i0}(x^i, x^{-i})$ of the actor i with respect to the decision variables x^i of this actor. Such sequences are generated for each actor. These sequences may or may not actually converge to the solution of respective optimization problems due to the dependence of the objectives not only on the decision variables of respective actors but also on the decision variables of all actors. Still, depending on the coordination between the step sizes used for their generation, the sequences may converge to the solution of optimization problems in a certain sense. In other cases, they will converge to the solutions of equilibrium problems, defined on the upper level of the problem hierarchy. For this reason, we put min or max from (23) in quotation marks. In the case of a single actor the problem (23)–(25) reduces to a proper optimization problem. We drop the index i then and the problem becomes the following:

$$\min_{x \in X}(\max) F_0(x) \tag{26}$$

subject to

$$F_j(x) \le 0,\ j = 1:J \tag{27}$$

The problem (23)–(25) looks similar to deterministic nonlinear optimization (or equilibrium) problems. This similarity is deceitful, however, because functions $F_{ij}(x^i, x^{-i})$, usually cannot be computed with the precision required the for application of deterministic nonlinear programming techniques. This is due to the presence of the expectation operator in the definition of these functions (8)–(10), which cannot be computed precisely for the problems of realistic dimensions and complexity. Therefore, the problem (23)–(25) belongs to the family of stochastic optimization (or equilibrium) problems (Birge and Louveaux (2011)). Other reasons are the long computing times necessary for precise simulations, when these functions are obtained through simulation models, and the transient dynamic behavior in the case of the steady state optimization (10) discussed in more detail in Section 3.

The SQG generates the sequences x^{is} following the stochastic gradient iteration similar to (2):

$$x^{i,s+1} = \Pi_{X_i}\left(x^{is} - \rho_{is}\left(\xi_{i0}^s + \sum_{j=1}^{m_i} C_{ij}\eta_{ij}^s \xi_{ij}^s\right)\right) \tag{28}$$

$$\eta_{ij}^s = \max\left(0, F_{ij}^s\right) \vee \mathbf{1}_{\{F_{ij}^s > 0\}} \tag{29}$$

Observe that in the case of a single actor, $i = 1$ and the absence of expectation type constraints, $m_i = 0$, the sequence (28), (29) coincides with the basic SQG process (2). In the case of several actors, the step size ρ_{is} is individual for actor i. This is important for the solution of bilevel stochastic optimization problems and Stackelberg games. The second term in the internal parenthesis in (28) is dedicated to the processing of expectation constraints. It makes (28) to be a basic SQG process of type (2) for minimization of function $F_{i0}(x^i, x^{-i})$

with added penalty term, which penalizes the violation of constraints (24). The positive constant C_{ij} is the penalty coefficient, vector ξ^s_{ij} is the stochastic gradient of constraint satisfying (18) and η^s_{ij} from (29) indicates the violation of constraint. Since exact values of $F_{ij}(x^i, x^{-i})$ are unknown, η^s_{ij} substitutes them with the estimates F^s_{ij} provided by the estimation level of the problem hierarchy. The two alternatives for η^s_{ij} shown in (29) correspond to two types of penalties. The first alternative is the classical quadratic penalty, while the second one is the non-smooth linear penalty. In the case of maximization, the signs in (28) are changed to the opposite ones.

2.3 Upper level: equilibrium problems (Q level)

This level constructs specific problem types and their solutions from the solution sequences x^{is} and function estimates F^s_{ij} provided by the two lower levels E and O, see Figure 1. We consider the following four problem types.

3a. Stochastic optimization (SO): This is a simple type of SQG problem with the empty third level of SQG hierarchy. In this case only single actor is present, $i = 1$. The problem formulation is shown in (26)–(27) where $F_j(x)$ can belong to one of the types (8)–(10). Observe that also the problems of this type can be quite complex because they may include expectation constraints (27) and the observations of functions $F_j(x)$ can be obtained by complex simulations.

3b. Bilevel stochastic optimization (BSO): These problems are known otherwise as leader-follower games or Stackelberg games with two actors (Stackelberg 1952). We describe them here in the interpretation of the leader-follower game. In this case, there are two actors, $i = 2$. The actor 1 called the *leader* announces his decision $\bar{x}^1$ to actor 2 called the *follower*. Knowing the decision of the leader the follower chooses his decision $\bar{x}^2$ solving a stochastic optimization problem

$$\min_{x^2 \in X_2} F_{20}\left(x^2, \bar{x}^1\right) \tag{30}$$

$$F_{2j}\left(x^2, \bar{x}^1\right) \le 0,\ j = 1:J \tag{31}$$

for fixed $x^1 = \bar{x}^1$. In this way the decision $\bar{x}^2$ will depend on $\bar{x}^1$: $\bar{x}^2 = \bar{x}^2(\bar{x}^1)$. Knowing this, the leader chooses his decision solving the problem

$$\min_{x^1 \in X_1} F_{10}\left(x^1, \bar{x}^2\left(x^1\right)\right) \tag{32}$$

$$F_{1j}\left(x^1, \bar{x}^2\left(x^1\right)\right) \le 0,\ j = 1:J \tag{33}$$

Bilevel optimization problems are quite challenging numerically even in the deterministic linear case. The SQG solves them by running two concurrent SQG processes (28)–(29) with the asymptotically vanishing ratio between the leader and follower step sizes: $\rho_{1s}/\rho_{2s} \to 0$.

3c. Stochastic Nash equilibrium (SNE): In this case, we have $I > 1$ actors with payoffs $F_{i0}(x^i, x^{-i})$ having one of the types (8)–(10) and possible constraints (24). The SQG runs I concurrent SQG processes (28)–(29), which under additional technical assumptions converge to the Nash equilibrium (Nash (1950)) if it exists.

3d. Bilevel stochastic Nash equilibrium (BSNE): This is the combination of problems 3b and 3c. There are the leader and follower levels as in (30)–(33), but there are $I_1 \geq 1$ leaders and $I_2 \geq 1$ followers, $I = I_1 + I_2$. Suppose that the actors' $i = 1{:}I_1$ are the leaders and the actors' $i = I_1 + 1{:}I$ are the followers. Then SQG runs I concurrent SQG processes (28)–(29) with asymptotically vanishing ratios between leader step sizes and follower step sizes: $\rho_{is}/\rho_{js} \to 0$ if $1 \leq i \leq I_1$ and $I_1 + 1 \leq j \leq I$.

We emphasize here again that in all the problem types mentioned above the payoff and constraint functions can originate from complex simulations.

3. Concurrent optimization and simulation with SQG

Here we discuss in more detail the issues regarding the integration of simulation and optimization using SQG. As we have mentioned above, the functions $f_{ij}(x^i, x^{-i}, \omega)$ or $f_{ij}(x^i, x^{-i}, y^{it}, \omega)$ from (8)–(10) can result from running of complex simulation model implemented by the user inside the function `fun` required by SQG. Such simulation can be quite complex, including in its body *yes* or *no* decisions of actors based on reaching certain triggering quantities certain thresholds, like acceptance or rejection of an investment project based on the predictions of profit. Or, the decision to release a certain amount of water from a reservoir based on the level of water in it. It can also include solutions of simpler equilibrium or optimization problems.

To be more specific, let us consider optimization of the steady state function $F_{ij}(x^i, x^{-i})$ on the infinite horizon (10), operating in discrete time $t = 1,\ldots,$ where we assume that the state transformation from y^{it} to $y^{i,t+1}$ is performed by the simulation model. We describe here how the optimal or equilibrium values of decision variables x^i can be obtained using $I + \sum_i n_i$ parallel simulation runs. In this case one simulation time step t coincides with one SQG iteration s from (28), so we use t also for indexing of SQG iterations.

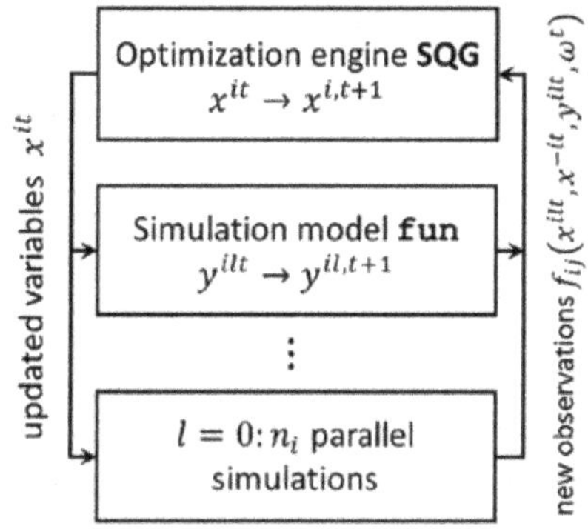

Figure 2. Concurrent optimization and simulation on the infinite horizon.

The concurrent simulation and optimization process is presented in Figure 2. It is performed as follows below.

1. *Initialization*: At the start the initial values x^{i0} are selected as the starting points for SQG iterations (28). Simulation processes Λ^{li}, $i = 1{:}I$, $l = 0{:}n_i$ are initialized. The points x^{i0} are sent to simulation processes Λ^{0i} and points x^{il0}, $l = 1{:}n_i$ obtained as in (17) for $s = 0$ are sent to processes Λ^{li}, $l = 1{:}n_i$. The initial values of simulation state variables y^{il0} for processes Λ^{li} are selected: $y^{il0} = y^{i0}$ for $l = 0{:}n_i$.
2. *Generic step*: By the beginning of simulation step t the current approximation to the optimal values of decision variables are x^{it} and the current states of the simulation processes are y^{ilt}. The following actions are performed at step t.

   2a. *SQG sends decision variables to simulation processes*: The points x^{it} are sent to simulation processes Λ^{0i} and points x^{ilt}, obtained as in (17) for $s = t$ are sent to processes Λ^{li}, $l = 1{:}n_i$.

   2b. *Observation of random parameters*: The new observation ω^t of random parameters Λ^{li} common to all simulation processes is made.

   2c. *Simulation step*: Simulation processes Λ^{li} obtain the observations $f_{ij}(x^i, x^{-i}, y^{it}, \omega)$ and update the respective state variables y^{ilt} to $y^{il,t+1}$ with equation (7), where $x^i = x^{ilt}$, $x^{-i} = x^{-it}$, $y^{it} = y^{ilt}$, $\omega = \omega^t$ for process Λ^{li}.

   2d. *Simulation processes send function observations to SQG*: The observations $f_{ij}(x^{ilt}, x^{-it}, y^{it}, \omega^t)$ are sent to SQG.

   2e. *SQG updates the decision variables*: SQG uses observations of $f_{ij}(x^{ilt}, x^{-it}, y^{it}, \omega^t)$ obtained from the simulation processes to update the function estimates F_{ij}^t, F_{ij}^{lt} as in (11), compute stochastic gradients ξ_{ij}^t as in (22) and obtain the update of decision variables $x^{i,t+1}$ as in (28).

2f. *Checking of stopping criterion*: The stopping criterion is checked and if satisfied the value of $x^{i,t+1}$ or its average over all or part of preceding iterations is taken as the solution of the problem, see (Nemirovski et al., 2009), Otherwise, the computations proceed with step $t + 1$.

In this manner, the optimal or equilibrium values of decision parameters will be obtained during a single concurrent run of $I + \sum_i n_i$ simulations. A finite simulation horizon as in (9) one has two choices. If T is small enough then the updates of decision parameters as in (28) can be made after the whole simulation run is performed and the values of the averaged observations as in the right-hand side of (9) are sent to the SQG. If T is large then the updates can be performed after each simulated time step as in the case of the infinite time horizon described above.

In the case of the finite time horizon, alternative methods like nonlinear programming algorithms (NLP) or genetic algorithms encounter additional pitfalls related to the *sample average approximation* (Shapiro and Xu 2008). This approach generates a large sample M of observations ω^m and substitutes the original problem with expectation functions $F_{ij}(x^i, x^{-i})$ from (9) by sample average approximation $F_{ij}^M(x^i, x^{-i})$ of these functions:

$$F_{ij}^M\left(x^i, x^{-i}\right) = \frac{1}{M}\sum_{m=1}^{M}\frac{1}{T}\sum_{t=1}^{T} f_{ij}\left(x^i, x^{-i}, y^{itm}, \omega^m\right). \tag{34}$$

The optimal or equilibrium values of the decision variables are obtained then by solving the appropriate problem with functions $F_{ij}^M(x^i, x^{-i})$ using a variety of methods from nonlinear optimization approaches to genetic algorithms. This approach encounters substantial difficulties in the very common case when the simulation model includes yes or no decisions. In such cases, the approximate functions $F_{ij}^M(x^i, x^{-i})$ will be discontinuous for any finite M even when the limiting functions $F_{ij}(x^i, x^{-i})$ exhibit smooth and unimodal behavior. Figures 3 and 4 show one example of an equilibrium problem in a simulation model of the energy market. The decision variables there are the thresholds, which trigger the acceptance of investment projects in the energy sector: following the industry practice, the project is accepted when the forecasted Return On Investment (ROI) exceeds the given threshold. It is necessary to find here the equilibrium values of the acceptance thresholds. Figures 3 and 4 show that the functions $F_{ij}^M(x^i, x^{-i})$ are discontinuous and piecewise constant for any finite M, which provides insurmountable difficulties for the NLP methods and even for genetic algorithms. The stochastic gradients succeed in finding the solution here because they address directly the expectation functions $F_{ij}(x^i, x^{-i})$, which is smooth and well behaved.

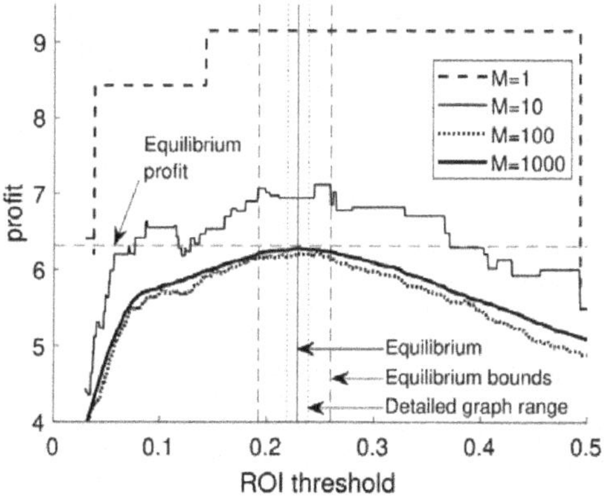

Figure 3. Sample average approximation of payoff function.

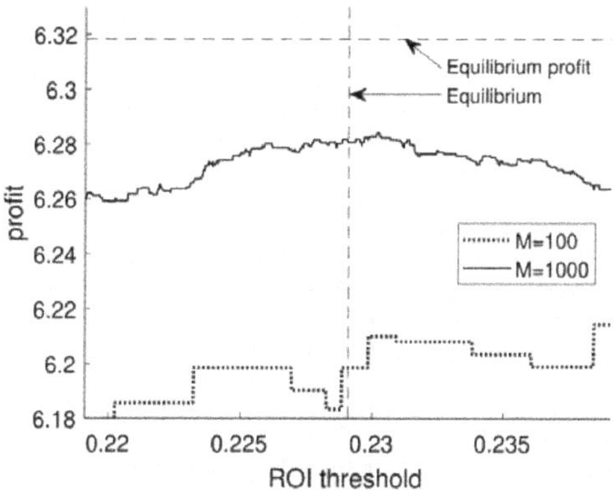

Figure 4. Sample average approximation of payoff function, magnified around the equilibrium.

4. Examples

The SQG was applied to solution of complex optimization and equilibrium problems on simulation models: pricing of services on social networks (Becker and Gaivoronski 2014), optimization of water resources networks (Gaivoronski and Zuddas 2012), maritime transportation (Zuddas and Gaivoronski 2013), energy markets (Gaivoronski and Harbo 2017).

5. Implementation

SQG runs on top of Matlab using descriptive language for defining the problem hierarchy (Section 2) and the solver parameters. Matlab Parallel Computing Toolbox is used for parallelization.

6. Conclusion

SQG is a powerful tool for solving a variety of stochastic optimization and equilibrium problems on realistic simulation models.

References

Becker, D.M. and A.A. Gaivoronski. 2014. Stochastic optimization on social networks with application to service pricing. Computational Management Science, 11(4): 531–562.

Birge, J.R. and F. Louveaux. 2011. Introduction to Stochastic Programming. Second Edition. Springer, New York.

Ermoliev, Y. 1983. Stochastic quasigradient methods and their application to system optimization. Stochastics, 9: 1–36.

Gaivoronski, A.A. 1978. Nonstationary stochastic optimization problems. Cybernetics and Systems Analysis, 14(4): 575–579.

Gaivoronski, A.A. 2005. SQG: Stochastic programming software environment. *In*: Wallace, S. and W. Ziemba (eds.). Applications of Stochastic Programming. SIAM.

Gaivoronski, A.A., G. Sechi and P. Zuddas. 2012. Balancing cost-risk in management optimization of water resource systems under uncertainty. Physics and Chemistry of the Earth, Parts A/B/C, 42-44(1): 98–107.

Gaivoronski, A., Z. Su and S. Harbo. 2017. Modelling investment in energy markets: stochastic equilibrium problem defined on simulation model. In ECSO 2017, Book of Abstracts.

Nash, J. 1950. Equilibrium points in n-person games. Proceedings of the National Academy of Sciences, 36(1): 48–49.

Nemirovski, A., G.L. Juditsky, G. Lan and A. Shapiro. 2009. Robust stochastic approximation approach to stochastic programming. SIAM Journal on Optimization, 19(4): 1574–1609.

Pug, G.C. 1996. Optimization of Stochastic Models. The Interface Between Simulation and Optimization. Kluwer, Boston.

Shapiro, A. and H. Xu. 2008. Stochastic mathematical programs with equilibrium constraints, modelling and sample average approximation. Optimization, 57(3): 395–418.

Stackelberg, H.V. 1952. The Theory of Market Economy. Oxford University Press, London.

Zenios, S.A. 2007. Practical Financial Optimization: Decision Making for Financial Engineers. Blackwell Publishing.

Zuddas, P., M. Di Francesco and A.A. Gaivoronski. 2013. Container transportation problem under uncertain demand and weather conditions. In 13th International Conference on Stochastic Programming, Book of Abstracts.

CHAPTER 2

Linking Catastrophe Modeling and Stochastic Optimization Techniques for Integrated Catastrophe Risk Analysis and Management

Tatiana Ermolieva,[1,*] *Yuri M Ermoliev,*[1] *Nadejda Komendantova,*[1] *Vladimir I Norkin,*[2] *Pavel S Knopov*[2] and *Vasyl M Gorbachuk*[2]

1. Introduction

The increasing vulnerability of modern society is an alarming tendency. Losses from natural and human-made catastrophes are rapidly increasing. Catastrophes destroy communication systems, electricity supply and irrigation, affecting agricultural and energy production and provision systems. They affect consumption, savings, and investments. In the EU, from 1980 to 2020, natural hazards affected nearly 50 million people and cost Member States on average approximately €12 billion per year (See European Commission 2021). In the US, the year 2021 brough very high share of natural disaster losses (roughly US$ 145 bn), of which some US$ 85 bn were insured, the overall and insured losses were significantly higher than in the two previous years (Overall losses 2020: US$ 100 bn, 2019: US$ 52 bn; insured losses

[1] International Institute for Applied Systems Analysis, Laxenburg, Austria.
[2] Glushkov's Institute of Cybernetics, Kyiv, Ukraine.
* Corresponding author: ermol@iiasa.ac.at

2020: US$ 67 bn, 2019: US$ 26 bn)—MunichRe 2022. Climate change is bringing along more extreme weather events worldwide (See Sigma 2022, Sigma 2021).

The primary reason for the increasing losses from catastrophes is the ignorance of risks leading to the concentration of industries, infrastructure, wealth, people in risk prone areas as well as the creation of new risk prone areas. This trend becomes alarming especially in the conditions of climate changes, increasing variability and frequency of natural hazards. Climate changes manifest in alteration of seasonal precipitation and temperature patterns and intensification of natural disasters. The impacts of climate changes can be magnified by the growing complexity of systemic interdependencies between economic sectors, introduction of new policies and technologies, rising demands, escalating frequency and severity of floods, hurricanes, storms, droughts, landslides, prolonged heatwaves. Climate changes put stress on water availability and quality, affect agricultural production, energy usage and production, thereby threatening the water-food-energy (WFE) security.

Continued urbanization and development in hazardous areas have been putting more people and wealth under risk. Urbanization and population concentration magnify the impact of hurricanes, windstorms, floods, heatwaves, epidemics, and other natural catastrophes, which can be further exacerbated by the systemic impacts stemming from the increasing interconnectedness between various socio-economic and environmental systems: "It is not just the nature of major catastrophe risks that seems to be changing, but also the context within which they appear and society's capacity to manage them" (OECD 2003).

This alarming tendency due to the combination of natural and human-induced risks calls for new risk-based approaches to economic development and catastrophe management. The economy can be considered a complex network of interdependent food, energy, water, and environmental systems, constantly facing shocks and changes, in particular, with catastrophic consequences. Economic assessment models involved in the risk analysis and impact assessment are primarily deterministic based on common utility-maximizing principles (See Ackerman et al. 2009, Boyle 2002, Chapman and Khanna 2000). They are not able to properly account for uncertainties, increasing variability and the frequency of extreme events, and catastrophic risks, inherent to climate change (See Fisher and Narain 2003, Goodess et al. 2003, Wright and Erickson 2003). Although climate change modelers recognize that the impact will be caused by extreme events along with changes in the patterns of variability, the ways to represent these variables in climate change assessment models are very limited (See Ackerman et al.

2009, Boyle 2002, Chapman and Khanna 2000, Fisher and Narain 2003, Goodess et al. 2003, Wright and Erickson 2003).

In traditional economic theory there is no special problem of catastrophic risks (See Arrow 1996, Chapman and Khanna 2000, Chichilinskii 1997). The modeling of economic behavior under uncertainty is often based, in fact, on rather strong assumptions of certainty. It is assumed that all economic agents know all possible shocks (states of the world), i.e., they know when, how often, and what may happen to each of them. Therefore, they can easily organize "markets", where everyone insures everyone by pooling resources available in any state of the entire society, i.e., a catastrophe becomes small on a world-wide scale. In reality, this pool does not exist, which calls for more realistic models with explicit representation of uncertainties and risks associated with decisions under uncertainties, safety constraints, and risk-based criteria. The advanced computational approaches allow us today to deal with large-scale decision-making problems in the presence of multidimensional mutually dependent random variables.

This chapter analyzes some methodological challenges involved in catastrophe risk modeling and management. In a sense, it summarizes and extends the discussions in papers Amendola et al. 2013, Amendola et al. 2000, Amendola et al. 2000, Baranov et al. 2002, Compton et al. 2008, Ekenberg et al. 2003, Ermoliev and Hordijk 2003, Ermoliev et al. 2018, Ermolieva et al. 2016, Ermolieva and Obersteiner 2004, Ermoliev et al. 2018, Ermoliev et al. 2000a, Ermoliev et al. 2000b, Ermoliev et al. 2001, Ermoliev and Norkin 1997, Ermolieva and Ermoliev 2005, Ermolieva et al. 2003, Ermolieva 1997, Ermoliev et al. 1998, Ermolieva 1997, Ermoliev et al. 2010. We argue that catastrophe risk analysis and management have to be addressed with an Integrated Assessment and Management modeling framework (IAMM) linking catastrophe risk models (CRM) with stochastic optimization (STO) techniques for the design of optimal and robust mitigation and adaptation strategies for risks of all kinds. Section 2 outlines the main features of catastrophe risk management: endogenous risks, highly mutually dependent losses, the lack of information, the need for long-term perspectives and geographically explicit models, the involvement of various agents (such as individuals, farmers, producers, consumers, governments, insurers, investors), safety and security requirements, and the need for robust decisions (See Ermoliev and Hordijk 2003). Section 3 discusses the main features of an integrated catastrophe management model, which can be considered as an IAMM. It may consist of several representative modules or submodels of various involved systems. Many authors (See, for instance, Boyle 2002, Clark 2002, Compton et al. 2008, Embrechts et al. 2000, Walker 1997) define three modules in CRM:

A scientific or hazard module comprising of an event generator and a local intensity calculation; an engineering module for structural damage estimation; and an insurance coverage module for insured loss calculation. The modules for damage estimation can also include agricultural, energy, water submodels for the assessment of potential damages in the systems and possible dependent systemic risks due to the interconnectedness and synergies among them (OECD 2003). The damage submodels are spatially detailed utilizing GIS data.

A Monte Carlo simulation of a hazard module and a loss calculation module produce a scenario-by-scenario analysis and assessment of feasible catastrophe scenarios and respective scenario-dependent management strategies. However, such an analysis can easily lead to an infinite number of scenarios', combinations of hazard loss and respective decisions. This analysis provides only a scenario-dependent evaluation without giving a clue what management options (decisions) are good (robust) with respect to all (or a percentile) of all catastrophe/hazard scenarios. Section 3 sketches out an appropriate decision-making model, which is an IAMM linking a hazard (loss calculation) model and an STO multiagent (multicriteria), spatial and dynamic model. This model emphasizes the cooperation of various involved agents in dealing with catastrophes, direct and indirect (negative and positive) effects, and the coexistence of anticipative (risk-averse) long-term ex-ante policies with adaptive (risk-taking) short-term ex-post policies. Sections 4 and 5 discuss in more detail the long-term economic effects of shocks. It points out that shocks which persist in time implicitly modify even sustained exponential economic growth towards stagnation. Proper injection of capital may be necessary only at certain stages of growth. Section 6 concludes the chapter.

2. Catastrophic risks: main challenges

There is a number of methodological challenges involved in catastrophic risk analysis and management, which we outline in this section.

2.1 Complex interdependencies

Catastrophes produce severe consequences, characterized by mutually dependent in space and time socio-economic and environmental impacts, structural and agricultural damages, losses of human lives, failures in energy and water production and supply systems, etc. The multivariate distribution of these losses is in general analytically intractable due to systemic interdependencies and interactions among various systems and policies. The distribution depends on the clustering of values and activities in the region and the patterns of catastrophes. Besides, it may dramatically depend on policy variables. For example, a dam fundamentally modifies the flood conditions downstream and along the site. This creates favorable conditions for new developments, infrastructure investments, land-use transformations, and insurance activities. On the other hand, a failure of the dam may lead to rare but more devastating losses in the protected area. Such interdependencies of decisions and risks restrict the straightforward "one-by-one" evaluations of feasible risk management decisions and regional development strategies. The so-called "if-then" analysis can quickly run into an extremely high number of alternatives. For example, with only 10 feasible decisions regarding insurance coverage, say 10%, 20%, ..., 100% loss coverage for a particular site (household, agricultural land, energy production facility, etc.) and 10 possible heights of a flood protection infrastructure (e.g., dam), the number of possible "if-then" combinations to be evaluated is 10^{10}.

The main idea in dealing with the problem of large dimensionality is to avoid exact evaluations of all possible alternatives and concentrate attention on the most promising directions. From a formal point of view, this is equivalent to the design of special search techniques (in the space of decision variables), making use of random simulations of catastrophes (catastrophe generator), respective losses/impacts (loss calculator), and iterative revision of feasible decisions towards the improvement of specific indicators (iterative STO procedure). This is a task of iterative stochastic optimization, e.g., stochastic quasigradients (SQG) methods (Ermolieva 1997, Ermolieva et al. 2017, Ermoliev and Wets 1988). The SQG iterative algorithms define a "searching" process, which resembles a sequential adaptive learning and improvement of decisions from data and simulations, i.e., the so-called Adaptive Monte Carlo optimization. The SQG methods are applicable in cases when traditional stochastic approximation, gradient or stochastic gradient methods do not work, in particular, to general two-stage problems with implicitly defined goals and constraints functions, nonsmooth and possibly discontinuous performance indicators, risk and uncertainties shaped by decision of various agents. The SQG-based procedures generate feedback to policy variables after each simulation and automatically drive them towards desirable combinations without going into exhausting "it-then" scenario analyses.

2.2 Rare events

The principal problem with the management of rare catastrophe risks is the lack of historical data on losses at particular locations, although rich data may exist on an aggregate regional level. Historical data are relevant to old policies and may have very limited value for new policies.

Models have to play a key role for generating plausible synthetic data and designing new *robust* policies (Ermoliev and Hordijk 2003, Ermolieva et al. 2016, Ermolieva and Obersteiner 2004, Ermoliev et al. 2018, Ermoliev et al. 2000a, Ermoliev et al. 2000b, Ermoliev et al. 2001, Ermoliev and Norkin 1997, Ermolieva and Ermoliev 2005, Ermolieva et al. 2003, Ermolieva 1997, Ermoliev et al. 1998, Ermolieva et al. 1997, Ermoliev et al. 2010).

Catastrophes may be of quite different in nature from episode to episode, exhibiting a wide spectrum of impact on infrastructures, public health, agricultural-energy-water and environment systems. Each of these episodes seems to be improbable and may be simply ignored in the so-called "practical approaches" or "scenario thinking". This may lead to rather frequent "improbable" catastrophes: although each of N scenarios (episodes) has a negligible probability p, the probability of one of them increases exponentially in N as $1 - (1 - p)^N = 1 - exp\{N \ln(1 - p)\}$. In other words, the integrated analysis of all possible, although rare, scenarios is essential.

2.3 Long-term perspectives

The proper assessment and management of rare risks requires long-term perspectives. The occurrence of a catastrophe within a small interval Δt is often evaluated by a negligible probability $\lambda\Delta t$, but the probability of a catastrophe in an interval $[0, T]$ increases as $1 - (1 - \lambda\Delta t)^{T/\Delta t} \approx 1 - e^{-\lambda T}$. Purely adaptive "learning-by-doing" or "learning-by-catastrophe" approaches may be extremely expensive. There are uncertainties regarding the real occurrence of rare events. In fact, a 1000-year flood may occur next year. For example, floods across Central Europe in 2002 were classified as 1000-, 500-, 250-, and 100-year events. Another tendency in risk assessment is to evaluate potential losses by using so-called annualization (average annual evaluation), i.e., by spreading damages, fatalities, and compensations from a potential, say 50-year flood, equally over 50 years (Atoyev et al. 2020, Dantzig 1979). In this case, roughly speaking, potential 50-year losses incurred, e.g., with a dam break, are evaluated as a sequence of independent annual events (failures), one meter of the breach in the first year, another—in the second year, and so on, until the final crash of the whole flood defense system in the 50th year. The main conclusion from this type of deterministic analysis is that catastrophe is not significant. Here, the dramatic confusion is from the wrong representation of spatio-temporal aspects of catastrophes. Catastrophes do not occur on average with average patterns. They hit a "spike" in space and time requiring immediate financial support and adequate emergency actions. In other words, the distributional aspects, i.e., temporal and spatial distributions of values and risks, are key issues to capture the main sources of vulnerability for designing robust policies.

2.4 Spatial aspects

Catastrophes have different spatial patterns and affect locations quite differently. For example, the location of properties or infrastructure with respect to the center of an earthquake is an extremely important piece of information. Together with the regional geology and the soil conditions, the location influences the degree of shaking, and, hence, damage incurred at the location. Deforestation at a particular location modifies the flood conditions only downstream and affects losses and respective risk management decisions, e.g., flood protection measures and insurance claims, only from specific locations. In other words, management of complex interdependencies among catastrophic risks, losses and decisions is possible only within a geographically explicit framework. Unfortunately, a general approach in risk assessment is to use so-called average hazard maps, i.e., maps showing average catastrophe patterns that will never be observed as a result of a real episode, as the map is the average image of all possible patterns that catastrophic events may have. Accordingly, social losses in affected regions are evaluated as the sum of individual losses computed on a location-by-location rather than pattern-by-pattern basis w.r.t. joint probability distributions. This highly underestimates the real socio-economic impacts of catastrophes, as the following simple example shows.

Consider the dilemma between the social and individual losses. In a sense, this example shows that $100 >> \overbrace{1+1+1+\ldots+1}^{100}$. Assume that each of 100 locations has an asset of the same type. An extreme event destroys all of them at once with probability 1/100. Consider also a situation without the extreme event, but with each asset still being destroyed independently with the same probability 1/100. From an individual point of view, these two situations are identical: an asset is destroyed with probability 1/100, i.e., individual losses and expected losses are also the same. Collective (social) losses are dramatically different. In the first case 100 assets are destroyed with probability 1/100, whereas in the second case 100 assets are destroyed only with probability 100^{-100}, which is practically 0. This example also illustrates the potential exponential growth of vulnerability from increasing network-interdependencies.

2.5 Assessment vs. robust solutions

Uncertainty is associated with every facet of catastrophe risk assessment. The exact evaluation of all interdependencies between catastrophe patterns, potential strategies, and related outcomes is impossible. It may easily run into an extremely high number of alternatives. Also, a strategy optimal for one scenario may not be optimal against multiple scenarios. In this situation the most important task seems to be the design of management strategies robust with respect to all potential scenarios. The straightforward assessment is never exact, while the preference structure among different decisions may be rather stable to errors. This is similar to the situation with two parcels: to find out their weights is a much more difficult task than to determine the heavier parcel. This simple observation, in fact, is the basic idea of the integrated management model and the stochastic optimization approaches discussed in Section 3: the evaluation of the robust optimal decisions is achieved without exact evaluation of all possible alternatives.

The underlying assumption of the robustness accounts for safety, flexibility, and optimality criteria of all agents against multiple potential scenarios of catastrophic events (See Ermoliev and Hordijk 2003, Ermoliev et al. 2018, Ermolieva et al. 2016, Ermolieva and Obersteiner 2004, Ermoliev et al. 2018, Ermoliev et al. 2000a, Ermoliev et al. 2000b, Ermoliev et al. 2001, Ermoliev and Norkin 1997, Ermolieva and Ermoliev 2005, Ermolieva et al. 2003, Ermolieva 1997, Ermoliev et al. 1998, Ermolieva et al. 1997, Ermoliev et al. 2010). Foremost, the robustness is associated with the safety criteria (See ANCOLD 1998, Atoyev et al. 2020, Bowles 2007, CETS 1985, Dams and Development 2000, Daykin et al. 1994, Embrechts et al. 2000, Ermoliev and Hordijk 2003, Ermoliev et al. 2018, Ermolieva et al. 2016, Ermolieva and Obersteiner 2004, Ermoliev et al. 2018, Ermoliev et al. 2000a, Ermoliev et al. 2000b, Ermoliev et al. 2001, Ermoliev and Norkin 1997, Ermolieva and Ermoliev 2005, Galambos 2003, 1997, Ermoliev et al. 1998, Ermolieva et al. 1997, Ermoliev et al. 2010, Ermolieva et al. 2016, Ermolieva et al. 2021, Gorbachuk et al. 2019, IAEA 1992), which deal with the Value-at-Risk considerations (See Artzner et al. 1999, Embrechts et al. 2000, Ermoliev and Hordijk 2003, Ermoliev et al. 2018, Ermolieva et al. 2016, Ermolieva and Obersteiner 2004, Ermoliev et al. 2018, Ermoliev et al. 2000a, Ermoliev et al. 2000b, Ermoliev et al. 2001, Ermoliev and Norkin 1997, Ermolieva and Ermoliev 2005, Ermolieva et al. 2003, Ermolieva 1997, Ermoliev et al. 1998, Ermolieva et al. 1997, Ermoliev et al. 2010, Rockafellar and Uryasev 2000). These type of risk indicators is a key for the regulation of low probability-high consequences risks. The introduction of safety constraints induces risk aversion among the agents and identifies a trade-off between ex-ante precautionary and ex-post adaptive measures. These concerns, in particular, the trade-off between structural and financial measures, i.e., between expenditures for dam reinforcement, values and production reallocation, insurance coverage in the region exposed to floods or a dam break events.

The balance between the precautionary and adaptive decisions depends on the financial capacities of the agents: how much they can invest now into ex-ante risk reduction measures, say, reinforcement of dams, improving building quality. Or for how much insurance can they buy; and how much they can be allowed to spend for covering losses in the case of a catastrophe occurring. As mentioned by many insurance authorities, the dam break failure may cause losses that not every government can fund, insurance or reinsurance will be able to absorb.

The future losses highly depend on currently implemented strategies. The ex-post decisions may turn to be much costlier and, obviously, these costs often come unexpectedly at the wrong time and place. The need for coexistence of both measures is dictated by the potential disastrous losses (see discussion in Section 4). This decision-making framework implies, in particular, that the capacity for adaptive ex-post decision making has to be created in an ex-ante manner. In the discussed approaches, specific attention is paid to the modeling of endogenous extreme events (scenarios) and unknown catastrophic risks, i.e., events and risks induced by new decisions for the analysis of which there is no real observations available.

The exact evaluation of all complex interdependencies is impossible and thus risk assessment will yield poor estimates. In this situation the most important task it the design of robust management strategies. This simple observation, in fact, is the basic idea of the stochastic optimization approaches proposed in Amendola, Ermolieva et al. 2013, Amendola et al. 2000, Amendola et al. 2000, Ermoliev and Hordijk 2003, Ermoliev et al. 2018, Ermolieva et al. 2016, Ermolieva and Obersteiner 2004, Ermoliev et al. 2018, Ermoliev et al. 2000a, Ermoliev et al. 2000b, Ermoliev et al. 2001, Ermoliev and Norkin 1997, Ermolieva and Ermoliev 2005, Ermolieva et al. 2003, Ermolieva 1997, Ermoliev et al. 1998, Ermolieva et al. 1997, Ermoliev et al. 2010, Gorbachuk et al. 2019, the evaluation of the optimal decisions is achieved without exact evaluation of all possible alternatives.

2.6 Multiagent aspects

The high consequences of catastrophes call for the cooperation of various agents such as governments, farmers, producers, consumers, insurers, investors, and individuals. This often leads to multi-objective stochastic optimization problems and game-theoretical models with stochastic uncertainties. The issues of ethics relate to possible disagreements between the agents and the level of their risk perception, awareness, and involvement in catastrophe management. These primarily stem from the individual unique assumptions about the facts associated with the risks, the lack of knowledge and perception of risks, and from the differences in their evaluation criteria. The disagreement in views also leads to discordance in actions.

For example, dams associate with long-term developments. Their construction is evaluated from maximization of the intergenerational utility (See ANCOLD 1998, CETS 1985, Dams and Development 2000). Different views over the benefits of dams arise when individuals rate their instantaneous goals higher than the common wealth, which immediately results into a dissent about the schedule of actions to be taken (evaluations and the respective actions to be taken). Many recognize the benefits and potential risks associated with dam breaks. Yet, it may still be unclear how the losses associated with dam breaks can be shared among the concerned agents in a fair way. As estimated by many insurance companies, the losses from major dam failures cannot be borne by insurance companies or reinsurance companies alone. There is a need for appropriate balance between risk mitigation and risk sharing structural and financial instruments involving the main concerned agents. In this case, the model provides a tool to develop proper perception based on learning from modeling and simulation.

For all these reasons models become essential for catastrophic risks management. The occurrence of various episodes (scenarios) and dependent losses in the region can be simulated on a computer in the same way as the episode may happen in reality. The stochastic optimization techniques can utilize this information for designing robust management strategies.

2.7 Safety constraints

The occurrence of a disaster for each agent is often associated with the likelihood of some processes abruptly passing individual "vital" thresholds. The design of risk management strategies therefore requires introduction and regulation of the safety constraints of the agents. For example, in insurance industry, the vital risk process is defined by flows of premiums and claims, whereas thresholds are defined by insolvency constraints (See Amendola et al. 2013, Amendola et al. 2000, Amendola et al. 2000, Compton et al. 2008, Ermoliev and Hordijk 2003, Ermoliev et al. 2018, Ermolieva et al. 2016, Ermolieva and Obersteiner 2004, Ermoliev et al. 2018, Ermoliev et al. 2000a, Ermoliev et al. 2000b, Ermoliev et al. 2001, Ermoliev and Norkin 1997, Ermolieva and Ermoliev 2005, Ermolieva et al. 2003, Ermolieva 1997, Ermoliev et al. 1998, Ermolieva et al. 1997, Ermoliev et al. 2010). A similar situation arises in the control of environmental targets, in private incomes and losses, in the design of disaster management programs (See Amendola, Ermolieva et al. 2013, Amendola et al. 2000, Amendola et al. 2000, Ermolieva et al. 2016, Ermolieva and Obersteiner 2004, Ermoliev et al. 2018, Ermoliev et al. 2000b, Ermolieva et al. 2003, Ermoliev et al. 1998). Formally, in the model it is represented as following. Assume that there is a random process R_t and the threshold is defined by a random ρ_t. Assume, R_t represents insurer's risk reserve (or "wealth" of a household). In general, it can be written for $t = 0,1,\ldots,T-1$:

$$R_{\square}^{t+1} = R_{\square}^{t} + I_{\square}^{t} + \sum_{j=1}^{m}[\pi_{\square}^{t}(q^t) - c_{\square}^{t}(q^t)] - O_{\square}^{t} - \sum_{j\in\varepsilon_t}(\omega_t)L_{\square}^{t}(\omega_t)q_{\square}^{t},$$

where $R_{\square}^{0}$ is initial risk reserve (initial wealth), $I_{\square}^{t}$ are various inflows, $O_{\square}^{t}$ – stands for expenditures, q_j^t is the coverage of a company in location j at time t, $\pi_j^t(q_{\square}^t)$ is the premium from contracts characterized by coverages $\{q_j^t\}$. Full coverages of losses correspond to $q_j^t = 1$, $q^t = \{q_j^t, j = \overline{1,m}\}$. In spatial multiagent modeling, R_t and ρ_t can be large-dimensional vectors reflecting the overall situation in different locations of a region.

Let us define the stopping time τ as the first-time moment t when R_t is below ρ_t. By introducing appropriate risk management decisions x it is often possible to affect R_t and ρ_t in order to ensure the safety constraints $P[R_t \geq \rho_t] \geq \gamma$, for some safety level γ, $t = 0,1,2,\ldots$ (similar to model in Section 3).

The use of this type of safety constraints is a rather standard approach for coping with risks in insurance, finance, and nuclear industries. For example, the safety regulations of nuclear plants assume that the violation of safety constraints may occur only once in 10^7 years, i.e., $\gamma = 1 - 10^{-7}$. It is remarkable that the use of stopping time τ has strong connections with the dynamic safety constraints and dynamic versions of static CVaR risk measures (See Ermoliev and Hordijk 2003, Ermolieva and Obersteiner 2004, Ermolieva and Ermoliev 2005, Ermolieva et al. 2003).

The ethical question about losers and winners concerns not only the economic evaluation of the benefits and costs associated with the operation of critical infrastructure (e.g., dams), it also relates to human and environmental values, which are often difficult to appraise in monetary terms. The safety constraints allow it to control the actions within admissible norms on risky installation, environmental pollution, wellbeing, historical values, and cultural preferences, in particular, restrict the growth of the wealth in risk prone areas.

Therefore, within the model ethical issues can be resolved by evaluating the overall safety coherent with spatio-temporal goals, constraints, and indicators of involved agents, whether these are households, farmers, governments, the business supplying the water, those living downstream of a large dam, or insurance covering the losses. The proper representation of safety constraints is a crucial aspect of catastrophe management problem. Catastrophic losses often have multi-mode distributions and therefore the use of standard mean value indicators (e.g., expected costs and profits) may be dramatically misleading.

2.8 Discounting

One of the fundamental ethical parameters in long term projects evaluation (e.g., critical infrastructure against floods) is the discount rate (see e.g., ANCOLD 1998, Chichilinskii 1997, Dams and Development 2000, OECD 2003, Ramsey 1928, Weitzman 1999 and references therein). In particular, the social discount rate reflects the level to which we discount the value of future generations' well-being in relation to our own. A social discount rate of 0, for example, means we value future generations' well-being equally to our own. Ramsey 1928 argued that applying a positive discount rate r to discount values across generations is unethical. Koopman (see Weitzman 1999), contrary to Ramsey, claimed that zero discount rate r would imply an unacceptably low level of current consumption.

There are several aspects of discounting in relation to critical infrastructure (e.g., dams) discussion. Traditional approaches to evaluation of infrastructure efficiency and safeness often use principles of the so-called net present value (NPV) or modified net present value (MNPV) to justify an infrastructure project. In essence, both rely on an assumption that the project is associated with an expected stream of positive or negative cash flows $V_0, V_1,..., V_T$, $V_t = Ev_t$ over a time horizon $T \leq \infty$. These flows traditionally account for engineering variables and reflect the cash flows of infrastructure owners, e.g., governments. For example, the sum may comprise of several years negative cash values reflecting the costs of construction and commissioning, followed by positive cash flow during the years of operating life without essential maintenance costs, when the infrastructure is producing positive cash flows. Finally, a period of expenditures on reconstruction and maintenance. Spatio-temporal profiles of the benefits and the potential infrastructure-induced (dam-induced) losses are not explicitly included in the evaluation. Assume that r is a constant prevailing market interest rate, then alternative dam projects are compared $V = V_0 + d_1V_1 +...+ d_TV_T$, where $d_t = d^t$, $d = (1 + r)^{-1}$, $t = 0,1,...,T$, is the discount factor and V denotes NPV. If the NPV is positive, the project has positive expected benefits and, therefore, is justifiable for implementation.

The time horizon $T \leq \infty$ and the choice of a discount rate r may dramatically affect the evaluation of the project. Diverse assumptions about the discount rate may lead to dramatically different policy recommendations and management strategies, which may provoke additional catastrophes and significantly contribute to increasing vulnerability of the society.

Behind the choice of the discount factor and the evaluation model are not really the fact but the ethical and intergenerational considerations (See Ramsey 1928). A lower discount rate emphasizes the role of distant costs and benefits, i.e., shifts the emphasis to future time periods. For example, a flat discount rate of 5–6% (traditionally used in dam projects) orients the analysis on a 20–30-year time horizon. Meanwhile, the explicit treatment of a potential 200 year disaster would require at least the discount rate of 0.5% instead of 5%. Otherwise, it can be easily illustrated that damages do not matter (they are discounted almost to 0). For the evaluation of truly long-term projects, e.g., "catastrophic" or economic developments projects, say, long-term investments into a dam or a dike system, the discount factors have to be relevant to expected horizons of potential catastrophes. The expected duration of projects evaluated with standard discount rate obtained from traditional capital markets does not exceed a few decades and, as such, these rates cannot match properly evaluations of projects oriented on 1000-, 500-, 250-, 100-year catastrophes.

Disadvantages of the standard NPV criterion are well analyzed. In particular, the NPV critically depends on some average interest rate, the application of which for evaluation of a practical project may not be easily implementable. For example, the problem that arises from the use of the expected value Er and the discount factor $(1 + Er)^{-t}$ implies additional significant reduction of future values in contrast to the real expected discount factor $E(1 + r)^{-t}$, since $E(1 + r)^{-t} >> (1 + Er)^{-t}$.

In addition, the NPV does not reveal the temporal variability of cash flow streams. Two alternative streams may easily have the same NPV despite the fact that in one of them all the cash is clustered within a few periods, but in another it is spread out evenly over time. This type of temporal heterogeneity is critically important for dealing with catastrophic losses, which occur suddenly as a "spike" in time and space.

3. Catastrophe modeling and stochastic optimization as integrated catastrophe assessment and management model

3.1 Integrated catastrophe modeling

Models for the analysis and management of catastrophic risks "link" several modules or submodels. Many authors (for instance, Baranov et al. 2002, Boyle 2002, Clark 2002, Walker 1997) consider three main modules in catastrophe modeling: A hazard module comprising an event generator; an engineering module for damage estimation; an insurance business module for the analysis of insured losses vs. insurance premiums.

In more general models, in addition to an engineering module for structural (infrastructure) damage estimation, sectorial agricultural, energy, water production and provision modules can be incorporated for calculating plausible impacts from natural disasters to food, energy, and water (FEW) production and provision systems, thereby representing the aspects of food, energy, water security. Agricultural impacts can be estimated by land use and agriculture production planning models (See Bielza Diaz-Caneja et al. 2009, Glauber 2015, Ermolieva et al. 2016, Ermolieva et al. 2021). These models incorporate the main crops and livestock production and management systems, characterized by systems-specific production costs, water and fertilizer requirements, emission factors, and other parameters. Catastrophic damages in agricultural systems can be caused by natural extreme events such as prolonged droughts, heat waves, seasonal variations in precipitation, wind storms, heavy precipitation, etc. The events affect crop and livestock yields, destroy agricultural land, diminish water infrastructure and supply, and cause production depletion, price volatility, market instability, trade bans, and tariff increases, to offset the impact of increasing world prices and cope with production shortages. In addition to exogenous shocks, endogenous policy-driven impacts in the agricultural sector can be caused by catastrophic damages (failures) in other sectors. Linkages between the agricultural and energy markets are tightened because of climate change concerns and the rapid energy sector transition towards renewable energy sources. Agricultural commodities become important energy resources because of biofuel mandates. Increasing frequency and variability of natural disasters, the vulnerability of crop yields, grain demand and price volatility influence, directly and indirectly, markets for transportation fuels and transportation costs. At the same time, crude oil, gas and electricity markets and prices have indirect effects on agricultural production costs and prices. Additional interactions between FEW systems emerge due to the uncertainties inherent to the introduction of new technologies, e.g., intermittent renewables, advanced irrigation, hydrogen production, water desalination, etc. For example, water desalination for agricultural water provision requires vast amounts of electricity. Therefore, a failure in the energy system due to an extreme event causes propagation of losses in interdependent FEW systems.

Energy impact can be estimated by energy production and supply planning models. These models include the main stages of energy flows from resources to demands: energy extraction from energy resources, primary energy conversion into secondary energy forms, transport and distribution of energy to the point of end, and conversion into products for end users to fulfill specific demands. The model can incorporate various energy resources as, e.g., coal, gas, crude oil, and renewables; primary energy sources include coal, crude oil, gas, solar, wind, etc.; secondary energy sources are fuel oil, methanol, hydrogen, electricity, ammonia, etc.; final energy products are coal, fuel oil, gas, hydrogen, ammonia, methanol, electricity, etc. Demands for useful energy products come from economic sectors: industrial, residential, transport, agricultural, water, and energy. Each technology is characterized by unit costs, efficiency, lifetime, emissions, etc. Catastrophic damages (failures) in energy systems can be induced by exogenous natural hazards such as, for example, windstorms, hurricanes, floods, and earthquakes, and by endogenous systemic failures, which can be triggered beyond the energy system. For example, intensive water pumping by farmers in water-scarce regions can cause water shortages for energy system cooling and lead to enforced energy production termination (Abrar 2016).

3.2 Catastrophic systemic risks

Interdependent agricultural, energy, and water sectors are connected through joint cross-sectoral relationships (balances) capturing, e.g., the requirements and limitations on the natural resource use and availability, and investments. Incorporation of joint food, energy, and water security constraints requires satisfaction of food, energy, and water demand by users in all potential scenarios of catastrophic shocks. In general, joint constraints impose restrictions on total production, resource use, and emissions by all sectors/regions. The constraints can establish supply-demand relationships between the systems. In addition, the balances can impose limitations (quotas) on total energy production by the energy sector (electricity, gas, diesel, etc.) and land use sector (biodiesel, methanol); total energy use by energy and agricultural sectors; total agricultural production by distributed farmers/regions, etc.

Increasing interdependencies among FEW systems involving interactions between nature, humans, and technology resemble a complex chain network connected through supply-demand relations. Disruption of such networks triggered, for example, by a natural disaster, induces systemic risks associated with critical imbalances and exceedances of vital thresholds, affecting FEW security at various levels with possible global spillovers. Risks of disruptions and failures in such systems may be unlike anything, which has been experienced in the past. These risks can be induced by human decisions in combination with natural shocks. For example, an extra load in a power grid triggered by a power plant or a transmission line failure can cause cascading failures with catastrophic systemic outages (Abrar 2016). In financial networks, an event at a company level can lead to severe instability or even to a crisis similar to the global financial crisis of 2008. A hurricane in combination with inappropriate dams maintenance and land use management can result in human and economic losses, similar to those induced by Hurricane Katrina. Another example relates to an increase of biofuels production, which affects crops and food prices, destabilizes food and water provision, and worsens environmental conditions. Systemic risks in interdependent systems can be defined as the risks of a subsystem (a part of the system) threatening the sustainable performance and achievement of FEW security goals. Thus, a shock in a peripheral subsystem induced (intentionally or unintentionally) by an endogenous or exogenous event, can trigger systemic risks propagation with impacts, i.e., instability or even a collapse, at various levels. The risks may have quite different policy-driven dependent spatial and temporal patterns. While standard risks analysis and assessment can rely on historical data, systemic cascading risks in FEW systems are implicitly defined by the whole structure and the interactions among the systems, in particular costs, production and processing technologies, prices, trade flows, risk exposure, FEW security constraints, risk measures, decisions of agents.

Prediction of systemic risks in integrated natural and anthropogenic policy-driven FEW systems is a rather tedious task. The main issue in this case is robust management of the risks, which can be achieved by equipping the systems with precautionary and adaptive strategies enabling the systems sufficient flexibility and robustness to maintain sustainable performance and fulfill joint FEW security goals independently of what systemic shock occurs (Ermolieva et al. 2016, Ermolieva et al. 2021).

Therefore, catastrophe analysis and management in interdependent food, energy, water systems require an integrated approach to understand and deal with the numerous interactions between the systems and potential catastrophic losses due to exogenous and endogenous systemic risks. This approach, compared to independent analysis, contributes immensely to sustainable development within and across sectors and scales.

Systemic losses in interdependent food, energy, water systems requires proper approaches to the design of respective insurance and loss coverage programs. Thus, for example, agricultural insurance covers mainly production risks related to weather conditions, pests and diseases, market conditions, liberalization policies, climate change. Energy sector insurance covers risks typical for energy sector from installation to operational phases, including business interruption, photovoltaic insurance, insurance for wind turbines, etc.

3.3 Catastrophe generators

The lack of historical data on catastrophic losses and the absence of analytical forms of joint distribution create a rapidly increasing demand for integrated catastrophe modeling (See Amendola et al. 2013, Amendola et al. 2000, Amendola et al. 2000, Baranov et al. 2002, Boyle 2002, Clark 2002, Ermolieva et al. 2003) incorporating natural catastrophe (hazards) models and socio-economic sectoral models. GIS-based computer catastrophe modeling attempts to simulate samples (scenarios) of mutually dependent catastrophe losses on the levels of a household, a city, a region, or a sector, from various natural hazards, e.g., floods, droughts, earthquakes, and hurricanes. These models are used as a tool for planning, emergency systems, lifeline analyses, and estimation of losses.

Most of the models involved in the analysis produce outputs that are distributional. That is, the results of catastrophe simulations are typically not simply an expected loss, but rather a set of geographically detailed loss scenarios enabling it to obtain of a multidimensional probability distribution, which may be analytically intractable and not follow a particular statistical distribution.

Natural catastrophe models have been developed for a wide range of catastrophic risks and geographic territories worldwide. All major natural hazards are modeled, including earthquakes, hurricanes, winter storms, tornadoes, hailstorms, and floods. The seismic hazard module simulates actual earthquake shaking. It uses or simulates locations and magnitudes of earthquakes. This module often comprises other physical phenomena associated with an earthquake including subsequent fires, and landslides. The movement of the seismic waves through the soil is modeled by attenuation equations. Seismic effects at a site depend on earthquake magnitude, distance from the source, and site characteristics, such as regional geology and soil types. The vulnerability module relates seismic shaking to structural and property damage. It determines the extent of damages to buildings and content at a site. In the general case, it calculates sectoral damages. The financial module assigns a cost to these damages and calculates the maximum potential and/or expected losses for either individual sites or regions. It calculates losses to structural damage, damage to property and content, agricultural losses, losses due to business interruption for example due to energy system failure, etc. This module includes data on building locations, building types, building contents, power sector values at risk, agricultural properties, etc.

The development of insurance instruments/products and programs can be an important risk mitigation strategy for power generation and agricultural activities (See Bielza Diaz-Caneja et al. 2009, Glauber 2015). Loss estimates can be presented, e.g., either in percentage terms (of the total value) or as a monetary value.

From Monte Carlo simulation of natural catastrophe scenarios, histograms of aggregate losses for a single location, a particular catastrophe zone, a sector, a country or worldwide can be derived from the catastrophe modeling. But catastrophe modeling has only marginal benefits when it is used in a traditional scenario-by-scenario manner for obtaining estimates of aggregate losses.

Although catastrophe modeling is considered to be a decision-making tool by many researchers and insurance companies, however, the decision variables are not explicitly incorporated in the existing catastrophe models. Following Amendola, Ermolieva et al. 2013, Amendola et al. 2000, Amendola et al. 2000, Baranov et al. 2002, Ermoliev et al. 2018, Ermolieva et al. 2016, Ermolieva and Obersteiner 2004, Ermoliev et al. 2018, Ermoliev et al. 2000a, Ermoliev et al. 2000b, Ermoliev et al. 2001, Ermoliev and Norkin 1997, Ermolieva and Ermoliev 2005, Ermolieva et al. 2003, Ermolieva 1997, Ermoliev et al. 1998, Ermolieva et al. 1997, Ermoliev et al. 2010, we can admit that the currently existing form of catastrophe modeling can only be a necessary element/subset of more advanced extensive decision-support models enabling simultaneous integrated analysis of the catastrophic risk portfolios/scenarios and the optimization of catastrophe management policies within the same modeling framework. The explicit introduction of decisions in such models opens up a possibility for integrated catastrophic risks management based on the contribution of individual risks to aggregate losses.

3.3.1 Example: A catastrophe flood model

Let us consider an example of a catastrophe model developed for the analysis of flood losses in case studies (see Ermoliev et al. 2018, Ermolieva et al. 2016, Ermoliev et al. 2018, Ermolieva et al. 2003 and references therein), i.e., a catastrophe flood model. The aim of the model is to generate potential samples of mutually dependent losses for a given vector of policy variables. For example, when there is a lack of historical data, models can estimate distributions of losses and gains for different locations, households, farmers, energy providers, water authorities, insurers, and governments. This is critically important in the case of rare events or new policies that have never been implemented in practice.

As is shown in Figure 1, the model for catastrophe flood analysis links, in general, five submodels (modules): the "River" module, the "Inundation" module, the "Vulnerability" module, the "Multi-Agent Accounting System", and the "Variability" module.

The "River" module calculates the volume of discharged water to the pilot region from different river sections for given heights of dikes, given scenarios of their failures or removals, and rains. The latter are modeled by upstream discharge curves. Thus, formally, the "River" module maps an upstream discharge curve into the volume of water released to the region from various sections. The underlying submodel is able to estimate the discharged volume of the water into the region under different conditions, for example, if the rain patterns change, if the dikes are heightened, or if they are strengthened or removed.

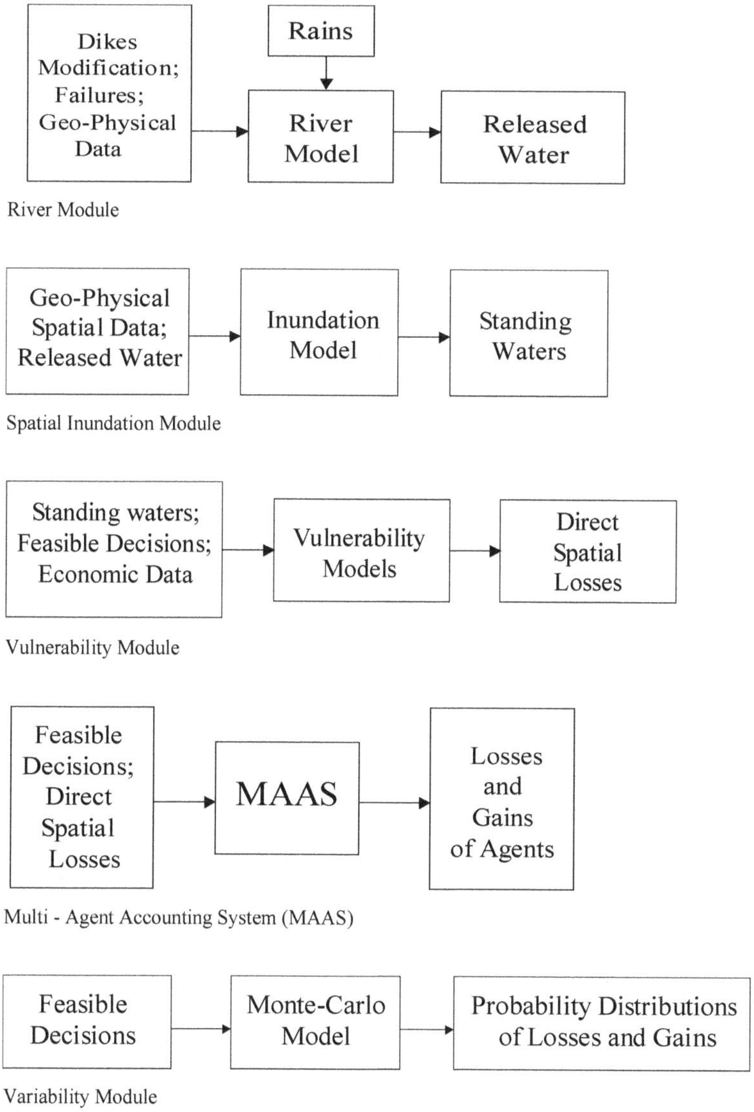

Figure 1. Modules of the Catastrophe Model.

The next module is the spatial GIS-based "Inundation" submodel. This module maps water released from the river into levels of standing water at fine resolutions (of, e.g., 10 sq. m) and thus it can estimate the area of the region affected by different decisions.

The "Vulnerability" module maps spatial patterns of released water into economic losses. This module calculates direct losses and may include possible cascading effects, such as floods causing fire and its consequences. It may also include loss reduction measures, e.g., new land-use modifications and flood preparedness measures. This module is able to indicate changes in economic losses from changes in risk reduction measures.

The "Multi-Agent Accounting System" module maps spatial economic losses into losses and gains of the involved agents. These agents are central and/or local governments, catastrophe insurance funds, investors, and "individuals" (cells) representing households, farmers, producers, consumers, etc.

Given sufficient data, the above-mentioned submodels can generate scenarios of losses and gains at different locations for specific scenarios of rains, dam failures, risk reduction measures and risk-spreading schemes. But there are significant uncertainties and considerable variability in the simulated losses and gains. A 50-year flood may occur in 5 days or in 70 years, which may induce considerably different losses depending on regional economic growth scenarios. Governments, mutual catastrophe funds, and insurers are especially concerned about the variability since they may not have the capacity to cover large losses. To maintain their solvency, they may charge higher premiums and taxes, which may result in overpayments by the insured (households, farmers, producers, consumers, etc.). Alternatively, they may undercharge contracts.

There may be alternative opinions about loss-reduction measures. A higher dike may fail and cause more damage in comparison to a dike without modification. The "Variability" module, a Monte Carlo model, transforms spatial scenarios of losses and gains among agents into histograms of probability distributions. For example, it derives histograms of direct losses at a location or a subregion. It also calculates histograms of overpayments and underpayments for different agents.

3.4 Linking stochastic optimization and catastrophe models into integrated assessment and catastrophe management framework

3.4.1 Adaptive Monte Carlo procedures

Catastrophe modeling discussed in previous sections opens up the possibility for "if-then" scenario analyses, which allows the evaluation of a finite number of policy alternatives. These analyses may run quickly into an infinite number of possible combinations. For example, an insurer in the region can have different policies regarding the extent of coverage that it offers, say 0%, 10%, 20%,..., 100%, i.e., altogether 11 alternatives. For 10 locations the number of possible combinations is 10^{11}. With 100 locations, the straightforward "if-then" analysis runs into an enormous number of scenarios leading to the "curse of dimensionality". The same computational complexity arises in dealing with location-specific land use management decisions, flood protection measures, dikes, channels, improvement of building codes, premiums, investments in different critical infrastructures, etc.

The fundamental question concerns the evaluation of a desirable policy without the evaluation of all the options. The complexity of this task is due to the analytical intractability of stochastic catastrophe models, often precluding the use of standard optimization methods, e.g., genetic algorithms. Therefore, in general cases we have to rely on the stochastic optimization methods Ermoliev and Hordijk 2003, Ermoliev et al. 2018, Ermolieva et al. 2016, Ermolieva and Obersteiner 2004, Ermoliev et al. 2018, Ermoliev et al. 2000a, Ermoliev et al. 2000b, Ermoliev et al. 2001, Ermoliev and Norkin 1997, Ermolieva and Ermoliev 2005, Ermolieva et al. 2003, Ermolieva 1997, Ermoliev et al. 1998, in particular, on the so-called Adaptive Monte Carlo Optimization. "Adaptive Monte Carlo" (See Ermolieva 1997, Ermoliev et al. 1998, Ermolieva et al. 1997, Pugh 1966) means a technique that makes on-line use of sampling information to sequentially improve the efficiency of the sampling itself.

We use "Adaptive Monte Carlo Optimization" in a rather broad sense, i.e., the efficiency of the sampling procedure is considered as a part of more general improvements with respect to different decision and goals. The "Adaptive Monte Carlo Optimization" model consists of three main interacting blocks: "Feasible Decisions", the "Monte Carlo Catastrophe Model", and "Indicators" (Figure 2).

For flood management, the block "Feasible Decisions" represents all feasible policies for coping with floods. They may include feasible heights of dikes, capacity of reservoirs, water channels, insurance coverage, land use modifications, etc. These variables affect performance indicators such as losses, premiums of funds and insurers, underpayments or overpayments by the insured, costs, insolvency, and stability indicators.

The essential feature of the "Adaptive Monte Carlo Optimization" model is the feedback mechanism updating decisions towards specific goals and the improvement of insolvency and stability indicators. The updating procedure

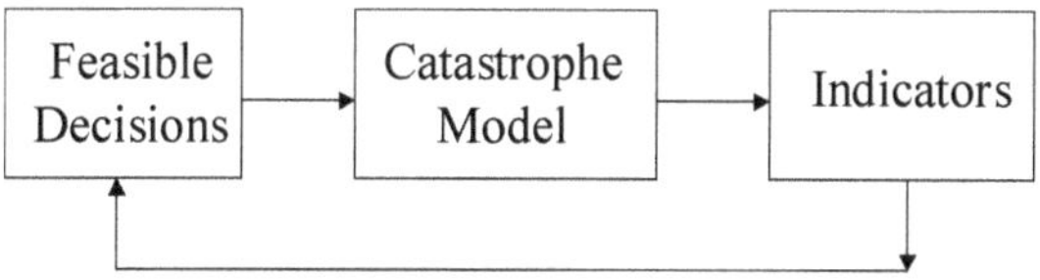

Figure 2. The Adaptive Monte Carlo Optimization Model.

relies on Stochastic Quasigradient (SQG) optimization techniques as discussed in Section 3.4.2. Losses are simulated by the catastrophe model, causing an iterative revision of the decision variables after each simulation run. In a sense, the Adaptive Monte Carlo optimization simulates in a remarkably simple and evolutionary manner the learning and adaptation process on the basis of the simulated reversible history of catastrophic events. This technique is unavoidable when the outcomes of the catastrophe model do not have a well-defined analytical structure.

3.4.2 General stochastic optimization model

Stochastic optimization provides a framework for the iterative revision of decisions embedded in the catastrophe model. These decisions influence the contribution of location-specific risks on the overall catastrophe losses. Let us outline the approach developed for a number of catastrophe risks: seismic, catastrophe floods, hurricanes, wind storms, livestock epidemics.

The main idea is based on subdividing the study region into grid cells or homogenous simulation units (homogenous response units) $j = 1,2,\dots, m$. These units may correspond to a collection of households at a certain site, a collection of grids (zones) with a similar land-use structure, or an administrative district, or a grid with a segment of a gas pipeline or energy production facility. The choice of cell representation and resolution provides a desirable representation of losses. In our case, the cells consist of the value of the physical structures. Catastrophes, which are simulated by the catastrophe model, affect at random different cells and produce mutually dependent at time t losses L_j^t. These losses can be modified by various decision variables. Some of the decisions reduce losses, say a dike, whereas others spread them on a regional, national, and international level, e.g., insurance contracts, catastrophe securities, credits, and financial aid. If $x = (x_1, x_2,\dots, x_n)$ is the vector of the decision variables, then losses L_j^t to a cell j at time t are transformed into $L_j^t(x)$. For example, we can think of $L_j^t(x)$ as L_j^t being affected by the decisions of the insurance to cover losses from a layer $[x_{j1}, x_{j2}]$ at a cell j in the case of a flood disaster at time t:

$$L_j^t(x) = L_j^t - max\{x_{j1}, min[x_{j2}, L_j^t]\} + x_{j1} + \pi_j^t, \tag{1}$$

where $max\{x_{j1}, min[x_{j2}, L_j^t]\} - x_{j1}$ are retained by insurance losses, and π_j^t is a premium function. The variable x_{j1} defines the deductible part ("trigger") of the contract and x_{j2} defines its "cap".

In the most general of cases, vector x comprises decision variables of different agents, including governmental decisions, such as the height of a new dike or a public compensation scheme defined by a fraction of total losses $\sum_{j=1}^m L_j^t$. The insurance decisions concern premiums paid by individuals and the payments of claims in the case of catastrophe. There are complex interdependencies among these decisions, which call for the cooperation of agents. For example, the partial compensation of catastrophe losses by the government enforces decisions on loss reductions by individuals and, hence, increases the insurability of risks, and helps the insurance to avoid insolvency. On the other hand, the insurance combined with individual and governmental risk-reduction measures can reduce losses, compensations and government debt and stabilize the economic growth of the region and the wealth of individuals.

3.4.3 Example: Design of a robust insurance program for catastrophe flood loss coverage

Let us now consider an example of designing a potential insurance system (a mutual catastrophe fund) and introduce some important indicators. In the following we do not, for the simplicity of notation, consider the most general situation, e.g., we consider only the proportional compensation by the government, proportional insurance coverages, and we do not use discount factors.

In this application the system is modeled until a first catastrophic flood, which occurs within a given time horizon. We define this moment as the stopping time. This event can be associated with, e.g., a dike break that may occur only after a 100-year, 150- or 1000-year flood. They are characterized by upstream discharge curves and the probability of breaking each of the dikes. The timing of the first catastrophic flood significantly affects the accumulation of risk reserves by the insurance, and total payments of individuals; for example, a 100-year flood with the break of a dike may occur in two years.

Let τ be a random (stopping) time to a first catastrophe within a time interval $[0, T]$, where T is some planning horizon, say, of 10 or 50 years. If no catastrophe occurs, then $\tau = T$. Since τ is associated with the break of a dike, the probability distribution of τ is, in general, affected by some components of vector x, e.g., by decisions on dike modifications, the land use changes, building reservoirs, etc. In this paper we discuss only the case when τ does not depend on x.

Let L_j^τ be random losses at location j at time $t = \tau$. In our analysis we evaluate the capacity of the catastrophe insurance in the upper Tisza region only with respect to financial loss-spreading decisions. Let us use a special notation for their components such as π_j, ϕ_j, v, q, y. If π_j is the premium rate paid by location j to the mandatory insurance, then the accumulated mutual catastrophe fund at time τ together with the proportional compensation $v \sum_j L_j^\tau$ by the government is equal to $\tau \sum_j \pi_j + v \sum_j L_j^\tau - \sum_j \phi_j L_j^\tau$, where $0 \le \phi_j \le 1$, is the insurance coverage for cell j. Thus, in this model we assume that the compensation to victims by the government is paid through the mandatory insurance.

The stability of the insurance program depends on whether the accumulated mutual fund together with the governmental compensation is able to cover claims, i.e., on the probability of event:

$$e_1 = \tau \sum_j \pi_j + v \sum_j L_j^\tau - \sum_j \phi_j L_j^\tau \geq 0. \quad (2)$$

The stability also depends on the willingness of individuals to accept premiums, i.e., with the probability of overpayments:

$$e_2 = \tau\pi_j - \phi_j L_j^\tau \geq 0, j = 1,..., m. \quad (3)$$

Apart from the compensation $v\sum_j L_j^\tau(x)$ the government arranges a contingent credit y with a fee q to improve the stability of the mandatory insurance by transforming event (2) into (4):

$$e_3 = \tau \sum_j \pi_j + v \sum_j L_j^\tau - \sum_j \phi_j L_j^\tau + y - \tau q y \geq 0. \quad (4)$$

Here we assume that the mandatory insurance pays the fee $\tau q y$ and receives the credit y, whereas the government pays back the credit with the interest rate γy, $\gamma > 1$.

The difference between compensation $v \sum_j L_j^\tau$ and contingent credit y is significant: the outflow of fees is smooth, whereas the compensation of claims has a sudden impact at time τ, and without y it may require a higher government compensation (greater v) possibly exceeding the available budget. Therefore, without ex-ante contingent injections of capital y the diversion of capital from other governmental needs may occur.

Let us note that the budget constraint, which is not considered explicitly in this model, raises a general question on the optimal dynamic management of the available budget in order to increase the stability of the mandatory insurance and its efficiency. For example, besides the contingent credit, a reasonable option may also be to invest some money in liquid assets. The main aim of our analysis is narrower: the evaluation of the mandatory insurance capacity and the demand for contingent credit.

Inequalities (3)–(4) define important events, which constrain the choice of the decision variables specifying the insurance program, i.e., the compensation rate v by the government, coverages by the insurance company ϕ_j, premiums π_j, and credit y with fee q. The likelihood of events (3)–(4) and values e_2, e_3 determine the stability (resilience) of the program. In a rough way this can be expressed in terms of the probabilistic constraint

$$P[e_2 > 0, e_3 < 0] \leq p, \quad (5)$$

where p is a desirable probability of the program's default, say a default that occurs only once in 100 years.

Constraint (5) is similar to the so-called insolvency constraint (See Amendola et al. 2013, Amendola et al. 2000, Amendola et al. 2000, Ermoliev and Hordijk 2003, Ermoliev et al. 2018, Ermolieva et al. 2016, Ermolieva and Obersteiner 2004, Ermoliev et al. 2018, Ermoliev et al. 2000a, Ermoliev et al. 2000b, Ermoliev et al. 2001, Ermoliev and Norkin 1997, Ermolieva and Ermoliev 2005, Ermolieva et al. 2003, Ermolieva 1997, Ermoliev et al. 1998, Ermolieva et al. 1997, Ermoliev et al. 2010), a standard for regulations of the insurance business. In the stochastic optimization constraint (5) is known as the so-called chance constraint. Unfortunately, this constraint does not account for the values e_2, e_3, what is important for the government, since it cannot walk away from the region in a distress. The main goal can now be formulated as the minimization of expected total losses $F(x) = E\sum_j(1-\phi_j)\, L_j^\tau + \gamma y$ including uncovered (uninsured) losses by the insurance contracts and the cost of credit γy, subject to the chance constraint (5), where vector x consists of the components π_j, ϕ_j, y.

Constraint (5) imposes significant methodological challenges even in cases when $\tau(x)$ does not depend on x and events (3)–(4) are defined by linear functions of decision variables. This constraint is of "black-and-white" character, i.e., it accounts only for a violation of (3)–(4) and not for its size. There are important connections between the minimization of $F(x)$ subject to highly non-linear and possibly discontinuous chance constraints (5) and the minimization of convex functions, which have important economic interpretation. Consider the following function

$$G(x) = F(x) + \alpha E\, max\left\{0, \sum_j \phi_j L_j^\tau - v\sum_j L_j^\tau - \tau\sum_j \pi_j - y + \tau q y\right\} + \beta E \Sigma_j\, max\left\{0, \tau\pi_j - \phi_j L_j^\tau\right\}, \tag{6}$$

where α, β are positive parameters.

It is possible to show (see, e.g., results in Ermoliev et al. 2000a, Ermoliev et al. 2000b, Ermoliev et al. 2001, Ermoliev and Norkin 1997) that for large enough α, β a minimization of function $G(x)$ generates solutions x with $F(x)$ approaching the minimum of $F(x)$ subject to (5) for any given level p.

The minimization of $G(x)$ defined by (6) has a simple economic interpretation. Function $F(x)$ comprises expected direct losses associated with the insurance program. The second term includes the expected shortfall of the program to fulfill the obligations; it can be viewed as the expected amount of needed for this purpose ex-post borrowing with a fee α. Similarly, the third term can be interpreted as the expected ex-post borrowing with a fee β needed to compensate overpayments. Obviously that large enough fees α, β will tend to preclude the violation of (3)–(4). Thus, the ex-post borrowing with large enough fees allows for a control of the insolvency constraints (5). It is easy to see that the use of the ex-post borrowing (expected shortfall) in the second term of $G(x)$ in combination with the optimal ex-ante contingent credit y controls the CVaR type risk measures. Indeed, the minimization of $G(x)$ is an example of stochastic minimax problems (see Ermoliev and Wets 1988, Chapter 22).

By using standard optimality conditions for these problems we can derive the optimality conditions for the contingent credit y. For example, assuming continuous differentiability of $G(x)$ which follows in particular from the continuity of underlying probability distributions, it is easy to see that the optimal level of the credit $y > 0$ satisfies the equation

$$\frac{\partial G}{\partial y} = \gamma - \alpha P\left[\sum_j \phi_j L_j^\tau - v\sum_j L_j^\tau - \tau\sum_j \pi_j > y\right] = 0. \tag{7}$$

Thus, the optimal amount of the contingent credit is defined as a quantile of the random variable $\sum_j \phi_j L_j^\tau - v \sum_j L_j^\tau - \tau \sum_j \pi_j$ specified by the ratio γ/α, which has to be not greater than 1. Hence, the expectation in the second term of $G(x)$ for optimal y is taken under the condition that y is the quantile of $\sum_j \phi_j L_j^\tau - v \sum_j L_j^\tau - \tau \sum_j \pi_j$. This is in accordance with the definition of CVaR. More general risk measures emerge from the optimality conditions of $G(x)$ with respect to premiums π_j, ϕ_j.

The importance of such an economically sound risk measure as expected shortfall was emphasized by many authors (see Artzner et al. 1999, Embrechts et al. 2000, Ermoliev and Hordijk 2003, Ermolieva et al. 2016, Ermoliev et al. 2018, Ermoliev et al. 2000a, Ermoliev et al. 2000b, Ermoliev et al. 2001, Rockafellar and Uryasev 2000, Yang 2000). Important connections of CVaR with the linear programs were discussed in Ermolieva et al. 2016, Ermoliev et al. 2018, Ermoliev et al. 2000a, Ermoliev et al. 2000b, Ermoliev et al. 2001, Rockafellar and Uryasev 2000. Let us note that $G(x)$ is a convex function in the case when τ and L_j^τ do not depend on x. In this case the stochastic minimax problem (6) can be approximately solved by linear programming methods (see general discussion in Ermolieva 1997). The main challenge is concerned with the case when τ and L^τ are implicit functions of x. Then we can only use the Adaptive Monte Carlo optimization. Let us outline only the main idea of these techniques. More details and further references can be found in Amendola et al. 2013, Amendola et al. 2000, Amendola et al. 2000, Ermoliev and Hordijk 2003, Ermoliev et al. 2018, Ermolieva et al. 2016, Ermolieva and Obersteiner 2004, Ermoliev et al. 2018, Ermoliev et al. 2000a, Ermoliev et al. 2000b, Ermoliev et al. 2001, Ermoliev and Norkin 1997, Ermolieva and Ermoliev 2005, Ermolieva et al. 2003, Ermolieva 1997.

3.4.4 Adaptive Monte Carlo optimization

Assume that vector x incorporates not only risk management decision variables but also includes components affecting the efficiency of the sampling itself (for more detail see Ermolieva 1997, Pugh 1966). An adaptive Monte Carlo procedure searching for a solution minimizing $G(x)$ of type (6) starts at any reasonable guess x^0. It updates the solution sequentially at steps $k = 0,1,...,$ by the rule $x^{k+1} = x^k - \rho_k \xi^k$, where numbers $\rho_k > 0$ are predetermined step-sizes satisfying the condition $\sum_{k=0}^{\infty} \rho_k = \infty$, $\sum_{k=0}^{\infty} \rho_k^2 = \infty$. For example, the specification $\rho_k = 1/k + 1$ would suit. Random vector ξ^k is an estimate of the gradient $G_x(x)$ or its analogs for nonsmooth function $G(x)$. This vector is easily computed from random observations of $G(x)$. For example, let G^k be a random observation of $G(x)$ at $x = x^k$ and $\widetilde{G}^k$ be a random observation of $G(x)$ at $x = x^k + \delta_k h^k$. The numbers δ_k are positive, $\delta_k \rightarrow 0$, $k \rightarrow \infty$, and h^k is an independent observation of the vector h with independent and uniformly distributed on $[-1, 1]$ components. Then ξ^k can be chosen as $\xi^k = [(\widetilde{G}^k - G^k)/\delta_k]h^k$. The formal analysis if this method, in particular, for discontinuous goal functions, is based on general ideas of the stochastic quasigradient (SQG) methods (see Ermoliev and Wets 1988 and further references in Amendola et al. 2013, Amendola et al. 2000, Amendola et al. 2000, Ermoliev and Hordijk 2003, Ermoliev et al. 2018, Ermolieva et al. 2016, Ermolieva and Obersteiner 2004, Ermoliev et al. 2018, Ermoliev et al. 2000a, Ermoliev et al. 2000b, Ermoliev et al. 2001, Ermoliev and Norkin 1997, Ermolieva and Ermoliev 2005, Ermolieva et al. 2003, Ermolieva 1997, Ermoliev et al. 1998, Ermolieva et al. 1997, Ermoliev et al. 2010).

4. The coexistence of ex-ante and ex-post decisions

The stochastic optimization model outlined in Section 3 includes long-term ex-ante (precautionary anticipative) and short-term ex-post (adaptive) decisions. The need for the coexistence of such decisions is especially evident for management of extreme situations and risks. Prediction of spatio-temporal patters of the risks is not possible. The main issue in this case is robust management of the risks, which can be achieved by equipping various involved agents and systems with precautionary and adaptive strategies enabling them sufficient flexibility and robustness to maintain safe and sustainable performance independently of what catastrophe scenario occurs. In the presence of uncertainties about catastrophe scenario, it is natural to employ two main types of strategies: the ex-ante strategic precautionary actions (engineering design, insurance contracts, mutual catastrophe funds, resource allocation, technological investments, water and grain reserves) and the ex-post adaptive adjustments (subsidies, prices, costs, ex-post borrowing) that are made after the occurrence of the scenario. The precautionary and adaptive measures can lessen chances of exceedances of vital thresholds and safety constraints, which could otherwise lead to systemic failures and the lack of security. A portfolio of robust interdependent ex-ante and ex-post strategies can be designed by using a two-stage stochastic optimization (STO) approach incorporating both types of decisions.

An ex-ante decision corresponds to risk-averse behavior and an ex-post decision models risk-prone behavior. The coexistence of these two types of behavior can be viewed as a rather flexible decision-making framework when we commit ourselves ex-ante only to a part of possible decisions and, at the same time, keep other options open until more information becomes available and can be effectively utilized by appropriate ex-post decisions. This type of model, the so-called two-stage stochastic optimization models, produces strong risk aversion even for linear utility functions. Consider a simple situation. Let us assume that there are two options to deal with catastrophe: to protect values against losses before the occurrence of a catastrophe or to borrow after its occurrence. We assume that the decision maker has a risk-neutral, linear disutility function $f(x, y) = cx + dy$, where c is the marginal cost for protection x, and $d(L)$ is the marginal cost for borrowing y in the case of losses L. The decision-making process has two stages. First of all, decision x is chosen before the observation of L. Decision y is chosen after the occurrence of the catastrophe, i.e., the observation of L. Therefore, we have (because of the linear disutility) $y = max\{0, L - x\}$. Thus, the decision x minimizes, in fact, the function $F(x) = cx + Ed(L)\, max\{0, L - x\} = cx + \int_x^{\bar{L}} d(l)(l - x)\phi(l)dl$. Assume that the probability distribution of L, $L \leq \bar{L}$, has a continuous density function ϕ. Then $F(x)$ is a strictly concave continuously differentiable function and, as it is easy to verify,

$$F'(x) = c - \int_x^{\bar{L}} d(l)\phi(l)dl.$$

The function $F'(x)$ is monotonically increasing for $x \to \bar{L}$. Therefore, if $c < Ed$, then there is a positive value $x = x^*$, $x^* \neq \bar{L}$, such that $F'(x^*) = 0$. Here $x^* = \bar{L}$ is excluded because $c > 0$. Hence, the minimization of the linear expected disutility function does not lead to the dominance of the preferable on an average ex-ante decision ($c < Eg$): both types of decisions coexist.

5. Indirect effects

Apart from the visible direct damage, catastrophes produce long-term indirect systemic effects. The cost of a damaged bridge may be incomparable to the cost of interrupted activities. A large catastrophic loss may absorb domestic savings and force the government into debt. The low-income countries lack the budgetary resources that would enable them to undertake the necessary growth adjustments. In Sections 2 and 3 it was discussed that the welfare growth (economic growth) can be represented by the growth in each location (cell). Inflows (investments, returns, etc.) are channeled through terms I_j^t, whereas outflows (expenditures, costs, etc.) are affected by L_j^t (in general, losses depend on decisions x). The trade-off between these forces defines the growth path of the location j, its possible stagnation and the contribution to the wealth of the region. Let us consider this issue by using a stylized model of sustainable economic growth. This model illustrates the importance of stochastic versus deterministic approaches to the omic growth.

5.1 Economic growth under shocks: stochastic vs deterministic model

Let us assume that the wealth (output) of the economy is defined by two factors: "capital" and "labor", and constant returns to scale. Then the output can be characterized in terms of capital-to-labor ratio k, and output-to-labor ratio y, $y = f(k)$, where $f(k)$ is the production function. Assume that output is subdivided into consumption and savings, and savings are equal to investments. If investments are simply a fraction s, $0 < s < 1$ of the output, and productivity of capital $f(k)/k$ is constant θ, then this leads to the very influential in growth planning Harrod-Domar model Ray 1998 with an exponential growth rate defined by the linear function

$$ln\, y = ln\, y_0 + (s\theta - \gamma - \delta)t,\ t > 0, \tag{8}$$

where γ is an exponential population growth rate, and δ is the capital depreciation rate. Shocks occur at some random time moments T_1, T_2,... and transform the linear function (8) into a highly nonlinear jumping random function (Figure 3)

$$ln\, y(t) = y_0 + (s\theta - y - \delta)t - L_1 - L_2 - ... - L_{N(t)}, \tag{9}$$

where $N(t)$ is the (random) number of shocks in the interval $[0, t]$. Let us ignore for a moment that shocks L_1, L_2,... may depend on the state of the economy, which is essentially important in the case of catastrophes. In our economy, it is $y(t)$. Let us also assume that random sizes of shocks L_1, L_2,... are independent, and identically distributed with a mathematical expectation μ. They are also independent of the inter-shock times $T_{k+1} - T_k$, and the inter-shock times have a stationary distribution with mathematical expectation λ. Then the expected exponential growth is still characterized by the linear in t function (Figure 3)

$$E\, ln\, y(t) = y_0 + (s\theta - \gamma - \delta - \lambda\mu)t. \tag{10}$$

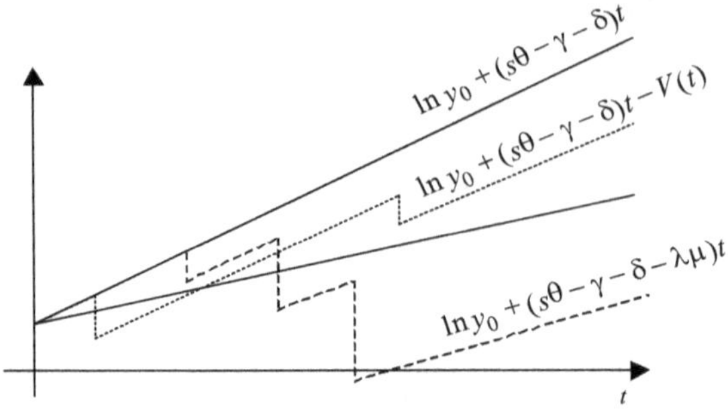

Figure 3. Expected and real growth rates.

From the strong law of large numbers, it follows that $ln\ y(t)/t$ approaches $s\theta - \gamma - \delta - \lambda\mu$ for large enough t and for each random path $y(t)$. In other words, the sustainable exponential growth "takes off" only after a long random time $T = T(y)$ specific for each path: $ln\ y(t) \approx y_0 + (s\theta - \gamma - \delta - \lambda\mu)t$, for $t > T$. On the way to sustained growth for $t < T$ the economy may stagnate as Figure 3 shows. The ignorance of risk, in this case, is equivalent to the substitution of the complex jumping process $ln\ y(t)$ by the linear function (10). This function still shows exponential growth, although with a large depreciation rate $\delta + \lambda\mu$, but it ignores possible stagnation of the economy in the interval $[0, T]$. The uncertainty analysis, which is usually recommended after such substitutions, can not reveal the possibility of stagnation either, since in our case it is equivalent to the turning around of the same linear function (10).

The behavior of $ln\ y(t)$ is radically changed under endogenously generated shocks. In fact, not only shocks $L_1, L_2, ...$ but also all parameters are affected by the growth $y(t)$. The saving rate s may critically depend on the overall level y and its distribution in the economy. Obviously, at low-income levels, saving rates are small. In this case, a shock may further reduce them even to negative values (borrowing). Let us make a rather optimistic assumption that only shocks depend on y, and shocks do not modify the production function of the economy. A shock at time t reduces current output y to a level $\tilde{y}$, $0 < \tilde{y} \leq y$. The probability distribution of $\tilde{y}$ is concentrated in the interval $[0, y]$. Assume that it has a density function Ψ, i.e., the probability of losses in the interval $[h, h + dh]$ is $\Psi(y, h)dh$. Then the expected losses μ depend on y and has the form $\mu(y) = \int_0^y h\Psi(y, h)dh$. It is easy to see that the second derivative $\mu''(y)$ involves the derivative $d\Psi(y, y)/dy$, which may be positive or negative at different levels of y. Therefore $\mu''(y)$ may also have an oscillating character, as is shown in Figure 4.

From (9) follows that for the shocks independent of the inter-shock times, the change of a given $y(t)$ to a random value $y(t + \Delta t)$ in the interval $[t, t + \Delta t]$

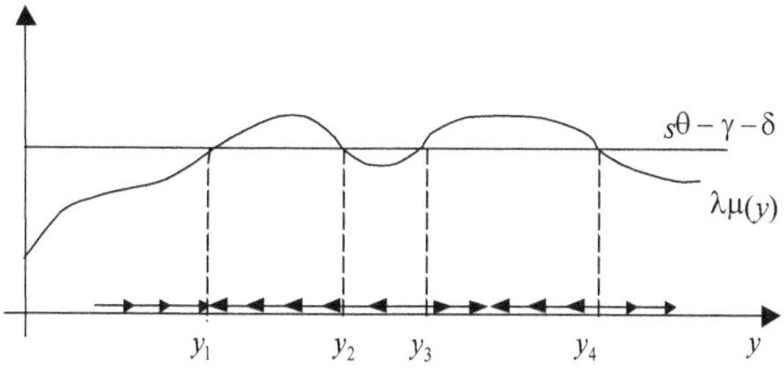

Figure 4. Traps and thresholds.

is such that the conditional expectation $E\{y(t+\Delta t)|y(t)\}$ of $y(t+\Delta t)$ for fixed $y(t)$ changes approximately as follows

$$\frac{E[y(t+\Delta t)\,|\,y(t)]-y(t)}{\Delta t}\approx y(t)[s\theta-\gamma-\delta-\lambda\mu(y(t))].$$

As we can see from Figure 3, the value $s\theta-\gamma-\delta-\lambda\mu(y)$ is positive in the interval $(0, y_1)$. Therefore, the economy is expected to grow in $(0, y_1)$. If the economy is in interval (y_1, y_2), the value $y(t+\Delta t)$ is expected to decrease towards y_1. The state y_1 is expected to be a trap, i.e., the economy stagnates around y_1. If the economy is pushed up to cross threshold y_2, it enters a path of sustained growth until it reaches the next trap. These conclusions concern only expectations. Nevertheless, it is possible to show that in our economy every random path $y(t)$ converges to a trap with probability 1. In other words, starting from the same initial conditions, the economy may end up (without assistance to growth) at different traps and stagnate within these traps thereafter. This simple model illustrates that assistance to growth is very important in such intervals as (y_1, y_2), (y_3, y_4). From the formal point of view, the assistance is equivalent to the increasing of $s\theta-\gamma-\delta$ or decreasing $\lambda\mu(y)$.

6. Concluding remarks

In this chapter, we argue that integrated modeling linking catastrophe models and stochastic optimization procedures is necessary for the analysis and management of exogenous (natural) disasters and anthropogenic (policy-driven) systemic risks emerging due to interdependencies between economic and natural systems. The existing catastrophe modeling involves three main modules: A hazard module, a module for loss calculation, and a module for loss coverage and spread between various involved agents such as insurers, governments, investors, and individuals. The catastrophe model is a Monte Carlo simulation model producing a scenario-by-scenario analysis and assessment of alternative catastrophe scenarios and respective feasible catastrophe management strategies. This analysis may result in an infinite number of scenario combinations: hazard loss and respective management decision(s), without providing robust management options (decisions), which leave in better-off conditions independently of what catastrophe/hazard scenario occurs. For the design of robust decisions, the linkage of catastrophe models and STO procedures is necessary. Section 3 sketches out an appropriate decision-making model, which is a multiagent (multicriteria), spatial and dynamic stochastic optimization model linked with a disaster simulation model. The model has been used in a number of case studies.

The Integrated Catastrophe Simulation and Stochastic Optimization model outlined in Section 3 can be used for risk management with the so-called rolling horizon. The model includes a time horizon τ, which may be a random variable, say the time of the first catastrophe from the given initial state at $t = 0$. This generates a sequence of decisions for $t = 0,1,\ldots$. After implementing decisions at $t = 0$, new information on catastrophe occurrence, losses, constraints satisfaction, etc. becomes available. This new information can trigger the revision of decisions if, e.g., safety constraints are no longer fulfilled or losses exceed some critical level. At the next modeling time period $t = 1$ the model is updated, a new sequence of decisions for $t = 1,2,\ldots$ with a new time horizon is obtained, decisions for $t = 1$ are implemented, and so on. Several data-intensive numerical experiments with the type of models outlined in this paper are discussed in Amendola et al. 2013, Amendola et al. 2000, Amendola et al. 2000, Compton et al. 2008, Ermolieva et al. 2016, Ermolieva and Obersteiner 2004, Ermolieva et al. 2003. These case studies used data on earthquakes from Russia and Italy and data on floods from Hungary, Ukraine, the Netherlands, and Austria.

Acknowledgements

The development of the presented methodologies and models was supported by a joint project between the International Institute for Applied Systems Analysis (IIASA) and National Academy of Sciences of Ukraine (NASU) on “Integrated robust modeling and management of food-energy-water-land use nexus for sustainable development”. The work has received partial support from the Ukrainian National Fund for Strategic Research, grant No. 2020.02/0121, and project CPEA-LT-2016/10003 jointly with Norwegian University for Science and technology. The paper contributes to EU PARATUS (CL3-2021-DRS-01-03, SEP-210784020) project on “Promoting disaster preparedness and resilience by co-developing stakeholder support tools for managing systemic risk of compounding disasters”.

References

Abrar, M. 2016. Power cut off and power blackout in India a major threat—An overview. Int. Journal of Advancements in Research and Technology, 5(7): 8–15.

Ackerman, F., S.J. DeCanio, R.B. Howarth and K. Sheeran. 2009. Limitations of integrated assessment models of climate change. J. Clim. Change, 95(3–4): 297–315.

Amendola, A., T. Ermolieva, J. Linnerooth-Bayer and R. Mechler. 2013. Integrated Catastrophe Risk Modeling: Supporting Policy Processes; Dordrecht: Springer Verlag, Dordrecht, Germany. ISBN 978-94-007-2225-5.

Amendola, Y., T. Ermoliev, V. Ermolieva, G. Gitits, J. Koff and A. Linnerooth-Bayer. 2000. Systems approach to modeling catastrophic risk and insurability. Natural Hazards Journal, Vol. 21, Issue 2/3.

Amendola, A., Y. Ermoliev and T. Ermolieva. 2000. Earthquake risk management: a case study for an Italian Region. *In*: Proceedings of the Second EuroConference on Global Change and Catastrophe Risk Management: Earthquake Risks in Europe IIASA, Laxenburg, Austria.

ANCOLD. 1998. Guidelines on Risk Assessment. Working Group on Risk Assessment, Australian National Committee on Large Dams, Sydney, New South Wales, Australia.

Arrow, K.J. 1996. The theory of risk-bearing: Small and great risks. Journal of Risk and Uncertainty, 12: 103–111.

Artzner, P., F. Delbaen, J.-M. Eber and D. Heath. 1999. Coherent measures of risk. Mathematical Finance, 9(3): 203–228.

Atoyev, K.L., A.N. Golodnikov, V.M. Gorbachuk, T. Ermolieva, Y. Ermoliev, V.S. Kiriljuk, P.S. Knopov and T.V. Pepeljaeva. 2020. Food, energy and water nexus: methodology of modeling and risk management. *In*: Zagorodny, A.G., Yu.M. Ermoliev, V.L. Bogdanov and T. Ermolieva (eds.). Integrated Modeling and Management of Food-Energy-Water NEXUS for Sustainable Development. Kiev, Ukraine: Committee for Systems Analysis at Presidium of National Academy of Sciences of Ukraine—National Member Organization of Ukraine in International Institute for Applied Systems Analysis (IIASA), 2020, ISBN 978-966-02-9344-1.

Baranov, S., B. Digas, T. Ermolieva and V. Rozenberg. 2002. Earthquake Risk Management: A Scenario Generator, IIASA Interim Report, IR-02-025.

Bielza Diaz-Caneja, M., C. Conte, C. Dittmann, F. Gallego Pinilla, J. Stroblmair and R. Catenaro. 2009. Risk management and agricultural insurance schemes in Europe. EUR 23943 EN. Luxembourg (Luxembourg): OPOCE; 2009. JRC51982.

Bowles, D. 2007. Tolerable Risk or Dams: How Safe is Safe Enough? Dams Annual Conference, March 2007, Philadelphia, Pennsylvania.

Boyle, C. 2002. Catastrophe modeling: Feeding the risk transfer food chain. Insurance Journal.

Chapman, D. and N. Khanna. 2000. Crying no wolf: Why economists don't worry about climate change, and should. Clim. Change, 47: 225–232.

Chichilinskii, G. 1997. What is Sustainable Development? Land Economics, 73/4: 467–491.

Clark, K.M. 2002. The use of computer modeling in estimating and managing future catastrophe losses. The Geneva Papers on Risk and Insurance, 27: 181–195.

Commissions on Engineering and Technical Systems (CETS). 1985. Safety of Dams: Flood and Earthquake Criteria (at http://www.nap.edu/openbook.php?record_id= 288&page=11).

Compton, K.L., R. Faber, T. Ermolieva, J. Linnerooth-Bayer and H.-P. Nachtnebel. 2008. Development of a catastrophe model for managing the risks of urban flash flooding in Vienna. IIASA Interim Report IR-08-001.

Dams and Development: A New framework for decision-making. 2000. The Report of World Commission on Dams. Earthscan Publications Ltd., London and Sterling, VA.

Dantzig, G.B. 1979. The role of models in determining policy for transition to a more resilient technological society. IIASA distinguished lecture series.

Daykin, C., T. Pentikainen and M. Pesonen. 1994. Practical Risk Theory for Actuaries: Monographs on Statistics and Applied Probability. Vol. 53, Chapman and Hall Ltd.

Ekenberg, L., L. Brouwers, M. Danielson, K. Hansson, J. Johannson, A. Riabacke and A. Vari. 2003. Flood risk management policy in the Upper Tisza Basin: A system analytical approach (IR-03-003). International Institute for Applied Systems Analysis, Laxenburg, Austria.

Embrechts, P., C. Klueppelberg and T. Mikosch. 2000. Modeling extremal events for insurance and finance. Applications of Mathematics, Stochastic Modeling and Applied Probability; Springer Verlag, Heidelberg, Germany.

Ermoliev, Y. and R. Wets (eds.). 1988. Numerical techniques of stochastic optimization. Computational Mathematics, Springer-Verlag.

Ermoliev, Y. and V. Norkin. 1997. On nonsmooth and discontinuous problems of stochastic systems optimization. European Journal of Operations Research.

Ermoliev, Y.M., T. Ermolieva, G.J. MacDonald and V.I. Norkin. 1998. On the design of catastrophic risk portfolios, IIASA Interim Report IR-98-056.

Ermoliev, Y., T. Ermolieva, G. MacDonald and V. Norkin. 2000a. Insurability of catastrophic risks: the stochastic optimization model. Optimization Journal, 47: 251–265.

Ermoliev, Y., T. Ermolieva, G. MacDonald and V. Norkin. 2000b. Stochastic optimization of insurance portfolios for managing exposure to catastrophic risks. Annals of Operations Research, 99: 207–225.

Ermoliev, Y., T. Ermolieva, G. MacDonald and V. Norkin. 2001. Problems on insurance of catastrophic risks. Cybernetics and Systems Analysis, 37(2).

Ermoliev, Y. and L. Hordijk. 2003. Global changes: Facets of robust decisions. *In*: Marti, K., Y. Ermoliev, M. Makowski and G. Pug (eds.). Coping with Uncertainty: Modeling and Policy Issue. Springer Verlag, Berlin, Germany.

Ermoliev, Y., T. Ermolieva, G. Fisher and M. Makowski. 2010. Extreme events, discounting and stochastic optimization. Annals of Operations Research, 177: 9–19.

Ermoliev, Y., T. Ermolieva, E. Rovenskaya, P.S. Knopov and V.M. Gorbachuk. 2018. Interdependencies, extreme events, systemic risks, and robust solutions. pp. 105–108. *In*: Proceedings of the 6-th International Conference on Mathematical Modeling, Optimization and Information Technologies. Kischinev, Moldova: Evrica.

Ermoliev, Y., S. Robinson, E. Rovenskaya and T. Ermolieva. 2018. Integrated catastrophic risk management: Robust balance between Ex-ante and Ex-post measures. SIAM News, 51(6): 4.

Ermolieva, T.Y. 1997. The design of optimal insurance decisions in the presence of catastrophic risks. IIASA Interim Report, IR-97-068; IIASA, Laxenburg, Austria.

Ermolieva, T., Y. Ermoliev and V. Norkin. 1997. Spatial stochastic model for optimization capacity of insurance networks under dependent catastrophic risks: Numerical experiments, IIASA Interim Report IR-97-028.

Ermolieva, T., Y. Ermoliev, G. Fischer and I. Galambos. 2003. The role of financial instruments in integrated catastrophic flood management. Multinational Finance Journal, 7/3&4: 207–230.

Ermolieva, T. and M. Obersteiner. 2004. Abrupt climate change: Lessons from integrated catastrophic risks management. World Resource Review, 16(1): 57–82.

Ermolieva, T. and Y. Ermoliev. 2005. Catastrophic risk management: Flood and seismic risks case studies. *In*: Wallace, S.W. and W.T. Ziemba (eds.). Applications of Stochastic Programming, MPS-SIAM Series on Optimization, Philadelphia, PA, USA.

Ermolieva, T., T. Filatova, Y. Ermoliev, M. Obersteiner, K.M. de Bruijn and A. Jeuken. 2016. Flood catastrophe model for designing optimal flood insurance program: Estimating location-specific premiums in the Netherlands. Risk Analysis, 1–17.

Ermolieva, T., P. Havlík, Y. Ermoliev, A. Mosnier, M. Obersteiner, D. Leclère, N. Khabarov, H. Valin and W. Reuter. 2016. Integrated management of land use systems under systemic risks and security targets: a stochastic Global Biosphere Management Model. Journal of Agricultural Economics, 67(3): 584–601.

Ermolieva, T., P. Havlik, Y. Ermoliev, N. Khabarov and M. Obersteiner. 2021. Robust management of systemic risks and Food-Water-Energy-Environmental security: Two-stage strategic-adaptive GLOBIOM model. Sustainability, 13(2): e857. 10.3390/su13020857.

European Commission. 2021. European disaster risk management, Factsheets (available at: https://ec.europa.eu/echo/what/civil-protection/european-disaster-risk-management_en).

Fisher, A.C. and U. Narain. 2003. Global warming, endogenous risk, and irreversibility. Environmental and Resource Economics, 25: 395–416.

Glauber, J.W. 2015. Agricultural insurance and the World Trade Organization. IFPRI Discussion Paper 1473. Washington, D.C.: International Food Policy Research Institute (IFPRI) (available online at http://ebrary.ifpri.org/cdm/singleitem/collection/p15738coll2/id/129733).

Goodess, C., C. Hanson, M. Hulme and T. Osborn. 2003. Representing climate and extreme weather events in integrated assessment models: a review of existing methods and options for development. Integr Assess, 4: 145–171.

Gorbachuk, V.M., Y. Ermoliev and T. Ermolieva. 2019. Two stage model of ecological and economic decisions. Odesa National University Herald, Series in Economics, 21(9): 142–147.

IAEA. 1992. The Role of probabilistic safety assessment and probabilistic safety criteria in nuclear power plant safety. Vienna: International Atomic Energy Agency (IAEA) (available at https://www.iaea.org/publications/3784/the-role-of-probabilistic-safety-assessment-and-probabilistic-safety-criteria-in-nuclear-power-plant-safety).

Khan, M., P. Montiel and N. Haque. 1990. Adjustments with growth: Relating the analytical approaches of the IMF and the World Bank. Journal of Development Economics, 32: 155–179.

MunichRe. 2022. Hurricanes, cold waves, tornadoes: Weather disasters in USA dominate natural disaster losses in 2021 (available at: https://www.munichre.com/en/company/media-relations/media-information-and-corporate-news/media-information/2022/natural-disaster-losses-2021.html).

Newel, R. and W. Pizer. 2000. Discounting the distant future: How much do uncertain rates increase valuations? Economics Technical Series, Pew Center on Global Climate Change, Wilson Blvd. Suite 550 Arlington, VA (USA) (Available at www.pewclimate.org).

OECD. 2003. Emerging systemic risks in the 21st century: An agenda for action. ISBN 92-64-19947-0–OECD.

Pugh, E.L. 1966. A gradient technique of Adaptive Monte Carlo. SIAM Review, 8(3): 346–355.

Ramsey, F. 1928. A Mathematical theory of savings. Economic Journal, 138: 543–559.

Ray, D. 1998. Development Economics. Princeton University Press, Princeton, New Jersey.

Rockafellar, T., and S. Uryasev. 2000. Optimization of conditional value-at-risk. The Journal of Risk, 2(3): 21–41.

Sigma. 2021. Natural catastrophes in 2020. Sigma, Swiss Re.

Sigma. 2022. Global insured catastrophe losses rise to USD 112 billion in 2021, Sigma, Swiss Re.

Walker, G. 1997. Current developments in catastrophe modelling. pp. 17–35. *In*: Britton, N.R. and J. Oliver (eds.). Financial Risks Management for Natural Catastrophes. Brisbane, Griffith University, Australia.

Weitzman, M. 1999. Just Keep on Discounting. But .... (Chapter 3). Discounting and intergenerational equity. Portney, P. and J. Weyant (eds.). Resources for the Future (RFF) Press, Washington DC (USA).

Wright, E.L. and J.D. Erickson. 2003. Incorporating catastrophes into integrated assessment: science, impacts, and adaptation. Climate Change, 57: 265–286.

Yang, H. 2000. An integrated risk management method: VaR Approach. Multinational Finance Journal, 4: 201–219.

Chapter 3

Authentication for Coalition Groups

Anatoly Anisimov[1,*] and *Andrey Novokshonov*[2]

1. Introduction

Modern technological systems fulfilling critically dependent tasks are marked by a rich variety of interaction types and exponentially growing complexity. These factors significantly affect the costs of risks and uncertainty in decision-making. Examples of such systems are various peer-to-peer networks, interactive human-machine and machine-to-machine complexes, military and corporative units, financial systems including cryptocurrencies and smart cards, the Internet of Things (IoT), and many others. Along with this, one can see the increasing activity of technologically elaborated adversarial attacks with the purpose of maliciously intruding into the work of such systems. In this regard, the speed and accuracy of entity authentication are of special importance. In many critical cases, establishing a well-founded identity of an object demands essential computational and managerial efforts, that may be inconsistent with the time-limited restrictions to an authentication query. Especially, since this problem is important due to the ubiquitous advent of IoT with hundreds and thousands of interacting "computationally weak" devices. Thus, in many critical cases, a quite complex problem arises. On the one hand, it is necessary to obtain a reasonable, secure, and mathematically grounded object identification. On the other hand, it should be done in minimal admissible time depending on specific applications.

A coalition group is a group of mutually trusted entities that act independently in an unreliable, possibly malicious, environment and

[1] Taras Shevchenko National University of Kyiv, 60 Volodymyrska Street, Kyiv, 01033, Ukraine.
[2] International Research and Training Center for Information Technologies and Systems under NAS and MES of Ukraine, 40 Ave. Glushkov, Kyiv, 03680, Ukraine.
Email: andrey.novokshonov@ukr.net
* Corresponding author: a.v.anisimov@knu.ua

communicate through unsecured open channels. The main feature of coalition authentication is the necessity for both parties to not only prove the knowledge of their own secret identifying information but also to demonstrate belonging to the same authorized group. This could be implemented by secure sharing some secret unique information. Coalition units assume the implicit or explicit impact of a trusted third party (TTP) that acts as a coalition controller.

Mutual authentication means a case when entities simultaneously identify each other by executing the same authenticating protocol. Mutual authentication originates from the network authentication client-server protocols first presented in the famous Needham-Schroeder protocol that later evolved into the well-known Kerberos system: Needham and Schroeder 1978. Mutual authentication is one of the most important parts of customized IoT systems that are marked by the huge variety of specific cases reflecting broad areas of IoT applications. See surveys Liu and Nepal 2017, Yang et al. 2019.

With the development of zero-knowledge interactive proofs, the authentication problem obtains a new lightening. In the challenge-response zero-knowledge protocols, authentication is carried out through a verification process where a claimant, called a prover, must corroborate to another party, called a verifier, the knowledge of some secret identifying information without revealing any part of it. Zero-knowledge property is a very strong security requirement. In many cases, such protocols are mathematically substantiated, but they are unilateral and cannot be automatically adapted to the mutual case. Some considerations of this issue are given in Menezes et al. 1996.

Interactive and non-interactive zero-knowledge protocols mainly concentrate on optimization proofs for general NP relations: Goldwasser et al. 1985, Blum et al. 1988, Goldreich et al. 1991, Goldreich and Oren 1994. Despite many impressive achievements in this direction, such protocols are still quite complex in implementation, mainly due to the form of an NP-problem representation, and are not well suited for the bilateral authentication response.

When applied to the coalition case, faster authenticating zero-knowledge protocols like the families of the Fiat-Shamir and the Schnorr protocols must maintain some extra interactions and setup computations: Fiat and Shamir 1987, Fiege et al. 1987, Guillou and Quisquater 1988, Schnorr 1991. The security of the Fiat-Shamir protocol relies on the strong RSA assumption, i.e., the difficulty of factoring the product of two large prime numbers. The security of the Schnorr protocol is backed by the complexity of the discrete logarithm problem in a group. Using the Fiat-Shamir heuristic these schemes can be converted into unilateral non-interactive protocols based on digital signatures that confirm the prover's identity. For mutual coalition authentication, such reduction demands additional interactions and/or setting procedures.

In this communication we consider two cases of coalition groups: (1) a coalition group is defined through the possession of a static shared secret value and in the other case: (2) parties demonstrate to each other the possibility of fast finding the solution to a given computationally hard problem like inverting a one-way function. The first case is mostly used, especially for the IoT. For demonstration purposes, for the second scenario, we choose the case when entities can find secret factors of an RSA-modulus independently created by TTP.

We start the description of the first scenario with a short survey of the family of Diffie-Hellman "*handshaking*" protocols establishing a shared secret key. The second case uses the implicit knowledge of a shared secret that in the considered case is a factor of a large RSA-number. These factors are created by TTP and can be easily computed by A and B only through their interactions.

2. Notations and definitions

We use standard notations that are commonly accepted in descriptions of cryptographic protocols: $PubK_A$ and $PrivK_A$ are public and private keys of an entity A, respectively; $\{m\}_K$ is a message m that encrypted or signed with the key K; notation $A \rightarrow B$: m_1, m_2 means the following: A sends to B messages m_1, m_2; $x \in_R \mathcal{M}$ denotes the random selection of x from the set $\mathcal{M}$; p is a prime number; Z_p^* denotes the multiplicative group of residues modulo p; $u \leftarrow$ *random* means the assignment to the variable u a random value taken from its domain; for this case we also use the notation $u \in_R \mathcal{D}$ when the domain $\mathcal{D}$ of u is specified.

Protocols are written with numerated lines. A set-up stage (key generation and distribution, basic agreements, etc.) is numerated with zeros.

For the reader's convenience, in number-theoretic protocols, we restrict our considerations only to the multiplication group of residues modulo a prime number p denoted as Z_p^*. Nevertheless, it is necessary to stress that all these protocols can be reformulated for cyclic subgroups of any other finite commutative groups with known orders or inverse operations. For example, popular groups of points of elliptic curves over finite fields can be used.

The central notion of public key cryptography is a *one-way function*, i.e., a function that is easy to compute but hard (computationally infeasible) to invert. There are some well-known candidates for one-way functions that use the widely believed hardness of inverse operations. Among them are integer factorization, discrete logarithm problems in a group, Diffie-Hellman problem, and modular square rooting. We consider a one-way function of a DL-form $f(x) = g^x$, where g is a given element of a group.

Hash functions. A cryptographic hash function h is a one-way function that maps Boolean strings of arbitrary size into Boolean strings of fixed size and satisfies the following conditions:

1. For any input x it is *easy* to compute $h(x)$.
2. (Preimage resistance). For any randomly chosen output value $y = h(x)$ it is computationally *infeasible* to find any x' such that $y = h(x')$.
3. (Collision resistance). It is computationally *infeasible* to find pairs x_1, x_2, $x_1 \neq x_2$, such that $h(x_1) = h(x_2)$.

The term "*computationally infeasible*" means that for any algorithm (PPT Turing machine) the probability to solve the problem is a negligible function depending on security parameters. The detailed mathematical specification of its definition can be found in any textbook on cryptography. Hereinafter, we do not specify intuitive terms "*easy to compute*" and "*computationally infeasible*" assuming by default the presence of necessary security restrictions.

Belare and Rogaway suggested considering an ideal hash function as an Oracle. At present, their approach is called the *standard Oracle model*. In the proposed model, Belare and Rogaway 1993, a hash function is considered as an Oracle which operates as follows: upon obtaining an input value x it outputs a uniformly distributed random value $y = h(x)$ from the image domain. This value y is fixed for all subsequent uses of $h(x)$. Many useful properties of cryptographic protocols have been proved in this model.

By a coalition group, we mean a group of mutually trusted entities that establish and maintain secure group communications in a possibly hostile environment. Usually, coalition principals establish beforehand a shared secret key or property that allows them to independently recognize each other and communicate through unsecure transmission channels. For technical details of secure group communication management, we address the book: by Zou et al. 2005. Coalition conditions assume an initial set-up with a trusted third party (TTP). We assume that after the set-up phase coalition members act independently and TTP does not take part in future coalition communications and cannot transfer any information about information of A and B to any other party.

3. "Handshaking" protocols

Many authentication requirements rely on sharing the same secret value among communicating parties. The first published famous solution for secure establishing a shared secret was suggested by Diffie and Hellman in 1976: Diffie and Hellman 1976.

DH-protocol. Entities A (Alice) and B (Bob) establish a mutual secret number K (a key).

begin DH

A

0.1. $p, g \in Z_p^*$.

0.2. $A \leftarrow B$: g, p – public keys.

1. $x_1 \in_R [1, p-1]$.

2. $y_1 \leftarrow a^{x1}\ mod\, p$.

3. $A \rightarrow B : y_1$.

B

4. $x_2 \in_R [1, p-1]$.

5. $\boldsymbol{y_2 \leftarrow a^{x_2}\ mod\, p}$.

6. $B \rightarrow A : y_2$.

7. $K \leftarrow y_1^{x2} mod\, p$

A

8. $K \leftarrow y_2^{x1}\, mod\, p$

end DH

It is easy to check that K computed at step 7 and K at step 8 are the same and equal to $g^{x1x2}\ mod\, p$.

Applying recursively the Diffie-Hellman construction to the group of n entities $A_1, A_2, ..., A_n$ with private keys $x_1, x_2, ..., x_n$, respectively, one can obtain a formula for a shared secret $K(A_1, A_2, ..., A_n)$ for the group of n entities, $K(A_1, A_2, ..., A_n) = g^{K(A_1, A_2, ..., A_{n-1})x_n}\ mod\ p$. Thus, for three entities the group key is equal to $g^{g^{x_1x_2}x_3}\ mod\ p$. This construction is known as the Tree DH-protocol (TDH).

Unfortunately, the DH-protocol and its modifications are vulnerable to the attack known as the "*man-in-the-middle*". In this attack, the adversary controlling communications between A and B can intercept, relay, and send their own messages in a way indistinguishable from communications between A and B.

Hereinafter, we omit notification *mod p* assuming computations in any appropriate commutative group G with group operation denoted as multiplication. For example, at present, groups of points of elliptic curves with fast addition are popular.

In 1985 T. ElGamal managed to adapt the DH-scheme to data transmissions: ElGamal 1985.

The ElGamal protocol for a message transfer. Like the DH-protocol, A and B generate a shared secret key g^{ab}. A uses g^a as a long-term public key, B operates with a one-time session key g^b. B sends to A a message $m \in G$.

A

Like the DH-protocol, A generates a static public key c,

0.1. $a \leftarrow random$, $c \leftarrow g^a$, a is a randomly chosen number.

0.2. $A \rightarrow B : c$.

B

1. $b \leftarrow random$.
2. $y \leftarrow g^b$, $z \leftarrow mc^b$.
3. $B \rightarrow A : y, z$.

A

4. $u \leftarrow y^{-a}$.
5. $m = uz$.

Correctness of the protocol follows from the consideration of the final expression : $uz = y^{-a} z = g^{b(-a)} mg^{ab} = m$.

This protocol assumes that in the group G one can effectively compute the inverse of any given element y (step 3). If Z_p^* is used, then $y^{-a} = y^{p-1-a}$. Also, inverse operation is easy in elliptic curves.

As was previously mentioned, all "hand-shaking" protocols that straightforwardly use the Diffie-Hellman approach are vulnerable to the "*man-in-the-middle*" attack. Nevertheless, in 1995 A. Meneses, M. Qu, and S. Vanstone suggested an original authentication protocol that, under some conditions, can resist that kind of attack: Meneses et al. 1995. The authors constructed a two-tier DH-protocol with two different pairs of public and private DH-keys for each principal. The first pair of public keys forms authorized static keys. A certification center is needed for their distribution. The second pair of private keys is used as session "*ephemeral*" keys.

It is assumed that elements of a group used in the MQV-protocol have a unified numerical representation. There is a parameter λ depending on this representation. For Z_p^* $\lambda = \left\lfloor \frac{p}{2} \right\rfloor$, which is half of an average message length. For elliptic curves, $\lambda = 80$.

MQV-protocol (basic version). A and B establish a shared secret.

Begin MQV

A

1. Choose randomly x_A, α.
   $x_A \leftarrow random$, $\alpha \leftarrow random$.
2. $y_A = Pub1K_A \leftarrow g^{x_A}$, $z_A = Pub2K_A \leftarrow g^{\alpha}$, $Priv1K_A \leftarrow x_A$, $Priv2K_A \leftarrow \alpha$.

B

3. Choose randomly x_B, β.
   $x_B \leftarrow random, \beta \leftarrow random.$
4. $y_B = Pub1K_B \leftarrow g^{x_B}$, $z_B = Pub2K_B \leftarrow g^{\beta}$, $Priv1K_B \leftarrow x_B$, $Priv2K_B \leftarrow \beta$.
5. *A* and *B* exchange public keys. As the result *A* and *B* knows, respectively:
   $A : y_A, z_A, y_B, z_B, x_A, \alpha;$
   $B : y_A, z_A, y_B, z_B, x_B, \beta.$

A

6. $S_A \leftarrow z_A\ mod\ 2^{\lambda} + 2^{\lambda}$, $T_A \leftarrow z_B\ mod\ 2^{\lambda} + 2^{\lambda}$, $U_A \leftarrow (\alpha + x_A S_A)$, $K = K_A = (z_B\, y_B^{TA})^{U_A}$.

B

7. $S_B \leftarrow z_B\ mod\ 2^{\lambda} + 2^{\lambda}$, $T_B \leftarrow z_A\ mod\ 2^{\lambda} + 2^{\lambda}$, $U_B \leftarrow (\beta + x_B S_B)$, $K = K_B \leftarrow (z_A y_A^{TB})^{U_B}$.

end MQV

Correctness of the MQV-protocol follows from equalities: $S_A = T_B$, $S_B = T_A$, $K_A = (z_B\, y_B^{TA})^{U_A} = g^{(\beta + x_B T_A)(\alpha + x_A S_A)} = g^{(\beta + x_B S_B)(\alpha + x_A T_B)} = (z_A y_A^{TB})^{U_B} = K_B = K$.

With the aim of the statistical alignment of numerical bits, the key should be set to the hash $h(K)$.

Since the shared MQV-key depends on two sets of public and private keys of each of communicating participants, the MQV-protocol resists the session "*man-in-the-middle*" attack.

There exist a few strengthening improvements for the MQV-protocol: HMQV, FHMQV, and IEEE standard P1363 for ECMQV (Elliptic Curve MQV).

4. Static mutually shared secret

Herein, we consider the case when membership to a coalition group is provided by proving the possession of a mutually shared secret number. This number is known to all members of the group a priori. In a secure way, *A* and *B* must check their belonging to the same group, i.e., to prove that they know the mutual secret number without revealing it (zero-knowledge).

Authentication with a hash function. Let *s* be a secret coalition number, *h* is a publicly known hash function. Consider the following protocols.

Protocol H1.
begin
A
   $x \leftarrow random.$
   $A \rightarrow B$: x, $h(s + x)$.
B
   $B \rightarrow A$: $h(s - x)$.
end

Another variant:

Protocol H2.
begin
A
  $x \leftarrow random.$
  $A \rightarrow B: y = h(s + x).$
B
  $B \rightarrow A: h(s - y).$
end

If both A and B know s and share a public hash function h, then they can easily check the correctness of messages. Properties of a hash function imply that for both parties the probability to create a correct message by guessing s is a negligible value.

There is another way to check the knowledge of s using the ElGamal construction.

Authentication with the ElGamal construction. In this protocol, both A and B send each other nonces encrypted as in the ElGamal encryption scheme. The only difference is in adding a secret coalition number to the power of a mutual DH-key.

Protocol ElGamal.
s is a shared secret. A proves to B the knowledge of s.
begin
  DH-part
    A
    $x \leftarrow random.$
    $A \rightarrow B: g^x.$
  B
  $y \leftarrow random.$
  $B \rightarrow A: g^y.$
  end of the DH-part
    A
    $u \leftarrow random.$
    $A \rightarrow B: ug^{xys}.$
  B
  $v \leftarrow random.$
  $B \rightarrow A: vg^{xys}, u.$
    A
    $A \rightarrow B: v.$
  end

Both A and B demonstrate the verifiable knowledge of s.

5. Splitted secret

We consider authentication procedures of the type "*friend or foe*". Like the Fiat-Shamir protocol, this condition is equivalent to the natural assumption that a prover A and a verifier B both had one-time access to the same mutually trusted center T that acts as a coalition controller. This party generates a secret that can be revealed and shared by A and B only after authenticating interactions.

We assume that coalition members operate in a malicious cyberspace, i.e., the probability of an adversary's attempts to learn an identifying coalition secret is high enough. Therefore, for security reasons communicating parties must keep their own independent secrets. These secrets became temporarily known only to the TTP during the set-up stage. Depending on this information the trusted third party creates some other independent secret information that is splitted between parties. It is included in the authentication process and allows parties to easily solve a hard problem $\mathcal{P}$ chosen and created by the TTP.

There are many ways to choose a coalition secret property $\mathcal{P}$ based on a hard algorithmic or NP-complete problem. For demonstration purposes, we choose the scenario when a coalition secret is the factor of a randomly chosen large RSA modulus. These factors can be revealed by A and B only through a three-round interaction using their own private keys and combined keys generated for them by T. Principals A and B must prove to each other the knowledge of their private keys without revealing them, and, additionally, find factors of the RSA modulus selected by T, thereby demonstrating the coalition membership. We emphasize that even with additional combined keys neither party can predetermine the coalition and the other party's secrets. Also, there is no need for them to publicly reveal factors of the RSA coalition number.

Symmetrically, for parties A and B roles "*prover/verifier*" can be exchanged. Thus, the proposed scheme provides mutual authentication. Also, it maintains forward secrecy and possesses all desirable properties of interactive proofs such as completeness, soundness, and mutual zero-knowledge.

In further detail: during the setup stage, parties A and B independently present to the trusted center T their private integer keys k_1 and k_2, respectively. The center T randomly generates a numerated list of RSA moduli $n = pq$ where p and q are prime numbers being coprime with k_1 and k_2. The list of these moduli could be public or at least should be known to A and B. Also, the center T creates and safely distributes between A and B corresponding additional session keys depending on multipliers of n, private keys k_1, k_2, and a random parameter r selected by T. Using these combined keys and their own private keys, through interactions A and B themselves can compute the shared RSA multipliers of n. This way they demonstrate the knowledge of the corresponding private keys verifying their identity and corroborating their belonging to the coalition group. The parameter r is a probabilistic seed of the verifier B which starts the interaction. After finishing the initial setup, the trusted center does not need to keep any secrets of A and B as well as prime factors of n.

We name the proposed scheme the $\alpha\beta$-protocol because it uses two basic simple interconnected procedures. We call them α and β. Procedure α is executed by the TTP. For A and B it generates an RSA modulus and additional session keys needed for easy finding factors of a coalition modulus. Procedure β is executed by A and B at runtime. It only consists of one modular multiplication of three integers and produces factors of the given RSA modulus. Compared with β, the procedure α is much more time-consuming mainly due to the necessity to find large prime numbers. For making this step easier, the TTP can compute the list of primes beforehand. Actually, the executing time of the procedure α does not affect the time of authentication because α is not used in interactions between A and B.

We describe two variants of the $\alpha\beta$-protocol: a simplified basic version and its extension. The basic protocol is only two-round. In the basic variant, A proves to B that she knows her private value k_1 and factors of the RSA number n without revealing k_1 and factors of n. The principal B only demonstrates the knowledge of the additional starting session key given by T. Thus, the protocol provides only partial mutual authentication. After an initial setup, the basic protocol uses only two interactions and fixed roles for A and B: A is a prover, and B is a verifier. Augmented with the third interaction, which is the additional response from B, the modified extended protocol provides the reciprocal authentication of both parties. For security reasons, in this second variant, the starting random seed r must be kept by B and transmitted to A in an encrypted form. Any suitable specific encoding might be used. In our implementation, we exploit the ElGamal protocol on elliptic curves. During other rounds, interacting parties execute only one modular multiplication of three integers and a hashing procedure.

The security of the protocol is based on the assumptions of the difficulty of RSA factorization and the existence of a cryptographic hash function. Additionally, in practice, both parties can easily evaluate the expected responding time of a counterparty and reject it in the case of exceeding this time. Zero-knowledge properties are proved in the random oracle model.

We consider the $\alpha\beta$-protocol as a three-factor authentication scheme because it uses three independent sets of unique factors: two private keys of A and B and the set of combined keys with uniform distribution needed for integer factorization of a number generated by the TTP.

6. The αβ-protocol

By default, we assume that all numbers considered hereinafter are positive natural integers.

Let $\mathbf{1}^t$ be a security parameter, $\boldsymbol{h}$ a public hash function acting on strings, $\boldsymbol{h}$: $\{\mathbf{0}, \mathbf{1}\}^* \rightarrow \{\mathbf{0}, \mathbf{1}\}^t$. For integers $\boldsymbol{x}, \boldsymbol{y}, \boldsymbol{z}$ given in binary, when writing $\boldsymbol{h}(\boldsymbol{x}, \boldsymbol{y}, \boldsymbol{z})$ we assume that arguments are concatenated.

In procedures α and β we use two interconnected functions $f_\alpha(x, y, z) = x^{-1}y^{-1} \; mod \; z$ and $f_\beta(x, u, z) = xu \; mod \; z$, respectively. For the first function arguments x, y should be coprime with z. We note that if u in $f_\beta(x, u, z)$ is a multiple of y, $u = yw$, and integers x, y, w are less than z then the product $f_\alpha(x, y, z) f_\beta(x, u, z) mod \, z$ returns w which is the factor of u. We use this property in the procedure β for finding factors of the RSA coalition number.

First, we describe procedure α.

Procedure $\alpha(x, y, t)$.
Input: x, y, t.
Output: r, n, a, c.
{
1. Find a random prime number p of length t.
2. Find a random prime number q of length t.
3. Generate a random number r of length t.
4. $n \leftarrow pq$.
5. $a \leftarrow f_\alpha(r, q, x) = r^{-1} \, q^{-1} \; mod \, x$.
6. $b \leftarrow h(r, p, x)$.
7. $c \leftarrow f_\alpha(b, p, y) = b^{-1} \, p^{-1} \; mod \, y$.
8. $result \leftarrow (r, n, a, c)$.
}

We assume that in the procedure α integers r, q are coprime with x, and b, p are coprime with y. This condition, if necessary, could be achieved by varying the parameter r.

Procedure $\beta(x, y, z, u)$.
Input: x, y, z, u.
Output: *result*.
$result \leftarrow xyz \; mod \; u$

The $\alpha\beta$-protocol consists of the initial setup interactions of A and B with the TTP T (steps marked by 0) and two subsequent authentication rounds between A and B.

Protocol $\alpha\beta(k_1, k_2, t)$.

0. *Initial setup*.

01. A and B randomly select their own long-term secret keys k_1 and k_2, respectively, of the bitlength $t + 1$. A and B send the corresponding keys k_1 and k_2 to T.

$A \rightarrow T: k_1, B \rightarrow T: k_2$.

02. T executes the procedure $\alpha(k_1, k_2, t)$ and obtains the resulting values r, n, a, c.

03. T publishes the number n (list of numbers).

T may create several such numbers n. For each n, several associated values r together with the depending on n values a, c may exist. It is convenient to consider all these values to be indexed by n through corresponding arrays.

04. T sends the number a to A, the pair (r, c) to B, and the number n to A and B.

$T \rightarrow A: a, T \rightarrow B: (r, c), T \rightarrow A, B: n$.

Parties A and B keep their numbers a and (r, c), respectively, as additional session keys. We assume that during the setup stage all parties A, B, and T behave honestly and cannot be impersonated by any other party.

1. *B starts the interaction with A*. From the list of the given RSA moduli, B randomly chooses a modulus $n = pq$ and starts the interaction pointing out which n he chooses and sending to A the corresponding value r associated with the chosen n.

$B \rightarrow A: r$.

2. (*Weak acceptance of B*). Upon obtaining r from B, A computes $f_\beta(r, a, n, k_1) = ran \bmod k_1$. Since $a = r^{-1}q^{-1} \bmod k_1$ and $n = pq$ this number should be a divisor of n, namely it equals to p. If so, A weakly accepts B and continues. Otherwise, A rejects.

At step 2 of the protocol, B demonstrates to A only the knowledge of the right integer r that matches a, n, and k_1 but neither the knowledge of his private secret k_2 nor the ability to factorize n. We consider this as *a weak identity proof* of B.

3. A computes $b \leftarrow h(r, p, k_1)$ and sends it to B.

$A \rightarrow B: b$.

4. (*Acceptance of A*). Upon receiving b, B calculates $f_\beta(b, c, n, k_2) = bcn \bmod k_2$ and accepts if and only if f_β returns a divisor of n. It should be q. Otherwise, B rejects the proof.

The protocol is successful if A weakly accepts B and B accepts A.

7. Properties of the αβ-protocol

While establishing properties of the developed ***αβ***-protocol, we use main properties of hash functions: (1) it is infeasible to compute the given image

$\boldsymbol{h}(z)$ without getting z; (2) having z, it is infeasible to find another $\tilde{z}$ such that $\boldsymbol{h}(z) = \boldsymbol{h}(\tilde{z})$; (3) $\boldsymbol{h}(z)$ can be considered as statistically independent of z and uniformly distributed. The last property is known as the cryptographic random oracle model. Infeasible means that the probability of the stated event is a negligible function. Thereafter, we use commonly accepted notions of interactive proofs such as completeness, soundness, and zero-knowledge.

Theorem 1. The αβ-protocol satisfies the following properties: completeness, soundness, and honest-verifier zero-knowledge.

Proof. We remind that in the *αβ*-protocol *A* is the prover and *B* is the verifier.

Completeness. If ***A*** and ***B*** are honest parties and follow the protocol, then ***B*** always accepts the proof with probability 1. Indeed, from the structure of $\boldsymbol{a}$, $\boldsymbol{a} = \boldsymbol{r}^{-1}\, \boldsymbol{q}^{-1}\ \boldsymbol{mod}\ \boldsymbol{k}_1$, it follows that $\boldsymbol{ran}\ \boldsymbol{mod}\ \boldsymbol{k}_1 = \boldsymbol{p}$. Analogous statement holds true for $\boldsymbol{b}$, $\boldsymbol{c}$, $\boldsymbol{n}$, $\boldsymbol{k}_2$ and $\boldsymbol{q}$, $\boldsymbol{bcn}\ \boldsymbol{mod}\ \boldsymbol{k}_2 = \boldsymbol{q}$, if and only if $\boldsymbol{c} = \boldsymbol{b}^{-1}\, \boldsymbol{p}^{-1}\ \ \boldsymbol{mod}\ \boldsymbol{k}_2$. If ***A*** and ***B*** are honest and follow the protocol then both ***A*** and ***B*** can easily determine factors of $\boldsymbol{n}$. Principal ***A*** computes $\boldsymbol{p}$ (step 2) using given $\boldsymbol{a}$, $\boldsymbol{n}$, and the incoming number $\boldsymbol{r}$ and ***B*** computes $\boldsymbol{q}$ (step 3) using given $\boldsymbol{c}$, $\boldsymbol{n}$, and the incoming $\boldsymbol{b}$. Also, ***A*** using the given key $\boldsymbol{a}$ and the derived factor $\boldsymbol{p}$ can compute the hash $\boldsymbol{b} = \boldsymbol{h}(\boldsymbol{a}, \boldsymbol{p}, \boldsymbol{k}_1)$. Thus, if ***A*** and ***B*** follow the protocol then ***A*** and ***B*** interchange only by right numbers specified in the protocol. It follows that ***A*** weakly accepts ***B***, and ***B*** accepts the proof of ***A***.

We stress that assuming the difficulty of the RSA factorization problem and because of a random choice of the starting number *r* without interactions neither *A* nor *B* can find factors of the coalition number *n*.

Soundness. Soundness means that any dishonest prover A^* (a PPT Turing machine) impersonating ***A*** will be unable to convince the honest verifier ***B*** that she knows the secret key $\boldsymbol{k}_1$ of ***A*** and factors of $\boldsymbol{n}$ except with negligible probability.

Assume that with non-negligible probability *B* accepts A^*. After obtaining the initial number *r*, A^* must respond by sending some b^* to *B*. The honest verifier *B* can accept if and only if $b^*\, cn\ mod\ k_2 = q$. Numbers *c* and *n* have been given to *B* by the trusted (honest) center *T*. This implies that (*c*, *n*) are true numbers created during the setup stage. It follows that $b^*\, cn\ mod\ k_2 \equiv b^*\, (b^{-1}\, p^{-1})pq\ mod\ k_2 = q$. In its turn, this implies equalities: $b^*b^{-1}\, p^{-1}p\ mod\ k_2 \equiv 1\ mod\ k_2$, $b^*b^{-1} \equiv 1\ mod\ k_2$. It follows that $b^* \equiv b\ mod\ k_2$. Since the bitlength of b^* should be less than that of k_2, we conclude that $b^* = b = h(r, p, k_1)$. Properties of a hash function imply that if A^* with non-negligible probability can compute $b = h(r, p, k_1)$, then with non-negligible probability during its activity A^* should get access to *a*, *p*, and k_1. This fact allows constructing a knowledge extractor simulating machine.

For extracting knowledge (r, p, k_1), we construct a simulating PPT machine *M* fed by the program of A^*. At the first round, *M* simply repeats actions of A^*

until A^* comes to the state s^* that sends b^* to B. Doing this, M simultaneously records on the tape internal coin tosses that leads A^* to the state s^*. Only the case $b^* = b = h(r, p, k_1)$ allows B to get the factor of n equal to q and come to the accepting state. The assumption of non-negligible probability of accepting A by B implies that M also reaches the state s^* with non-negligible probability. The function h is a hash function. This implies that A^* can calculate b only when she got arguments r, p, k_1 at preceding steps. It follows that M repeating initial steps of A^* also must obtain parameters r, p, and k_1 at preceding steps. Reaching the state s^*, machine M makes a copy of b and at the next round will use it for the search of arguments (r, p, k_1).

At the second round, M rewinds to the starting state of A^* and using the stored record of coin tosses deterministically repeats the behavior of A^* leading it to the state s^* and additionally makes the search for r, p. During this stage of execution, upon obtaining on the working tape any value v M checks if $h(v) = b$. After finding such v M uses it for extracting with non-negligible probability (the last $t + 1$ bits of v) and obtains p (the last t remaining prefix bits in the string v after extracting k_1) and r (the remaining prefix part).

Thus, this way with non-negligible probability M extracts the secret parameter k_1 of A and factors of n.

Honest-verifier zero-knowledge. Zero-knowledge means that there exists a PPT simulating Turing machine ***Sim*** that without knowledge of the value $\boldsymbol{k_1}$ and factors of $\boldsymbol{n}$ using the program of the verifier ***B*** generates the transcript of communications between ***A*** and ***B*** indistinguishable from the original interactions of ***A*** and ***B*** in the protocol. In our case, both ***A*** and ***B*** are assumed to be honest and a singular interaction transcript has the form $(\boldsymbol{r}, \boldsymbol{b})$, where $\boldsymbol{b} = \boldsymbol{h}(\boldsymbol{r}, \boldsymbol{p}, \boldsymbol{k_1})$. In the random oracle model, the value b is considered an independent random value. Thus, the pair $(\boldsymbol{r}, \boldsymbol{b})$ can be considered as a pair of two random numbers of the length $\boldsymbol{t}$. Such pairs can easily be generated by a PPT Turing machine. This ends the proof of Theorem 1.

In the αβ-protocol the value r is kept and transmitted to A in the plaintext form. To prevent many types of adversarial attacks impersonating B, we suggest keeping this value in an encrypted form. Let E_A be an encryption function with the decryption D_A known to A. Principal B obtains from TTP $E_A(r)$. For starting authentication, this encoded value is transmitted to A. After being decrypted, it allows A to find factors of n and to continue the protocol.

8. Mutual authentication

After accepting A, B can act as a prover similar to the considered $\alpha\beta$-procedure for verifier A. Thus, if the TTP adds one necessary additional parameter to A this makes the protocol to be symmetrical with respect to the role "prover-verifier". This gives mutual authentication with three communication rounds. We call this variant the extended $\alpha\beta$-protocol.

9. Performance evaluation

The implementation of the mutual $\alpha\beta$-protocol was done using Python programming language. The testing environment consists of AMD Athlon 64 X2 2,79 GHz CPU and 4 GB RAM. The bitlength of a is 128 bits; SHA-256 is used as a hash function h. Evaluation results are shown in Table 1.

Table 1. Performance of mutual αβ-protocol.

t	Trusted center (sec.)	Prover (sec.)	Verifier (sec.)
64	0.02218965	0.00001027	0.00000207
128	0.05343343	0.00001129	0.00000264
256	0.12412936	0.00001428	0.00000430
512	0.45839436	0.00003285	0.00000847
1024	2.25825591	0.00004635	0.00002024
2048	14.37915106	0.00028879	0.00009394

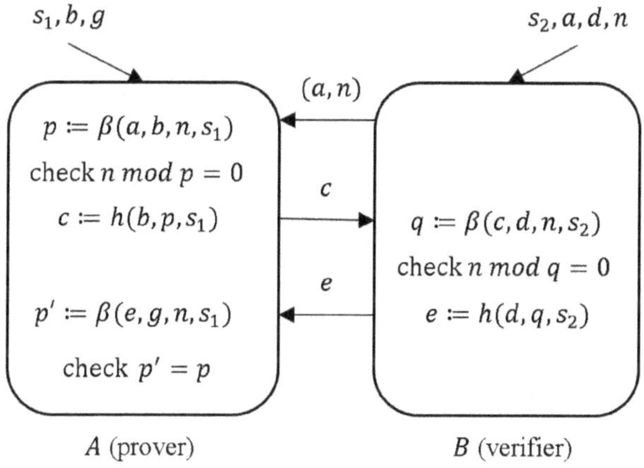

Figure 1. Mutual αβ-protocol.

References

Bellare, M. and P. Rogaway. 1993. Random oracles are practical: a paradigm for designing efficient protocols. In Proceedings of the 1st ACM conference on Computer and communications security (CCS '93). Association for Computing Machinery, New York, NY, USA, 62–73.

Blum, M., B. Feldman and T. Micali. 1988. Non-interactive zero-knowledge and its applications. Proceedings of the Twentieth Annual ACM Symposium on Theory of Computing (STOC '88), New York, NY, USA, 103–112.

Diffie, W. and M. Hellman. 1976. New directions in cryptography. IEEE Trans. Inf. Theor. 22(6): 644–654.

El Gamal, T. 1985. A public key cryptosystem and a signature scheme based on discrete logarithms. In Proceedings of CRYPTO 84 on Advances in Cryptology. Springer-Verlag, Berlin, Heidelberg, 10–18.

Fiat, A. and A. Shamir. 1986. How to prove yourself: practical solutions to identification and signature problems. Proceedings on Advances in Cryptology (CRYPTO '86). Santa Barbara, CA, USA: 186–194.

Fiege, U., A. Fiat and A. Shamir. 1987. Zero knowledge proofs of identity. Proceedings of the Nineteenth Annual ACM Symposium on Theory of Computing (STOC '87), New York, NY, USA: 210–217.

Goldwasser, S., S. Micali and C. Rackoff. 1985. The knowledge complexity of interactive proof-systems. Proceedings of the Seventeenth Annual ACM Symposium on Theory of Computing (STOC '85). New York, NY, USA: 291–304.

Goldreich, O., S. Micali and A. Wigderson. 1991. Proofs that yield nothing but their validity or all languages in NP have zero-knowledge proof systems. J. ACM 38(3): 690–728.

Goldreich, O. and Y. Oren. 1994. Definitions and properties of zero-knowledge proof systems. J. Cryptol. 7(1): 1–32.

Guillou, L. and J. Quisquater. 1988. A practical zero-knowledge protocol fitted to security microprocessor minimizing both transmission and memory. Lecture Notes in Computer Science on Advances in Cryptology (EUROCRYPT '88), Davos, Switzerland: 123–128.

Li, N., D. Liu and S. Nepal. 2017. Lightweight mutual authentication for IoT and its applications. IEEE Trans. Sustainable Computing 2(4): 359–370.

Menezes, A., M. Qu and S. Vanstone. 1995. Some new key agreement protocols providing mutual implicit authentication. Second Workshop on Selected Areas in Cryptography (SAC 95).

Menezes, A., S. Vanstone and P. Van Oorschot. 1996. Identification and Entity Authentication. in Handbook of Applied Cryptography. 1st ed. USA: CRC Press, Inc.: 385–424.

Needham, R. and M. Schroeder. 1978. Using encryption for authentication in large networks of computers. Commun. ACM 21(12): 993–999

Schnorr, C. 1991. Efficient signature generation by smart cards. J. Cryptol. 4(3): 161–174.

Yang, T., G. Zhang, L. Liu, Y. Yang, S. Zhao, H. Sun and W. Wang. 2019. New features of authentication scheme for the IoT: A survey. In Proceedings of the 2nd International ACM Workshop on Security and Privacy for the Internet-of-Things (IoT S&P'19), New York, USA: 44–49.

Zou, X., B. Ramamurthy and S.S. Magliveras. 2004. Secure Group Communications Over Data Networks. Springer-Verlag TELOS, Santa Clara, CA, USA.

CHAPTER 4

Robust Constructions of Risk Measures for Optimization under Uncertainty

Vladimir S Kirilyuk

1. Introduction

The main problem with applying classical models of stochastic programming (SP) is the presence of a wide uncertainty in practice. In such models, knowledge of the probability distribution of random variables (r.v.) is assumed, the average values (or other statistical quantities) of which are used as criteria or constraints for finding optimal solutions. However, as a rule, this assumption is not justified. Even long-term observations do not always allow for identifying a probability distribution. Moreover, knowledge of the distribution extracted from the data reflects their past history, and not the future, which contains new uncertainties. In this regard, the SP models should be modified taking into account such practical realities.

The sense of such modifications is that the optimization takes into account the mentioned criteria not for a known distribution, but for a set of distributions (probability measures) that correspond to our knowledge about the distribution. Since in this case the values of statistical criteria are used, which are the worst by the set of distributions, the optimal solutions obtained using them will be resistant to such uncertainty. This line of research is called distributionally robust optimization, see, for example, Delage et al. 2010, Wiesemann et al. 2014, and Shapiro 2017.

V. M. Glushkov Institute of Cybernetics, National Academy of Sciences of Ukraine.
Email: vlad00@ukr.net

In addition, the apparatus of coherent risk measures (CRM) introduced by Artzner et al. 1999 is now widely known. The dual representation of such risk measures quantifies the risk by the worst-case average loss when the probability measure varies over a set that depends on the initial probability measure and reflects the construction of the CRM.

In this paper, both of these approaches are used to assess risk under uncertainty. If a certain risk measure is used to assess risk for a known probability measure, then under uncertainty, where such probability measure is determined accurate to a set, the risk is estimated by the worst-case risk measure when the probability measure is varied over the set. Such an estimate is called a robust construction of the initial risk measure and is described by the worst-case average losses over a set of probability measures constructed in a special way.

We study such robust constructions of risk measures, their calculation, as well as optimal portfolio solutions on the reward-risk ratio, which is described with their use.

This chapter considers the apparatus of polyhedral coherent risk measures (PCRM), which allows one to represent a number of stochastic programming and robust optimization problems within a unified approach (Kirilyuk 2015). In addition, the use of PCRM and polyhedral uncertainty sets significantly simplifies calculations. Then the calculation of robust PCRM constructions with such uncertainty sets is reduced to linear programming problems (LPP). It is shown that problems of portfolio optimization on the reward-risk ratio using robust constructions of the average return and PCRM are reduced to the corresponding LPP as well.

Finally, we consider similar formulations of portfolio problems in the case of imprecise scenario estimates of r.v.

2. Dual representation of coherent risk measures

Let a probability space $(\Omega, \sum, P_0)$ and space $Z \ni X$ of r.v. be given, where $X: \Omega \to \overline{R}$ is a measurable function. Let r.v. $X(\omega)$ describes a value of random losses, which is characterized by preference "the less – the better".

A function $\rho: Z \to \overline{R}$ is called a ***coherent*** risk measure (CRM) by Artzner et al. 1999 if it has the following properties:

A0) normalized $\rho(0) = 0$;

A0a) lower semicontinuous (l.s.c.) and proper in sense of Rockafellar 1970;

A1) convex: $\rho(\lambda X_1 + (1 - \lambda X_2)) \leq \lambda\rho(X_1) + (1 - \lambda)\, \rho(X_2)$, $0 \leq \lambda \leq 1$;

A2) monotonous: $\rho(X) \leq \rho(Y)$ if $X \leq Y$ (on distribution);

A3) translation invariant: $\rho(X + a) = \rho(X) + a$, $a \in R$;

A4) positive homogeneous: $\rho(\lambda X) = \lambda\rho(X)$, $\lambda \geq 0$.

Recall that a risk measure $\rho(\cdot)$ is called ***convex*** by Föllmer and Schied 2002 if the positive homogeneity property A4) is not postulated.

Let the pair of dual spaces $Z = L_p(\Omega, \sum, P_0)$, $p \in [1, +\infty)$ and $Z^* = L_q(\Omega, \sum, P_0)$, $q \in (1, +\infty]$, $1/p + 1/q = 1$ be given. Action of a linear continuous operator $\zeta \in Z^*$ on $X \in Z$ has the following form

$$\langle \zeta, X \rangle = \int_\Omega \zeta(\omega) X(\omega) dP_0(\omega).$$

A risk measure $\rho(\cdot)$, by definition, is a convex proper and l.s.c. function. Recall that its conjugate function $\rho^*: Z^* \to \bar{R}$ is defined as

$$\rho^*(\zeta) = \sup_{X \in Z} \left[\langle \zeta, X \rangle - \rho(X) \right].$$

Using the Fenchel-Moreau theorem, $\rho(\cdot)$ can be presented by the following dual form (Shapiro et al. 2009, Theorem 6.4):

$$\rho(X) = \sup_{\zeta \in M} \left[\langle \zeta, X \rangle - \rho^*(\zeta) \right] = \sup_{\zeta \in M} \left[\int_\Omega \zeta(\omega) X(\omega) dP_0(\omega) - \rho^*(\zeta) \right], \quad (1)$$

where $M = \mathrm{dom}(\rho^*)$ is the domain of function $\rho^*(\cdot)$.

Remark 1. This representation is valid for the reflexive spaces $L_p(\Omega, \sum, P_0)$, $p \in (1, +\infty)$. For the pair of spaces $L_\infty(\Omega, \sum, P_0)$ and $L_1(\Omega, \sum, P_0)$, the first space is considered with the weak topology, and the second one – with the weak* topology (Shapiro 2017).

Note that for the CRM this representation (1) is greatly simplified:

$$\rho(X) = \sup_{\zeta \in M} \langle \zeta, X \rangle = \sup_{\zeta \in M} \int_\Omega \zeta(\omega) X(\omega) dP_0(\omega). \quad (2)$$

Remark 2. We can assume that such set M is convex and weakly* closed, since $\rho(\cdot)$, as a support function, is in one-to-one correspondence with its convex and weakly* closed hull.

Moreover, properties of monotonicity A2) and translation invariance A3) for $\rho(\cdot)$ are equivalent, respectively, to the following conditions (Shapiro et al. 2009, Theorem 6.4):

$$\zeta(.) \geq 0 \text{ a.s. and } \int_\Omega \zeta(\omega) dP_0(\omega) = 1. \quad (3)$$

Consequently, such $\zeta(\cdot)$ from (1) and (2) are some probability densities.

It is easy to see that under conditions (3)

$$Q(P_0) = \left\{ P(.) : P(A) = \int_A \zeta(\omega) dP_0(\omega), \zeta \in M \right\} \quad (4)$$

is a set of probability measures of form $dP = \zeta dP_0$ on $(\Omega, \sum)$, which is associated with set M of such densities (Radon-Nikodym derivatives).

Accordingly, equality (2) is then represented as

$$\rho(X) = \sup_{\zeta \in M} \int_\Omega \zeta(\omega) X(\omega) dP_0(\omega) = \sup_{P \in Q(P_0)} \int_\Omega X(\omega) dP(\omega) = \sup_{P \in Q(P_0)} E_P[X]. \quad (5)$$

This fact was first shown in the fundamental work of Artzner et al. 1999. Note that the convex risk measure from (1) is described by Föllmer and Schied 2002 in the following form:

$$\rho(X) = \sup_{P \in Q(P_0)} \left(E_P[X] - \alpha(P) \right), \quad (6)$$

where $\alpha(\cdot)$ is some penalty function (see (1)).

The set of probability measures $Q(P_0)$ depends on the initial measure P_0. Consequently, this is some multivalued mapping (m.m.) Q: $\Theta \to 2^\Theta$, where Θ is the set of probability measures, which represents the risk measure $\rho(\cdot)$ by an initial probability measure P_0 in form (5).

Definition 1. A risk measure $\rho(\cdot)$ of the form (2) will be called ***polyhedral*** CRM (PCRM) if M either is the set

$$M_0 = \left\{ \zeta(.) \geq 0 \ a.s., \int_\Omega \zeta(\omega) dP_0(\omega) = 1 \right\}, \quad (7)$$

or its intersection with any of the following two sets:

$$M_1 = \left\{ \int_\Omega \zeta(\omega) X_i(\omega) dP_0(\omega) \leq c_i, i = 1, \ldots, k \right\} \quad (8)$$

for some $\{X_i \in Z, c_i \in R, i = 1, \ldots, k\}$;

$$M_2 = \{\zeta(.) \leq \gamma \ a.s.\} \quad (9)$$

for some $\gamma > 0$. Here M_0 is called the standard part of the description of M.

Remark 3. In set M_1 from (8), we can take into account the restrictions on the first moments of some r.v. $X_i \in Z$.

Remark 4. In comparison with a similar definition from Kirilyuk 2004, this definition is more cumbersome, but it allows it to work with measurable spaces L_p (Ω, $\sum$, P_0) and exactly corresponds with the risk representation in form (2).

Depending on the type of M consider the following four cases:

C0) $M = M_0$; C1) $M = M_0 \cap M_1$; C2) $M = M_0 \cap M_2$; C3) $M = M_0 \cap M_1 \cap M_2$.

Examples:

E1) Maximum (worst) losses: $\operatorname{ess\,sup} X = \sup_{P \in \Theta} E_P[X]$, case C0),

$$M = M_0 = \left\{ \zeta(.) \geq 0 \ a.s., \int_\Omega \zeta(\omega) dP_0(\omega) = 1 \right\} \Rightarrow Q(P_0) = \Theta,$$

where Θ is the set of all possible probability measures of the form (4);

E2) Average losses: $E_{P_0}[X]$, case C2),

$$M = M_0 \cap M_2 = \left\{0 \le \zeta(.) \le 1 \; a.s., \int_\Omega \zeta(\omega)dP_0(\omega) = 1\right\} = \{\zeta(.) = 1 \; a.s.\} \Rightarrow Q(P_0) = \{P_0\};$$

E3) Conditional Value-at-Risk (CVaR) from Rockafellar and Uryasev 2000, case C2),

$$CVaR_\alpha(X) = \inf_{\eta \in R}\left[\eta + \frac{1}{1-\alpha}E(X-\eta)_+\right].$$

Its dual representation (2) is described by Shapiro et al. 2009, p. 272 as

$$CVaR_\alpha(X) = \sup_\zeta\left[\langle \zeta, X\rangle : 0 \le \zeta(.) \le \frac{1}{1-\alpha} \; a.s., \int_\Omega \zeta(\omega)dP_0(\omega) = 1\right].$$

Therefore, $Q(P_0)$ is represented in the form (4) for

$$M = M_0 \cap M_2 = \left\{0 \le \zeta(.) \le \frac{1}{1-\alpha} \; a.s., \int_\Omega \zeta(\omega)dP_0(\omega) = 1\right\},$$

where M_2 of form (9) for $\gamma = 1/(1-\alpha)$.

Discrete distributions. Let there be a finite set of elementary outcomes $\Omega = \{\omega_1, \dots, \omega_n\}$, consider a discretely distributed probability measure (a vector of scenario probabilities) $p_0 = (p_1^0, \dots, p_n^0)$, $\sum_{i=1}^n p_i^0 = 1, p_i^0 \ge 0$ on it. Then r.v. X (as a distribution on scenarios) is represented by a vector $x \in R^n$, and a coherent risk measure of the form (5) is:

$$\rho_{p_0}(x) = \sup_{p \in Q(p_0)} E_p[x] = \sup_{p \in Q(p_0)} \langle p, x\rangle, \tag{10}$$

where $Q(p_0)$ is a convex compact, $Q(p_0) \subseteq S^n$, where S^n is the unit simplex.

Accordingly, m.m. $Q(.)$ has the following forms for examples E1)–E3) respectively:

E1) maximum (worst) losses, case C0),

$$Q(p_0) = \left\{\zeta p_0^T : \sum_{i=1}^n \zeta_i p_i^0 = 1, \zeta p_0^T \ge 0\right\} = \left\{p : \sum_{i=1}^n p_i = 1, p \ge 0\right\} = S^n;$$

E2) average losses, case C2), $Q(p_0) = \{p_0 \zeta^T : \zeta^T = (1, \dots, 1)\} = \{p_0\}$;

E3) $CVaR_\alpha(\cdot)$, case C2),

$$Q(p_0) = \left\{p_0\zeta^T : 0 \le \zeta \le \frac{1}{1-\alpha}, \sum_{i=1}^n p_i^0 \zeta_i = 1\right\} = \left\{p : Ip \le \frac{1}{1-\alpha} p_0, p \ge 0, \sum_{i=1}^n p_i = 1\right\}.$$

For discrete distributions, set M_1 for PCRM from Definition 1 can be described by a matrix B and a vector c with dimensions $k \times n$ and k respectively:

$$M_1 = \left\{ \int_\Omega \zeta(\omega) X_i(\omega) dP_0(\omega) \le c_i, i = 1,\ldots,k \right\} = \left\{ \zeta p_0^T : B\zeta p_0^T \le c \right\} = \left\{ p : Bp \le c \right\},$$

where i-th row of matrix B describes scenario values of r.v. X_i.

For cases C1)–C3), sets $Q(p_0)$ are respectively of the following form:

$$\text{C1)}\ M = M_0 \cap M_1, Q(p_0) = \left\{ \zeta p_0^T : B\zeta p_0^T \le c, \zeta p_0^T \ge 0, \sum_{i=1}^n \zeta_i p_i^0 = 1 \right\} \Rightarrow$$

$$Q(p_0) = \left\{ p : Bp \le c, p \ge 0, \sum_{i=1}^n p_i = 1 \right\}; \tag{11}$$

$$\text{C2)}\ M = M_0 \cap M_2, Q(p_0) = \left\{ \zeta p_0^T = p_0 \zeta^T : 0 \le \zeta \le \gamma, \sum_{i=1}^n \zeta_i p_i^0 = 1 \right\} \Rightarrow$$

$$Q(p_0) = \left\{ p : Ip \le \gamma p_0, p \ge 0, \sum_{i=1}^n p_i = 1 \right\}; \tag{12}$$

$$\text{C3)}\ M = M_0 \cap M_1 \cap M_2, Q(p_0) = \left\{ \zeta p_0^T : B\zeta p_0^T \le c, 0 \le p_0 \zeta^T \le \gamma, \sum_{i=1}^n \zeta_i p_i^0 = 1 \right\} \Rightarrow$$

$$Q(P_0) = \left\{ p : Bp \le c, 0 \le p \le \gamma p_0, \sum_{i=1}^n p_i = 1 \right\} = \left\{ p : \begin{pmatrix} B \\ I \end{pmatrix} p \le \begin{pmatrix} c \\ \gamma p_0 \end{pmatrix}, p \ge 0, \sum_{i=1}^n p_i = 1 \right\}. \tag{13}$$

Set $Q(p_0)$ can be represented in the unified form as

$$Q(p_0) = \left\{ p : Bp \le c(p_0), p \ge 0, \sum_{i=1}^n p_i = 1 \right\}, \tag{14}$$

where for cases C1)–C3) the following designations are used:

C1) $B = B$, $c(p_0) = c$; C2) $B = I$, $c(p_0) = \gamma p_0$; C3) $B = \begin{pmatrix} B \\ I \end{pmatrix}, c(p_0) = \begin{pmatrix} c \\ \gamma p_0 \end{pmatrix}$.

Remark 5. Case C0) is considered in example E1).

Remark 6. In contrast to the definition from Kirilyuk 2004, for discrete distributions Definition 1 more narrowly represents set $Q(p_0)$, since in (14) matrix B does not depend on p_0, and $c(p_0)$ depends on a specific way described by cases C2)–C3).

For such set $Q(\cdot)$, PCRM in accordance with (5) has the form:

$$\rho_{p_0}(x) = \sup_{p \in Q(p_0)} E_p[x] = \sup_{p \in Q(p_0)} \langle p, x \rangle. \tag{15}$$

Remark 7. It is easy to see that nothing changes in the reasoning if the set of elementary outcomes $\Omega = \{\omega_1,\dots, \omega_n,\dots\}$ is countable and r.v. X is described by $x = \{x_1,\dots, x_n,\dots\} \in l_2$ (known Hilbert space). Then $\zeta \in l_2$ and the appropriate linear operator are considered instead of matrix B, $\langle\zeta, p\rangle$ instead of ζp^T and series $\sum_{i=1}^{\infty}$ instead of $\sum_{i=1}^{n}$.

3. Robust constructions of risk measures by uncertainty set

Suppose now that we cannot identify the initial probability measure P_0, and define it up to an uncertainty set U in form of $P_0 \in U$. Estimating the maximum (worst-case) risk in such a situation using measure (6), we obtain

$$\rho_U(X) = \sup_{P_0 \in U} \rho_{P_0}(X) = \sup_{P \in Q(P_0), P_0 \in U} (E_P[X] - \alpha(P)) = \sup_{P \in Q(U)} (E_P[X] - \alpha(P)). \quad (16)$$

If $\rho_{P0}(\cdot)$ is CRM, $\alpha(\cdot) \equiv 0$ (see (5)), and then

$$\rho_U(X) = \sup_{P_0 \in U} \rho_{P_0}(X) = \sup_{P \in Q(U)} E_P[X] = \sup_{P_0 \in U} \left\{ \int_{\Omega} \zeta(\omega) dP_0(\omega), \zeta \in M \right\} \quad (17)$$

Definition 2. Measure $\rho_U(\cdot)$ from (16), (17) will be called the ***robust construction*** of risk measure $\rho(\cdot)$ by uncertainty set U.

Remark 8. As is easy to see, for a coherent risk measure

$$\rho_U(X) = \sup_{P \in Q(U)} E_P[X] = \sup_{P \in \overline{co}Q(U)} E_P[X],$$

where $\overline{co}$ means convex weakly* closed hull of a set (see Remark 2).

Remark 9. Such constructions were called risk measures "induced by uncertainty sets" in Kirilyuk 2008, and by "robust constructions" – in Kirilyuk 2019.

It is easy to see that $\rho_U(\cdot)$ retains properties of the initial risk measure $\rho(\cdot)$. Indeed, let us denote the initial risk measure as $\rho_{P0}(\cdot)$, indicating its dependence on the probability measure P_0. Since each $\rho_P(\cdot)$, $P \in U$ is constructed in form (1) taking into account conditions (3), then it is convex and the l.s.c. function of X. Therefore, its epigraph $epi\rho_P = \{(X, r): \rho_P(X) \le r\}$ is a convex closed set. Then, as follows from (8),

$$epi\rho_U = \bigcap_{P \in U} epi\rho_P.$$

The intersection operation retains properties of closedness and convexity for sets. Therefore, set $epi\rho_U$ is also closed and convex, which implies convex and l.s.c. properties of the function $\rho_U(\cdot)$.

If $\rho_P(\cdot)$ are positively homogeneous, then $epi\rho_P$ are cones. Then their intersection is a cone as well. Therefore, $\rho_U(\cdot)$ is positively homogeneous.

Thus, the following elementary proposal is true.

Proposal 1. Robust construction $\rho_U(\cdot)$ of a convex (coherent) risk measure $\rho_{P_0}(\cdot)$ is also a convex (coherent) risk measure.

It is easy to see that robust construction (16), (17) is determined by two factors: uncertainty set U and m.m. $Q(\cdot)$ from (4). Set U is determined by available information on probability measure P_0, and m.m. $Q(\cdot)$—by the initial risk measure $\rho(\cdot)$.

Usually, the data available in the applications does not allow identifying the initial probability measure P_0, but permits estimating it by an uncertainty set as $P_0 \in U$. Therefore, the formulation of problems using such robust constructions ensures stability relatively to such uncertainty for the problem's solutions.

Definition 3. A set of probability measures will be called ***polyhedral*** if it is represented in the following form:

$$U = \left\{ P : P(A) = \int_A \zeta(\omega) dP_0(\omega),\ \zeta \in M_0 \cap M_1 \right\}, \tag{18}$$

where M_0, M_1 are described by relations (7) and (8), respectively.

Discrete distributions. In the case of discrete distributions, the polyhedral set U from (18) can be represented in form $Q(P_0)$ from (11) for case C1):

$$U = \left\{ p : B_u p \le c_u, \sum_{i=1}^{n} p_i = 1, p \ge 0 \right\}, \tag{19}$$

where B_u and c_u are a matrix and a vector of corresponding dimensions.

Remark 10. Set U from (19) does not depend on the probability measure P_0, which we do not know.

Consider a couple of elementary examples of such sets U.

***Example* 1.** Let a vector of scenario probabilities p_0 be estimated from below and from above, respectively, by vectors p_l and p_u, i.e., by the following uncertainty set U:

$$U = \left\{ p : p_l \le p \le p_u, p \ge 0, \sum_{i=1}^{n} p_i = 1 \right\}.$$

It is presented in form (19), where $B_u = \begin{pmatrix} -I \\ I \end{pmatrix}$, $c_u = \begin{pmatrix} -p_l \\ p_u \end{pmatrix}$.

***Example* 2.** Suppose now there is an urn with balls, say, red, black and yellow. It is known that there are 30 red balls, and the total number of black and yellow balls is not less than 60, but not more than 70. Let p_i, $i = 1,2,3$ be the probabilities of choosing a red, black and yellow ball from the urn, respectively. Then set U has the following form:

$$U = \left\{ p = (p_1, p_2, p_3) : 0.3 \le p_1 \le \frac{1}{3}, \frac{2}{3} \le p_2 + p_3 \le 0.7, \sum_{i=1}^{3} p_i = 1, p \ge 0 \right\}.$$

It is presented in form (19), where $B_u = \begin{pmatrix} -1 & 0 & 0 \\ 1 & 0 & 0 \\ 0 & -1 & -1 \\ 0 & 1 & 1 \end{pmatrix}$, $c_u = \begin{pmatrix} -0.3 \\ 1/3 \\ -2/3 \\ 0.7 \end{pmatrix}$.

Let us now consider a robust construction of a coherent risk measure (17) with a polyhedral uncertainty set of form (10). It is described as

$$\rho_U(x) = \sup_{p \in Q(U)} \langle p, x \rangle = \sup_{p \in \overline{co} Q(U)} \langle p, x \rangle, \tag{20}$$

where $\overline{co}$ is the convex closed hull of set $Q(U)$.

Then, for examples of risk measures E1)–E3), their robust constructions are described in the form (20), where sets $Q(U)$ to have the form:

E1) for maximum (worst) losses $Q(U) = \left\{ p : \sum_{i=1}^{n} p_i = 1, p \ge 0 \right\}$;

E2) for average losses $Q(U) = U = \left\{ p : B_u p \le c_u, \sum_{i=1}^{n} p_i = 1, p \ge 0 \right\}$;

E3) for $CVaR_\alpha(\cdot)$

$$Q(U) = \bigcup_{\left\{ B_u p_0 \le c_u, p_0 \ge 0, \sum_{i=1}^{n} p_i^0 = 1 \right\}} \left\{ p : Ip \le \frac{1}{1-\alpha} p_0, \sum_{i=1}^{n} p_i = 1, p \ge 0 \right\} =$$

$$= \left\{ p : (p, p_0) \in \left\{ B_u p_0 \le c_u, Ip \le \frac{1}{1-\alpha} p_0, \sum_{i=1}^{n} p_i = 1, \sum_{i=1}^{n} p_i^0 = 1, p, p_0 \ge 0 \right\} \right\}.$$

For PCRM presented in form (10), such set has the following form

$$Q(U) = \bigcup_{\left\{ B_u p_0 \le c_u, p_0 \ge 0, \sum_{i=1}^{n} p_i^0 = 1 \right\}} \left\{ p : Bp \le c(p_0), \sum_i^n p_i = 1, p \ge 0 \right\}.$$

Let us now write out robust constructions for the cases C1)–C3):

$$\text{C1)}\ \rho_U(X) = \sup_p \left\{ \langle x, p \rangle : Bp \le c, \sum_{i=1}^{n} p_i = 1, p \ge 0 \right\};$$

$$\text{C2)}\ \rho_U(X) = \sup_{(p,p_0)} \left\{ \langle x, p \rangle : B_u p_0 \le c_u, Ip \le \gamma p_0, \sum_{i=1}^{n} p_i = 1, \sum_{i=1}^{n} p_i^0 = 1, p, p_0 \ge 0 \right\};$$

$$\text{C3)}\ \rho_U(X) = \sup_{(p,p_0)} \left\{ \langle x, p \rangle : B_u p_0 \le c_u, Bp \le c, p \le \gamma p_0, \sum_{i=1}^{n} p_i = 1, \sum_{i=1}^{n} p_i^0 = 1, p, p_0 \ge 0 \right\}.$$

Remark 11. In case C1), set $Q(P_0)$ does not depend on the initial probability measure P_0; therefore, the risk measure coincides with its robust construction. For convenience of presentation, we introduce the following notation:

$$B_0 = \begin{pmatrix} 1 \ldots 1 \\ -1 \ldots -1 \end{pmatrix}, c_0 = \begin{pmatrix} 1 \\ -1 \end{pmatrix}, \tag{21}$$

which allow us to represent the standard normalization condition for probabilities $\sum_{i=1}^{n} p_i = 1$ in form $B_0 p \le c_0$, as $\sum_{i=1}^{n} p_i \le 1$ and $-\sum_{i=1}^{n} p_i \le -1$.

Taking into account the notation (21), the calculation of the robust constructions $\rho_U(X)$ for cases C1)–C3) reduces, respectively, to solve the following LP problems:

$$\max_p \ \langle x, p \rangle, \quad \begin{pmatrix} B \\ B_0 \end{pmatrix} p \le \begin{pmatrix} c \\ c_0 \end{pmatrix}, p \ge 0 \tag{22}$$

$$\max_{(p,p_0)} \langle x, p \rangle = \max_{(p,p_0)} \langle x, p \rangle, \tag{23}$$
$$\begin{pmatrix} I \\ B_0 \end{pmatrix} p \le \begin{pmatrix} \gamma p_0 \\ c_0 \end{pmatrix}, p \ge 0 \qquad \begin{pmatrix} I \\ B_0 \\ O_2 \\ O_1 \end{pmatrix} p + \begin{pmatrix} -\gamma I \\ O_1 \\ B_u \\ B_0 \end{pmatrix} p_0 \le \begin{pmatrix} o_1 \\ c_0 \\ c_u \\ c_0 \end{pmatrix}, p, p_0 \ge 0$$
$$\begin{pmatrix} B_u \\ B_0 \end{pmatrix} p_0 \le \begin{pmatrix} c_u \\ c_0 \end{pmatrix}, p_0 \ge 0$$

$$\max_{(p,p_0)} \langle x, p \rangle = \max_{(p,p_0)} \langle x, p \rangle, \tag{24}$$
$$\begin{pmatrix} B \\ I \\ B_0 \end{pmatrix} p \le \begin{pmatrix} c \\ \gamma p_0 \\ c_0 \end{pmatrix}, p \ge 0 \qquad \begin{pmatrix} B \\ I \\ B_0 \\ O_2 \\ O_1 \end{pmatrix} p - \begin{pmatrix} O_3 \\ -\gamma I \\ O_1 \\ B_u \\ B_0 \end{pmatrix} p_0 \le \begin{pmatrix} c \\ o_1 \\ c_0 \\ c_u \\ c_0 \end{pmatrix}, p, p_0 \ge 0$$
$$\begin{pmatrix} B_u \\ B_0 \end{pmatrix} p_0 \le \begin{pmatrix} c_u \\ c_0 \end{pmatrix}, p_0 \ge 0$$

where o_1, O_1, O_2, O_3 are zero vector and zero matrices of dimension n, $2 \times n$, $1 \times n$, $m \times n$, and l, m are the number of rows of matrices B_u and B, respectively. Let us formulate this.

Proposal 2. For cases C1)–C3), the calculation of the robust construction of risk measure is reduced to solving LPP of the form (22)–(24), respectively.

Corollary 1. For $CVaR_\alpha(\cdot)$, the calculation of the robust construction of risk measure $CVaR_{\alpha,U}(X)$ is reduced to solving LPP (23) for $\gamma = 1/(1-\alpha)$.

4. Optimal solutions on reward-risk ratio and portfolio optimization

4.1 Robust construction of reward

In financial applications, decisions are usually made on a reward-risk ratio. If a certain risk measure is used to assess the risk, then a reward function must also be determined. Even in the classical work Markowitz 1952, the average return was considered in this capacity, which requires clarification under uncertainty.

Let us also specify the sign of the argument for the risk measure. If we take a random financial flow X as an initial, then the value $(-X)$ describes its losses, therefore, we use them to construct a risk measure $\rho(-X)$. Accordingly, the "–" sign is added to argument of the risk measure:

$$\rho_{P_0}(-X) = \sup_{P\in Q(P_0)} \int_\Omega (-X(\omega))dP(\omega) = \sup_{P\in Q(P_0)} E_P[-X].$$

Let us consider the average return as a reward function $r(\cdot)$, i.e., $r_{P_0}(X) = E_{P_0}[X]$. For uncertainty set U, by analogy with the risk measure, we estimate the reward as the guaranteed (worst-case) reward on this set as

$$r_U(X) = \inf_{P_0\in U} r_{P_0}(X) = \inf_{P_0\in U} E_{P_0}[X]. \tag{25}$$

Definition 4. Function $r_U(\cdot)$ from (25) will be called the ***robust construction*** of reward $r_{P_0}(\cdot)$ (average return) with respect to uncertainty set U.

As it is easy to see, the function $r_U(\cdot)$ possesses properties close to those for the CRM. It is normalized; its analogs of axioms A1)–A4) have the form:

A1a) $r_U(X + a) = r_U(X) + a$;

A2a) $r_U(X_1 + X_2) \geq r_U(X_1) + r_U(X_2)$;

A3a) $r_U(\lambda X) = \lambda r_U(X)$, $\lambda \geq 0$;

A4a) $r_U(X_1) \geq r_U(X_2)$, $X_1 \geq X_2$.

These are quite rational properties. The concavity of the function means that asset diversification does not reduce the worst (guaranteed) reward.

4.2 Portfolio optimization on the reward-risk ratio

Let the distribution of the return of portfolio components r_j, $j = 1,...,k$ be described by a matrix H of dimension $n \times k$, where j-th column represents the distribution of j-th component. Vector $w = (w_1,...,w_k)$ describing the portfolio structure is considered as a variable, where $\sum_1^k w_i = 1, w_i \geq 0, i = 1,...,k$. It is necessary to find a portfolio structure w that optimizes its result in term of the reward-risk ratio.

In the case of a known vector of scenario probabilities $p_0 = (p_1^0,...,p_n^0)$, if the portfolio risk is estimated by PCRM $\rho_{p_0}(\cdot)$ of form (15), (11)–(13), and the reward is estimated by the average return, the following two related problems can be considered:

1) minimization of the portfolio risk measure under lower bounds on its average return μ_0;
2) maximization of the average portfolio return under upper bounds on its risk measure ρ_0.

These problems, respectively, have the following form:

$$\begin{array}{ll} \min & \rho_{p_0}(-Hw). \\ \sum_1^k w_i = 1, \ w \geq 0 & \\ E_{p_0}[Hw] \geq \mu_0 & \end{array} \qquad \begin{array}{ll} \max & E_{p_0}[Hw]. \\ \sum_1^k w_i = 1, \ w \geq 0 & \\ \rho_{p_0}(-Hw) \leq \rho_0 & \end{array}$$

As is known from Kirilyuk 2004, 2008, both of them are reduced to the appropriate LPP, therefore, they can be efficiently solved.

In the case when the vector of scenario probabilities is estimated by uncertainties set U in form $P_0 \in U$, similar problems are posed with their robust constructions:

$$\begin{array}{ll} \min & \rho_U(-Hw). \\ \sum_1^k w_i = 1, \ w \geq 0 & \\ r_U[Hw] \geq \mu_0 & \end{array} \tag{26}$$

$$\begin{array}{ll} \max & r_U[Hw]. \\ \sum_1^k w_i = 1, \ w \geq 0 & \\ \rho_U(-Hw) \leq \rho_0 & \end{array} \tag{27}$$

Let us show that for the case of a polyhedral set U, which is described in form (11), both problems are reduced to the appropriate LPP.

Theorem 1. If problems (26) and (27) with reward function $r_U(\cdot)$ (25) and risk measure $\rho_U(\cdot)$ for cases C1)–C3) in form (22)–(24) have solutions, then their optimal portfolios are, respectively, the components w of solutions (v, u, w) of the following LPP:

$$\begin{aligned} &\min_{(w,u,v)} \quad \langle (c, c_0), u\rangle \\ &(B^T B_0^T)u + Hw \geq 0 \\ &\langle (-c_u, -c_0), v\rangle \geq \mu_0 \\ &(-B_u^T - B_0^T)v - Hw \leq 0 \\ &\sum\nolimits_1^k w_i = 1, w \geq 0, u \geq 0, v \geq 0 \end{aligned} \tag{28}$$

$$\begin{aligned} &\max_{(w,u,v)} \quad \langle (-c_u, -c_0), v\rangle. \\ &\left(-B_u^T - B_0^T\right)v - Hw \leq 0 \\ &\langle (c, c_0), u\rangle \leq \rho_0 \\ &\left(B^T B_0^T\right)u + Hw \geq 0 \\ &\sum\nolimits_1^k w_i = 1, w \geq 0, u \geq 0, v \geq 0 \end{aligned} \tag{29}$$

$$\begin{aligned} &\min_{(w,u,v)} \quad \langle (o_1, c_0, c_u, c_0), u\rangle \\ &\begin{pmatrix} I & B_0^T & O_2^T & O_1^T \\ -\gamma I & O_1^T & B_u^T & B_0^T \end{pmatrix} u \geq \begin{pmatrix} -Hw \\ o_1 \end{pmatrix} \\ &\langle (-c_u, -c_0), v\rangle \geq \mu_0 \\ &(-B_u^T - B_0^T)v - Hw \leq 0 \\ &\sum\nolimits_1^k w_i = 1, w \geq 0, u \geq 0, v \geq 0 \end{aligned} \tag{30}$$

$$\begin{aligned} &\max_{(w,u,v)} \quad \langle (-c_u, -c_0), v\rangle. \\ &\left(-B_u^T - B_0^T\right)v - Hw \leq 0 \\ &\langle (o_1, c_0, c_u, c_0), u\rangle \leq \rho_0 \\ &\begin{pmatrix} I & B_0^T & O_2^T & O_1^T \\ -\gamma I & O_1^T & B_u^T & B_0^T \end{pmatrix} u \geq \begin{pmatrix} -Hw \\ o_1 \end{pmatrix} \\ &\sum\nolimits_1^k w_i = 1, w \geq 0, u \geq 0, v \geq 0 \end{aligned} \tag{31}$$

$$
\begin{aligned}
&\min_{(w,u,v)} \qquad \langle (c, o_1, c_0, c_u, c_0), u \rangle \\
&\begin{pmatrix} B^T & I & B_0^T & O_2^T & O_1^T \\ O_3^T & -\gamma I & O_1^T & B_u^T & B_0^T \end{pmatrix} u \geq \begin{pmatrix} -Hw \\ o_1 \end{pmatrix} \\
&\langle (-c_u, -c_0), v \rangle \geq \mu_0 \\
&(-B_u^T - B_0^T)v - Hw \leq 0 \\
&\textstyle\sum_1^k w_i = 1, w \geq 0, u \geq 0, v \geq 0
\end{aligned} \tag{32}
$$

$$
\begin{aligned}
&\max_{(w,u,v)} \qquad \langle (-c_u, -c_0), v \rangle. \\
&\left(-B_u^T - B_0^T\right)v - Hw \leq 0 \\
&\langle (c, o_1, c_0, c_u, c_0), u \rangle \leq \rho_0 \\
&\begin{pmatrix} B^T & I & B_0^T & O_2^T & O_1^T \\ O_3^T & -\gamma I & O_1^T & B_u^T & B_0^T \end{pmatrix} u \geq \begin{pmatrix} -Hw \\ o_1 \end{pmatrix} \\
&\textstyle\sum_1^k w_i = 1, w \geq 0, u \geq 0, v \geq 0
\end{aligned} \tag{33}
$$

where case C1) corresponds to problems (28)–(29), C2)–(30)–(31), C2)–(32)–(33).

Proof. Since the proof for all cases is carried out similarly, we present it only for case C2). Let us write down problems (26) and (27) for the reward function $r_U(\cdot)$ (22) and risk measure $\rho_U(\cdot)$ (23). Accordingly, they have the following form:

$$
\min_{\substack{\sum_1^k w_i = 1, w \geq 0 \\ \min_{p_0} \langle Hw, p_0 \rangle \geq \mu_0 \\ \begin{pmatrix} B_u \\ B_0 \end{pmatrix} p_0 \leq \begin{pmatrix} c_u \\ c_0 \end{pmatrix}, p_0 \geq 0}} \quad \max_{(p, p_0)} \quad \langle -Hw, p \rangle, \quad \begin{pmatrix} I \\ B_0 \\ O_2 \\ O_1 \end{pmatrix} p + \begin{pmatrix} -\gamma I \\ O_1 \\ B_u \\ B_0 \end{pmatrix} p_0 \leq \begin{pmatrix} o_1 \\ c_0 \\ c_u \\ c_0 \end{pmatrix}, p, p_0 \geq 0 \tag{34}
$$

$$
\max_{\substack{\sum_1^k w_i = 1, w \geq 0 \\ \max_{(p,p_0)} \langle -Hw, p \rangle \leq \rho_0 \\ \begin{pmatrix} I \\ B_0 \\ O_2 \\ O_1 \end{pmatrix} p + \begin{pmatrix} -\gamma I \\ O_1 \\ B_u \\ B_0 \end{pmatrix} p_0 \leq \begin{pmatrix} o_1 \\ c_0 \\ c_u \\ c_0 \end{pmatrix}, p, p_0 \geq 0}} \quad \min_{p_0} \quad \langle Hw, p_0 \rangle. \quad \begin{pmatrix} B_u \\ B_0 \end{pmatrix} p_0 \leq \begin{pmatrix} c_u \\ c_0 \end{pmatrix}, p_0 \geq 0 \tag{35}
$$

Since they have solutions, their internal LPP has solutions as well:

$$\max_{(p,p_0)} \quad \langle -Hw, p\rangle,$$
$$\begin{pmatrix} I \\ B_0 \\ O_2 \\ O_1 \end{pmatrix} p + \begin{pmatrix} -\gamma I \\ O_1 \\ B_u \\ B_0 \end{pmatrix} p_0 \le \begin{pmatrix} o_1 \\ c_0 \\ c_u \\ c_0 \end{pmatrix}, p, p_0 \ge 0 \qquad (36)$$

$$\min_{p_0} \quad \langle Hw, p_0\rangle.$$
$$\begin{pmatrix} B_u \\ B_0 \end{pmatrix} p_0 \le \begin{pmatrix} c_u \\ c_0 \end{pmatrix}, p_0 \ge 0 \qquad (37)$$

Moreover, due to specific constraints, they have finite solutions. Therefore, (36) and (37) are equivalent to their dual LPP, which, respectively, have the following form:

$$\min_u \quad \langle (o_1, c_0, c_u, c_0), u\rangle,$$
$$\begin{pmatrix} I & B_0^T & O_2^T & O_1^T \\ -\gamma I & O_1^T & B_u^T & B_0^T \end{pmatrix} u \ge \begin{pmatrix} -Hw \\ o_1 \end{pmatrix}, u \ge 0 \qquad (38)$$

$$\max_v \quad \langle (-c_u, -c_0), v\rangle.$$
$$\left(-B_u^T - B_0^T\right) v - Hw \le 0, \ v \ge 0 \qquad (39)$$

Replacing problems (36), (37) by (38), (39) in problems (34), (35), respectively, we obtain

$$\min_{\substack{w \\ \sum_1^k w_i = 1, w \ge 0 \\ \max_{v,\ (-B_u^T - B_0^T)v - Hw \le 0, v \ge 0} \langle (-c_u, -c_0), v\rangle \ge \mu_0}} \quad \min_{\substack{u \\ \begin{pmatrix} I & B_0^T & O_2^T & O_1^T \\ -\gamma I & O_1^T & B_u^T & B_0^T \end{pmatrix} u \ge \begin{pmatrix} -Hw \\ o_1 \end{pmatrix}, u \ge 0}} \quad \langle (o_1, c_0, c_u, c_0), u\rangle, \qquad (40)$$

$$\max_{\substack{w \\ \sum_1^k w_i = 1, w \ge 0 \\ \min_{u,\ \begin{pmatrix} I & B_0^T & O_2^T & O_1^T \\ -\gamma I & O_1^T & B_u^T & B_0^T \end{pmatrix} u \ge \begin{pmatrix} -Hw \\ o_1 \end{pmatrix}, u \ge 0} \langle (o_1, c_0, c_u, c_0), u\rangle \le \rho_0}} \quad \max_{\substack{v \\ \left(-B_u^T - B_0^T\right) v - Hw \le 0, \ v \ge 0}} \quad \langle (-c_u, -c_0), v\rangle. \qquad (41)$$

It is also easy to see the equivalence of the following conditions:

$$\begin{array}{c}\max_{v\geq 0} \quad \langle(-c_u,-c_0),v\rangle \geq \mu_0 \\ \left(-B_u^T - B_0^T\right)v - Hw \leq 0\end{array} \Leftrightarrow \begin{array}{c}\exists v \geq 0 : \langle(-c_u,-c_0),v\rangle \geq \mu_0, \\ \left(-B_u^T - B_0^T\right)v - Hw \leq 0\end{array} \tag{42}$$

$$\begin{array}{c}\min_{u\geq 0} \quad \langle(c_0,o_1,c_0,c_u),u\rangle \leq \rho_0 \\ \begin{pmatrix} I & B_0^T & O_2^T & O_1^T \\ -\gamma I & O_1^T & B_u^T & B_0^T \end{pmatrix} u \geq \begin{pmatrix} -Hw \\ o_1 \end{pmatrix}\end{array} \Leftrightarrow \begin{array}{c}\exists u \geq 0 : \langle(o_1,c_0,c_u,c_0),u\rangle \leq \rho_0. \\ \begin{pmatrix} I & B_0^T & O_2^T & O_1^T \\ -\gamma I & O_1^T & B_u^T & B_0^T \end{pmatrix} u \geq \begin{pmatrix} -Hw \\ o_1 \end{pmatrix}\end{array} \tag{43}$$

Replacing the left conditions of relations (42), (43) with their equivalent right conditions in problems (40), (41), respectively, we finally obtain

$$\begin{array}{c}\min_{(w,v)} \\ \sum_1^k w_i = 1, w \geq 0 \\ \langle(-c_u,-c_0),v\rangle \geq \mu_0 \\ \left(-B_u^T - B_0^T\right)v - Hw \leq 0, v \geq 0\end{array} \quad \begin{array}{c}\min_u \\ \begin{pmatrix} I & B_0^T & O_2^T & O_1^T \\ -\gamma I & O_1^T & B_u^T & B_0^T \end{pmatrix} u \geq \begin{pmatrix} -Hw \\ o_1 \end{pmatrix}, u \geq 0\end{array} \quad \langle(o_1,c_0,c_u,c_0),u\rangle, \tag{44}$$

$$\begin{array}{c}\max_w \\ \sum_1^k w_i = 1, w \geq 0 \\ \langle(o_1,c_0,c_u,c_0),u\rangle \leq \rho_0 \\ \begin{pmatrix} I & B_0^T & O_2^T & O_1^T \\ -\gamma I & O_1^T & B_u^T & B_0^T \end{pmatrix} u \geq \begin{pmatrix} -Hw \\ o_1 \end{pmatrix}, u \geq 0\end{array} \quad \begin{array}{c}\max_v \\ \left(-B_u^T - B_0^T\right)v - Hw \leq 0,\ v \geq 0\end{array} \quad \langle(-c_u,-c_0),v\rangle. \tag{45}$$

It is easy to see those problems (30) and (31) are nothing but problems (44) and (45), respectively, which proves the theorem for case C2).

Corollary 2. If $CVaR_\alpha(\cdot)$ is used as the initial risk measure, then portfolio optimization problems are reduced to solving LPP (30)–(31) for $\gamma = 1/(1-\alpha)$.

4.3 Portfolio optimization by Sharpe ratio

Let us now consider for a portfolio the maximization of Sharpe ratio with robust constructions of the reward function and the risk measure

$$ShR_U(X) = \frac{r_U(X)}{\rho_U(X)}.$$

We represent it as a minimization problem in the following form

$$\min_{\sum_1^k w_i=1, w\geq 0} \frac{1}{ShR_U(Hw)} = \min_{\sum_1^k w_i=1, w\geq 0} \frac{\rho_U(Hw)}{r_U(Hw)}. \tag{46}$$

As known, its classical analogue for a known vector of scenario probabilities p_0 is reduced to solving an LP problem (see, for example, Kirilyuk

2014). Let us show that problem (46) can also be reduced to an appropriate LP problem.

Theorem 2. If the reward function $r_U(\cdot)$ from (25) does not equal 0, then the optimal portfolios for the problem (46) with such $r_U(\cdot)$ and risk measures $\rho_U(\cdot)$ for cases C1)–C3) in form (22)–(24) are $w = \tilde{w}/t$ in the solutions $(\tilde{w}, t, \tilde{u}, \tilde{v})$ of the following LP problems, respectively:

$$\min_{(\tilde{w},t,\tilde{u},\tilde{v})} \quad \langle (c, c_0), \tilde{u} \rangle,$$
$$(B^T B_0^T)\tilde{u} + H\tilde{w} \geq 0$$
$$\langle (-c_u, -c_0), \tilde{v} \rangle = 1$$
$$(-B_u^T - B_0^T)\tilde{v} - H\tilde{w} \leq 0$$
$$\sum_1^k \tilde{w}_i = t, \tilde{w} \geq 0, \tilde{u} \geq 0, \tilde{v} \geq 0, t > 0$$

$$\min_{(\tilde{w},t,\tilde{u},\tilde{v})} \quad \langle (o_1, c_0, c_u, c_0), \tilde{u} \rangle,$$
$$\begin{pmatrix} I & B_0^T & O_2^T & O_1^T \\ -\gamma I & O_1^T & B_u^T & B_0^T \end{pmatrix} \tilde{u} \geq \begin{pmatrix} -H\tilde{w} \\ o_1 \end{pmatrix}$$
$$\langle (-c_u, -c_0), \tilde{v} \rangle = 1$$
$$(-B_u^T - B_0^T)\tilde{v} - H\tilde{w} \leq 0$$
$$\sum_1^k \tilde{w}_i = t, \tilde{w} \geq 0, \tilde{u} \geq 0, \tilde{v} \geq 0$$

$$\min_{(\tilde{w},t,\tilde{u},\tilde{v})} \quad \langle (c, o_1, c_0, c_u, c_0), \tilde{u} \rangle.$$
$$\begin{pmatrix} B^T & I & B_0^T & O_2^T & O_1^T \\ O_3^T & -\gamma I & O_1^T & B_u^T & B_0^T \end{pmatrix} \tilde{u} \geq \begin{pmatrix} -H\tilde{w} \\ o_1 \end{pmatrix}$$
$$\langle (-c_u, -c_0), \tilde{v} \rangle = 1$$
$$(-B_u^T - B_0^T)\tilde{v} - H\tilde{w} \leq 0$$
$$\sum_1^k \tilde{w}_i = t, \tilde{w} \geq 0, \tilde{u} \geq 0, \tilde{v} \geq 0$$

The theorem-proof can be easily obtained using arguments similar to those used in the proof of Theorem 6 from Kirilyuk 2014, or by applying Lemma 1 from Schaible 1974, taking into account the results of the previous theorem.

Remark 12. If $\rho_U(\cdot) \geq 0$, then, as follows from Schaible 1974, equality $\langle (-c_u, -c_0), \tilde{v} \rangle = 1$ in the problem constraints can be replaced by an inequality $\langle (-c_u, -c_0), \tilde{v} \rangle \geq 1$.

5. Risk measures for imprecise scenario values of random variables

Suppose that we know only the lower and upper bounds of scenario values of r.v. X, not their values. On probability space $(\Omega, \sum, P_0)$ r.v. X is described by its measurable (random) lower X_l and upper X_u bounds in the form

$$X_l(\cdot) \le X(\cdot) \le X_u(\cdot) \ a.s.$$

Such a situation was called in Kirilyuk 2018 the case of imprecise scenario estimates of r.v. X. In it the PCRM technique can be used as well.

So, then the risk measure for estimates X_l and X_u can be considered as their lower and upper estimates for risk measure of r.v. X as

$$\rho(X_l) \le \rho(X) \le \rho(X_u).$$

In this case, the concept of a robust construction of a risk measure becomes much more complicated, since it is necessary to consider a whole set of Pareto optimal values for the following problem

$$\begin{pmatrix} \rho_{p_0}(X_l) \\ \rho_{p_0}(X_u) \end{pmatrix} \to \max_{p_0 \in U}$$

In this case, various problem statements are possible, for example, one can make a convolution of such criteria, or use one of these criteria $\rho_U(X_l)$, $\rho_U(X_u)$ for minimization with restrictions on the second. This can be done under uncertainty for the reward function also.

So, let's turn to the problem of portfolio optimization. Suppose that we do not know the matrix H of return distributions of components of financial instruments by scenarios, but only its lower and upper estimates, represented by matrices H^l and H^u, respectively. Those, $H = \{r_{ij}\}_{i=1,j=1}^{n,k}$ is element-wise evaluated by matrices $H^l = \{r_{ij}^l\}_{i=1,j=1}^{n,k}$ and $H^u = \{r_{ij}^u\}_{i=1,j=1}^{n,k}$, where

$$r_{ij}^l \le r_{ij} \le r_{ij}^u, \ i = 1,\dots, n, j = 1,\dots, k.$$

Then analogs of problems (26)–(27) in such a situation are, respectively, the following problems

$$\min_{\substack{\sum_1^n w_i = 1, w_i \ge 0 \\ \rho_U(-H^l w) \le \rho_0^l \\ r_U(H^u w) \ge \mu_0^u \\ r_U(H^l w) \ge \mu_0^l}} \rho_U(-H^u w), \qquad \min_{\substack{\sum_1^n w_i = 1, w_i \ge 0 \\ \rho_U(-H^u w) \le \rho_0^u \\ r_U(H^u w) \ge \mu_0^u \\ r_U(H^l w) \ge \mu_0^l}} \rho_U(-H^l w),$$

$$\begin{array}{ll}
\max\limits_{\sum_1^n w_i = 1, w_i \geq 0} r_U(H^u w), & \max\limits_{\sum_1^n w_i = 1, w_i \geq 0} r_U(H^l w), \\
r_U(H^l w) \geq \mu_0^l & r_U(H^u w) \geq \mu_0^u \\
\rho_U(-H^l w) \leq \rho_0^l & \rho_U(-H^l w) \leq \rho_0^l \\
\rho_U(-H^u w) \leq \rho_0^u & \rho_U(-H^u w) \leq \rho_0^u
\end{array}$$

in which μ_0^j, ρ_0^j, $j = l, u$, respectively, are the lower and upper constraints on the reward function (from below) and the robust construction of the risk measure (from above).

If the listed problems with reward functions $r_U(\cdot)$ in the form (25), (19) and risk measures $\rho_U(\cdot)$ in the form (22)–(24) have solutions, then they, like problems (26) and (27), can be reduced using similar reasoning to the respective LPP.

At the same time, taking into account the number of restrictions in these problems, the obtained LPP will be of a significantly higher dimension compared to the problems described in Theorem 1. Nevertheless, such a reduction will greatly facilitate the solution of the original portfolio optimization problems.

6. Conclusion

For the case of uncertainty, when the initial probability measure is not known and is described by an uncertainty set U, the robust construction of risk measure is proposed. Under using PCRM and a polyhedral set U, the calculation of such robust construction of risk measure is reduced to solving appropriate LPP.

The problems of portfolio optimization by a reward-risk ratio are considered, where the reward and risk are described by the robust constructions of the average return and PCRM, respectively. It is shown how tasks of maximizing the reward under upper bounds on the risk and minimizing the risk under lower bounds on the reward are reduced to appropriate LPP. It is described how the problem of maximizing the Sharpe ratio for a portfolio with these robust constructions reduces to the corresponding LP problem as well.

Similar statements of portfolio optimization problems and their reduction to LPP for the case of imprecise scenario estimates of r.v. are discussed also.

References

Artzner, P., F. Delbaen, J.M. Eber and D. Heath. 1999. Coherent measures of risk. Mathematical Finance, 9(3): 203–228.

Delage, E. and Y. Ye. 2010. Distributionally robust optimization under moment uncertainty with application to Data-Driven Problems. Operations Research, 58(3): 595–612.

Föllmer, H. and A. Schied. 2002. Convex measures of risk and trading constraints. Finance Stochastics, 6(4): 429–447.

Kirilyuk, V.S. 2004. The class of polyhedral coherent risk measures. Cybernetics and System Analysis, 40(4): 599–609.

Kirilyuk, V.S. 2008. Polyhedral coherent risk measures and investment portfolio optimization. Cybernetics and System Analysis, 44(2): 250–260.

Kirilyuk, V.S. 2014. Polyhedral coherent risk measures and optimal portfolios on the reward-risk ratio. Cybernetics and System Analysis, 50(5): 724–740.

Kirilyuk, V.S. 2015. Risk measures in stochastic programming and robust optimization. Cybernetics and System Analysis, 51(6): 874–885.

Kirilyuk, V.S. 2018. Polyhedral coherent risk measures in the case of imprecise scenario estimates. Cybernetics and System Analysis, 54(3): 423–433.

Kirilyuk, V.S. 2019. Polyhedral coherent risk measures and robust optimization. Cybernetics and System Analysis, 55(6): 999–1008.

Markowitz, H.M. 1952. Portfolio selection. Journal of Finance, 7(1): 77–91.

Rockafellar, R.T. 1970. Convex Analysis. Princeton University Press, Princeton.

Rockafellar, R.T. and S. Uryasev. 2000. Optimization of conditional value-at-risk. Journal of Risk, 2(3): 21–41.

Schaible, S. 1974. Parameter-free convex equivalent and dual programs of fractional programming problems. Zeitschrift für Oper. Res. 18(5): 187–196.

Shapiro, A., D. Dentcheva and A. Ruszczynski. 2009. Lectures on Stochastic Programming. Modeling and Theory. SIAM, Philadelphia.

Shapiro, A. 2017. Distributionally robust stochastic programming. SIAM J. Optim. 27(4): 2258–2275.

Wiesemann, W., D. Kuhn and M. Sim. 2014. Distributionally robust convex optimization. Operations Research 62(6): 1358–1376.

Chapter 5

On Minimum Length Confidence Intervals

Pavel S Knopov,[1,*] *Arnold S Korkhin*[2] and *Liliia B Vovk*[3]

1. Introduction: Statement of the problem

In mathematical statistics, the determination of the interval estimate (confidence interval) of a certain parameter $\alpha^0 \in \mathbb{R}^1$ is a widespread problem, which consists in finding such quantities A and B for which the equality $P(A \leq \alpha^0 \leq B) = v$ is satisfied, where the probability v is a given value, A and B are the observation functions. Definition of A and B is based on equality

$$P(A \leq \alpha^0 \leq B) = P(a \leq X \leq b),$$

where X is a random variable (r. v.) with known probability density $f(x)$.

A confidence interval length is often determined by the equalities:

$$B - A = c(b - a) \tag{1}$$

or

$$B - A = c(a^{-1} - b^{-1}), \tag{2}$$

where c is some value that depends on the sample, but does not depend on a and b. Therefore, for a particular sample, c is a constant.

[1] V. M. Glushkov Institute of Cybernetics, National Academy of Sciences of Ukraine.
[2] Pridneprovskaya Academy of Construction and Architecture, Dnipro, Ukraine.
[3] National Technical University of Ukraine "Igor Sikorsky Kyiv Polytechnic Institute".
* Corresponding author: knopov1@yahoo.com

The question arises, how to choose a and b so that the length $l = (b - a)$ for case (1) or $l = (a^{-1} - b^{-1})$ for case (2) is minimal, which provides a minimum the length of the confidence interval for the unknown parameter α^0. Note that minimizing $l = (b - a)$ is the problem of the r. v. X probability interval $[a, b]$ length minimizing.

A number of articles are devoted to this problem, among which we will point out Tate et al. 1959, Levy et al. 1974, Ferentinos et al. 2006. In the first two papers, a and b are determined to minimize the confidence intervals for the variance and variance ratio, respectively. The values of a and b are given for several values of v, the number of degrees of freedom of the χ^2 distribution $2 \leq q \leq 29$ in Tate et al. 1959 and twelve values of two degrees of freedom of Fisher distribution in Levy et al. 1974 for $v = 0.95$. The article Ferentinos et al. 2006, which is a generalizing work, analyzes solutions to the problems of the confidence interval length minimizing for cases (1) and (2), when a probability density $f(x)$ is a differentiable function. However, the necessary and sufficient conditions for a minimum in relation to the case (1) are not explicitly defined, and for case (2) there are no sufficient minimum conditions. The question of multiextremality, which arises when $l = (a^{-1} - b^{-1})$, is not considered.

In this article, we present the necessary and sufficient conditions for minimum length confidence intervals existence for cases (1) and (2). For frequent cases when X has the χ^2 distribution or Fisher distribution, conditions are given when the length $l = (a^{-1} - b^{-1})$ minimization problem has one minimum. A simplification of this problem is proposed, which allows it to be reduced to the one-dimensional minimization problem on the interval [0,1].

This work also has a practical side: it describes the definition of intervals for the degrees of freedom of the χ^2 distribution and Fisher distribution, which provides a decrease of the confidence interval length by a given relative value, i.e., the question of the effectiveness of using the minimum length confidence intervals is being studied.

2. Minimization of a probability interval length – Case (1)

Let us present the necessary and sufficient conditions for the minimum length of the probability interval $l = (b - a)$, which coincides with the minimum of the confidence interval according to (1).

Theorem 1. Let a and b, respectively, be the left and right ends of some 100v% probability interval of a random variable X having a unimodal probability density $f(x)$ defined on the interval I. The probability v is given. Then, for $a = a^*$ and $b = b^*$ to be the ends of the interval of the probability of the minimum length, it is necessary and sufficient to satisfy the following equalities for three possible cases of the position of $f(x)$ mode.

1) Mode of $f(x)$ is laying in the point d – the left point of the interval I, $f(x)$ is strictly monotonically decreasing, then

$$a^* = d, b^* \text{is found from the equality} \int_d^{b^*} f(x)dx = v. \quad (3)$$

2) Mode of $f(x)$ is laying in the point e – the right point of the interval I, $f(x)$ is strictly monotonically increasing, then

$$b^* = e, a^* \text{is found from the equality} \int_{a^*}^{e} f(x)dx = v.$$

3) Mode of $f(x)$ is the inner point of the interval I (the whole number axis or non-negative half-axis); $f(x)$ is a differentiable function, then a^* and b^* are laying respectively to the left and to the right of the mode, and

$$f(a^*) = f(b^*), \int_{a^*}^{b^*} f(x)dx = v. \quad (4)$$

Proof. 1) Finding the minimum length 100v% probability interval is the following optimization problem.

It is necessary to find such quantities a and b for which

$$b - a \to \min \quad (5)$$

under constraints

$$\int_a^b f(x)dx - v = 0, a - d \geq 0. \quad (6)$$

Let us compose Lagrange function for the problem (5), (6).

$$L(z,\boldsymbol{\lambda}) = (b-a) + \lambda_1\left(\int_a^b f(x)dx - v\right) - \lambda_2(a-d), z = \begin{bmatrix} b \\ a \end{bmatrix}, \boldsymbol{\lambda} = \begin{bmatrix} \lambda_1 \\ \lambda_2 \end{bmatrix} \quad (7)$$

where λ_i, $i = 1,2$ are Lagrange multipliers.
The necessary minimum condition for the problem (5), (6) is

$$\nabla_z L(z,\boldsymbol{\lambda}) = \boldsymbol{O}_2; \int_a^b f(x)dx - v; -\lambda_2 a = 0, \lambda_2 \geq 0 \quad (8)$$

where $\nabla_z L(z,\boldsymbol{\lambda}) = \begin{bmatrix} \frac{\partial L}{\partial b} \frac{\partial L}{\partial a} \end{bmatrix}'$, $\boldsymbol{O}_2$ is a zero two-dimensional vector, stroke means transpose.

Condition (8) in expanded form is

$$\frac{\partial L}{\partial b} = 1 + \lambda_1 f(b) = 0, \frac{\partial L}{\partial a} = -1 - \lambda_1 f(a) - \lambda_2 = 0, -\lambda_2(a-d) = 0, \lambda_2 \geq 0. \quad (9)$$

Hereinafter $\frac{\partial f(c)}{\partial x}$ means the derivative of $f(x)$ at the point $x = c$.

Let $\lambda_2 = \lambda_2^* = 0$, where λ_2^* is a solution of the equalities and inequalities system (9). Then from the first two equations in (9) we obtain $f(a^*) = f(b^*)$. This equality contradicts the condition of the theorem assertion 1) on the probability density strictly monotone decrease. Hence, $\lambda_2 > 0$. From this and the third equation in (9), we obtain the first equality in (3) $a = a^* = d$. Then the second equality is (3) follows from $f(x)$ monotonic decrease.

It remains for us to prove that the solution of the system of equations (3) gives the minimum point in problem (5), (6). Let us denote the left-hand sides of the constraints $g_1(z) = \int_a^b f(x)dx - v = 0$, $g_2(z) = -a + d \leq 0$. Their gradients at the point $z = z^* = [a^*\ b^*]'$ are the problem (5), (6) solutions

$$\nabla g_1(z^*) = \begin{bmatrix} f(b^*) \\ -f(a^*) \end{bmatrix}, \nabla g_2(z^*) = \begin{bmatrix} 0 \\ -1 \end{bmatrix}. \tag{10}$$

Moreover, $g_1(z^*) = 0$, $g_2(z^*) = 0$.

We have $f(b^*) \neq 0$, since otherwise, due to the strictly monotonic decrease of $f(x)$ we obtain $\int_d^{b^*} f(x)dx = 1$, which contradicts the first restriction in (6). For $f(b^*) \neq 0$ vectors in (10) will be linearly independent, since the matrix $[\nabla g_1(z^*) \vdots \nabla g_2(z^*)]$ is non-degenerate.

So, at the point z^* of the optimization problem (5), (6) solution all constraints are satisfied, and their left sides gradients are linearly independent. Then, according to Poliak 1983 Ch. 9, Sec. 2, Theorem 4 the points a^* and b^* are the ends of the minimum length probability interval.

2) The proof of the theorem assertion 2) is similar to the proof of assertion 1).
3) In this case, no restrictions are imposed on the ends of the minimum length probability interval, therefore condition (9) takes the form

$$\frac{\partial L}{\partial b} = 1 + \lambda_1 f(b) = 0, \quad \frac{\partial L}{\partial a} = -1 - \lambda_1 f(a) = 0. \tag{11}$$

We have from (11) $\lambda_1 = -\dfrac{1}{f(a)} < 0$. Substituting this expression into the first equation in (11), we obtain the first equality in (4).

Let us show that condition (4) defines the minimum points of the problem

$$b - a \to \min, g_1(z) = \int_a^b f(x)dx - v = 0. \tag{12}$$

According to Poliak 1983 Ch. 8, §1, Theorem 5, a^* and b^* determine the problem (11) minimum point on the plane, if $\boldsymbol{s}'(\nabla_z^2 L(z^*, \lambda_1^*))\boldsymbol{s} > 0$ for all vectors $\boldsymbol{s} \in \mathbb{R}^2$ satisfying the equality $\boldsymbol{s}'\nabla g_1(z^*) = 0$. Here $\nabla_z^2 L(z^*, \lambda_1^*)$ is the matrix of second derivatives of Lagrange function with respect to a and b components of z calculated for $a = a^*$, $b = b^*$ and $\lambda_1 = \lambda_1^*$, $\mathbf{s}'(\nabla_z^2 L(z^*, \lambda_1^*))\mathbf{s}$ is the quadratic form of $\mathbf{s} = [s_1\ s_2]'$.

We have from (10) $\mathbf{s}'\nabla g_1(z^*) = s_1 f(b^*) - s_2 f(a^*) = 0$. Hence,

$$s_1 = \frac{f(a^*)}{f(b^*)} s_2. \tag{13}$$

We get from (11) $\nabla_z^2 L(z^*, \lambda_1^*) = \begin{bmatrix} \lambda_1^* \dfrac{\partial f(b^*)}{\partial b} & 0 \\ 0 & -\lambda_1^* \dfrac{\partial f(a^*)}{\partial a} \end{bmatrix}$. Taking into account (13) we get $\mathbf{s}'(\nabla_z^2 L(z^*, \lambda_1^*))\mathbf{s} = \lambda_1^* s_2^2 \left(\dfrac{\partial f(b^*)}{\partial b} \left(\dfrac{f(a^*)}{f(b^*)} \right)^2 - \dfrac{\partial f(a^*)}{\partial a} \right)$.

Since $\lambda_1^* < 0$, for this value to be positive it is sufficient if $\dfrac{\partial f(b^*)}{\partial b} < 0$ and $\dfrac{\partial f(a^*)}{\partial a} > 0$. These inequalities are satisfied when $f(x)$ mode is between the values a^* and b^*. Assertion 3) of the theorem is proved, which completes its proof.

Corollary 1. Let $f(x)$ be a symmetric probability density with respect to $x = 0$. Then from the first condition in (4) we obtain $f(-b^*) = f(b^*)$.

Theorem 1 combines the results obtained in Ferentinos et al. 2006 by another method.

Now let us compare the $100v\%$ probability interval of the minimum length with the interval, the boundaries of which are, respectively, the lower and upper $100p\%$ percentage points of the distribution under consideration, moreover, $p = \dfrac{1-v}{2}$. This interval is usually used in practical calculations. See, for instance, Ledermann et al. 1984. Table 1 shows the results of comparing the lengths of the probability interval for Fisher distribution. In it, q_1 and q_2 are the degrees of freedom of this distribution, a, b are the boundaries of the interval, l and L are respectively the lengths of the interval got with the traditional method and its minimum length. The length decrease of the probability interval was determined by the formula $(l - L)\, l^{-1}\, 100\%$.

The boundaries of the minimum length probability interval a and b were obtained as a result of solving the optimization problem (5), (6). To solve this, the Mathcad 14 software system was used. The point with coordinates $a = 2$, $b = 100$ was chosen as the iterative solution process starting point.

It follows from Table 1 that the minimum length interval differs significantly from the traditionally calculated interval for Fisher distribution. The discrepancy is especially large for the probability density corresponding to the degrees of freedom $q_1 = 20, q_2 = 2$, for which the mathematical expectation and higher order moments do not exist.

Let's look at Table 2, obtained similarly to Table 1 for χ^2 distribution. In it, q is the number of degrees of freedom of this distribution.

Table 1. Comparison of the minimum length probability interval and the traditional interval for Fisher distribution lengths (v = 0.95).

Degrees of freedom	Values							
q_1	15	20	20	20	20	30	30	3
q_2	12	2	5	12	15	15	10	50
Length decrease, %	13.7	50.5	26.3	13.3	11.2	10.8	15.1	15.9

Table 2. Comparison of the minimum length probability interval and the traditional interval for χ^2 distribution lengths (v = 0.95).

Degrees of freedom	Values									
q	2	4	8	10	20	30	40	50	60	80
Length decrease, %	18.2	11.4	5.8	4.6	2.3	1.5	1.11	0.90	0.745	0.582

It follows from Table 1 and Table 2 comparison that the gain from using the minimum length probability interval for χ^2 distribution is small and noticeably smaller than for Fisher distribution. According to the Table 2, the largest discrepancy between the lengths of intervals determined by different methods occurs for small samples.

Figure 1 shows the dependence of the interval of probabilities length decrease from Table 2 on the skew ratio $Sk = \sqrt{\frac{2}{q}}$ (for χ^2 distribution). For this distribution, the coefficient of skewness $\gamma_{as} = 2Sk$ (See Zellner 1971, Appendix A3). It can be seen from Figure 1 that the difference between the probability interval minimum length and its length calculated in the traditional way increases as the skew ratio increases (the number of degrees of freedom q decreases).

For Fisher distribution, such a clear dependence on the coefficients characterizing the distribution density asymmetry as in Figure 1 does not exist, which can be explained, in particular, by the absence of the distribution moments necessary to calculate these coefficients for some degrees of freedom.

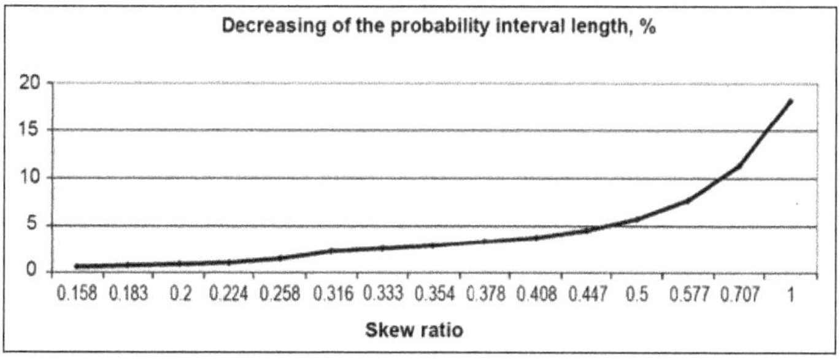

Figure 1. Influence of the skew ratio on the efficiency of using the minimum length probability interval for χ^2 distribution.

3. Minimization of a probability interval length – Case (2)

Theorem 2. Let a and b, respectively, be the left and right boundaries of some $100v\%$ probability interval of a random variable with a known probability density $f(x)$. The probability v is given. Then, for the value $a^{-1} - b^{-1} > 0$ to be minimal, for the solution $a = a^*$, $b = b^*$ of the problem

$$a^{-1} - b^{-1} \to \min, g_1(z) = \int_a^b f(x)dx - v = 0, z = \begin{bmatrix} b \\ a \end{bmatrix}. \tag{14}$$

it is necessary and sufficient to fulfill the following conditions:

$$1)\ (b^*)^2 f(b^*) = (a^*)^2 f(a^*), \int_{a^*}^{b^*} f(x)dx = v, \tag{15}$$

$$2)\ 2(a^*)^{-3}\left(1 - \frac{b^*}{a^*}\right) - \frac{1}{(a^*)^2 f(a^*)}\left(\frac{\partial f(b^*)}{\partial b}\left(\frac{b^*}{a^*}\right)^4 - \frac{\partial f(a^*)}{\partial a}\right) > 0. \tag{16}$$

Proof. Lagrange function for the problem (14)

$$L(z, \lambda_1) = (a^{-1} - b^{-1}) + \lambda_1\left(\int_a^b f(x)dx - v\right), z = \begin{bmatrix} b \\ a \end{bmatrix} \tag{17}$$

where λ_1 is Lagrange multiplier.

A necessary minimum condition for this problem

$$\nabla_z L(z, \lambda_1) = \boldsymbol{O}_2, \int_a^b f(x)dx = v. \tag{18}$$

From (18) we obtain equations for three unknowns a, b and λ_1:

$$\frac{\partial L}{\partial b} = b^{-2} + \lambda_1 f(b) = 0,\ \frac{\partial L}{\partial a} = -a^{-2} - \lambda_1 f(a) = 0,\ \int_a^b f(x)dx = v. \tag{19}$$

From the first two equations in (19) we obtain

$$\lambda_1 = \lambda_1^* = -\frac{(a^*)^{-2}}{f(a^*)} \tag{20}$$

and the theorem condition 1).

The sufficient condition for the minimum of problem (14) will be of the same form as in the proof of Theorem 1: $s'(\nabla_z^2 L(z^*, \lambda_1^*))s > 0$, where L is given by formula (17); $z^* = [b^* \ a^*]'$; the components $s = [s_1 \ s_2]'$ satisfy condition (13).

Using expressions for the derivatives of Lagrange function (19), we obtain

$$\nabla_z^2 L(z^*, \lambda_1^*) = \begin{bmatrix} -2(b^*)^{-3} + \lambda_1^* \dfrac{\partial f(b^*)}{\partial b} & 0 \\ 0 & 2(a^*)^{-3} - \lambda_1^* \dfrac{\partial f(a^*)}{\partial a} \end{bmatrix}.$$

Taking into account (13), we obtain

$$s'(\nabla_z^2 L(z^*, \lambda_1^*))s = 2s_2^2 \left(\frac{1}{(a^*)^3} - \frac{1}{(b^*)^3} \left(\frac{f(a^*)}{f(b^*)} \right)^2 \right) + \lambda_1^* s_2^2 \left(\frac{\partial f(b^*)}{\partial b} \left(\frac{f(a^*)}{f(b^*)} \right)^2 - \frac{\partial f(a^*)}{\partial a} \right).$$

Hence, using the first equality in (15), and, since $s_2^2 > 0$, we have (16). ◊

Corollary 2. The mode of $f(x)$ lays between the values a^* and b^* which are the solution to problem (14).

Proof. In (16), the first term is negative. For this inequality to be satisfied, the second term must be positive. Since according to (20) $\lambda_1^* < 0$, the second term in (16) will be positive if $\dfrac{\partial f(b^*)}{\partial b} < 0$, $\dfrac{\partial f(a^*)}{\partial a} > 0$.◊

Unlike the problem considered in Theorem 1, where there is one minimum, the fulfillment of Theorem 2 conditions allows us to check that the obtained problem (14) solution a^*, b^* is a local minimum point. Finding the global minimum in case when this problem has several local minima, is a separate difficult task. Therefore, it is important to know the cases when it is possible to guarantee the existence of one minimum in problem (14). The most common cases are when $f(x)$ in (14) is the probability density of χ^2 or Fisher distribution. The two statements below allow us to guarantee the uniqueness of solution (14) for the indicated cases.

First of all, let us simplify problem (14). Using the first condition in (15), we exclude one variable in (14). We have from (15) an equation connecting the sought variables a and b:

$$\frac{f(b)}{f(a)} = \frac{a^2}{b^2} = t^2 \tag{21}$$

where $t \in [0,1]$.

We put $a = a(t)$, $b = b(t)$. Moreover, according to (21)

$$a(t) = tb(t). \tag{22}$$

The solution of problem (14) $a^* = a(t^*)$, $b^* = b(t^*)$, where t^* is the solution of the equation with respect to t

$$\varphi(t) = \int_{a(t)}^{b(t)} f(x)dx = v \text{ for } 0 \le t \le 1, \tag{23}$$

must obey this equality.

The existence of a unique solution to equation (23) implies the existence of a unique solution to problem (14). Consider two statements that guarantee the uniqueness of the solution (23).

Statement 1. Let in (23) $f(x)$ be χ^2 distribution density: $f(x) = C(q)x^{(q-2)/2}\, e^{-1/2}$, $x > 0$; $f(x) = 0$, $x \le 0$, where q is a number of degrees of freedom, $C(q)$ is a constant not depending on x. Then the function $\varphi(t)$ is strictly monotonically decreasing on the interval (0,1].

Proof. From (21), formula for $f(x)$ and equality $a = tb$, we get $\frac{f(b)}{f(a)} = \frac{(1-t)exp\left(-\frac{b}{2}\right)}{t^m} = t^2$, where $m = \frac{q}{2} - 1$. Hence, we get

$$b = b(t) = -(q + 2) \ln t(1 - t)^{-1}. \tag{24}$$

The derivative of the integral (23) is

$$\frac{d\varphi(t)}{dt} = \frac{db(t)}{dt} f(b(t)) - \frac{da(t)}{dt} f(a(t)). \tag{25}$$

Let us define the values on the right-hand side of this equality. We have, respectively, from (22) and (23)

$$\frac{da(t)}{dt} = \frac{db(t)}{dt} t + b(t), \tag{26}$$

$$\frac{db(t)}{dt} = -(q+2)\left(\frac{1}{t(1-t)} + \frac{\ln t}{(1-t)^2}\right). \tag{27}$$

Taking into account (26) and (21) we get from (25)

$$\frac{d\varphi(t)}{dt} = f(a(t))\left[\frac{db(t)}{dt} t(t-1) - b(t)\right]. \tag{28}$$

Substituting into this formula from (24) and (27) the expressions for the function $b(t)$ and its derivative, we obtain $\frac{d\varphi(t)}{dt} = f(a(t))\left[1+\frac{\ln t}{1-t}(1+t)\right]$. Analysis of the function $\frac{\ln t}{1-t}$ shows that $\frac{\ln t}{1-t} \le -1$ for $0 \le t \le 1$. Hence, as $0 < a(1) < \infty$, according to (22) and (24), we have $\frac{d\varphi(t)}{dt}$ for $0 < t \le 1$. The statement is proven.

Statement 2. Let in (23) $f(x)$ be Fisher distribution density,

$$f(x) = C(q_1, q_2)x^{(q_1-2)/2}\,(1 + lx)^{-(q_1+q_2)/2},\ x > 0;\ f(x) = 0,\ x \le 0 \tag{29}$$

where q_1 and q_2 are the degrees of freedom, $l = \frac{q_1}{q_2}$, $C(q_1, q_2)$ be a constant not depending on x, at that, $q_2 > 2$. Then the function $\varphi(t)$ is strictly monotonically decreasing on the interval (0,1].

Proof. We have from (21) and (29) for $a = tb$

$$\frac{f(b)}{f(a)} = \frac{(1+lbt)^m}{(1+lb)^m t^k} = t^2 \tag{30}$$

where $m = \frac{(q_1+q_2)}{2}$, $k = \frac{(q_1-2)}{2}$.

Under the condition of the statement $q_2 > 2$, we obtain that $u = \frac{q_1+2}{q_1+q_2} < 1$. Then, having solved equation (30) with respect to b, we have

$$b = b(t) = \frac{(1-t^u)}{l(t^u - t)} \tag{31}$$

Hence,

$$\frac{db(t)}{dt} = -\frac{(1-t^u)ut^{u-1}}{l(t^u-t)} - \frac{(1-t^u)(ut^{u-1}-1)}{l(t^u-t)^2}.$$

Let us determine the expression $\frac{db(t)}{dt}t(t-1) - b(t)$, appearing in formula (28), as a function of t in explicit form. We have, according to the expressions obtained for the function $b(t)$ and its derivative, after transformations

$$V(t) = \frac{db(t)}{dt}t(t-1) - b(t) = -\frac{(1-t^u)}{l(t^u-t)}\left[1-(1-t)\frac{ut^u(t^u-t+1)-t}{t^u-t}\right]. \tag{32}$$

Since under the condition of this statement $u < 1$, then $\ln t^u = u \ln t > \ln t$, $\ln t^u < 0$, $0 < t < 1$. Hence,

$$1 > t^u > t \text{ for } 0 < t < 1;\ t^u = t \text{ for } t = 0,1. \tag{33}$$

The minimum of the function $\psi(t) = t^u - ut^u\,(t^u - t + 1)$ for $t \in [0,1]$ is equal to zero. Therefore, $t^u \geq ut^u\,(t^u - t + 1)$ for $t \in [0,1]$. Using this inequality, as well as (33), we have

$$(1-t)\frac{ut^u(t^u - t + 1) - t}{t^u - t} \leq (1-t)\frac{t^u - t}{t^u - t} < 1, 0 < t < 1. \quad (34)$$

We denote: $Q(t) = (1-t)\dfrac{ut^u(t^u - t + 1) - t}{t^u - t}$, the value in square brackets on the equality (32) right side as $S(t)$. Then $S(t) = 1 - Q(t)$.

If $Q(t) \geq 0$, $0 < t < 1$ then according to (34) $1 > Q(t) \geq 0$, $0 < t < 1$, what entails $S(t) > 0$ and from (32) we have $V(t) < 0$, $0 < t < 1$, whence, in accordance with (28), it follows

$$\frac{d\varphi(t)}{dt} < 0, 0 < t < 1. \quad (35)$$

If $Q(t) < 0$, $0 < t < 1$, then $S(t) > 0$, what entails (35) as well.

We obtain from (22), (31) $a(1) = \dfrac{u}{l(1-u)} > 0$, which gives from (29) $f(a(1)) > 0$. Analysis of the expression (32) shows that $V(t) \to -\dfrac{2u}{l(1-u)} < 0$ as $t \to 1$. Then we have from (28) $\dfrac{d\varphi(1)}{dt} < 0$. Statement 2 follows from this and (35).

The theorem below follows from the above two statements.

Theorem 3. If the length of the confidence interval for some distribution characteristic is proportional to the value $a^{-1} - b^{-1}$, where a and b are the ends of the probability interval of a random variable with the known probability density $f(x)$ of χ^2 or Fisher distribution, and Fisher distribution satisfies the conditions of Statement 2, then there exists a unique minimum length of the confidence interval being the solution to equation (23).

Instead of solving equation (23), you can solve the optimization problem, which follows from this equation:

$$O(t) = \left| \int_{a(t)}^{b(t)} f(x)dx - v \right| \to \min, 0 \leq t \leq 1, \quad (36)$$

where $b(t)$ is defined by formula (24) (for distribution χ^2) or (31) (for Fisher distribution), $a(t)$ is got from (22).

The objective function in problem (36) depends on a variable t given on the interval [0,1]. According to Theorem 3, this function is unimodal. Therefore, problem (36) can be simply solved by the Fibonacci or golden ratio methods, described in detail in the monograph Wilde 1964 and in a number of manuals on optimization methods, for example Minoux 1989, pp. 2.4, 2.5.

Let us determine the efficiency of using the confidence interval of the minimum length, determined by formula (2), depending on the degrees of freedom of Fisher and χ^2 distributions.

Tables 3 and 4 show the decrease in the length of such an interval in comparison with the traditional method for these distributions.

Table 3. Comparison of a minimum length confidence interval and a traditional interval for χ^2 distribution lengths ($v = 0.95$).

An interval length is determined by formula (2).

Degrees of freedom	**Values**									
q	1	2	3	4	5	9	11	15	20	30
Length decrease, %	75.0	50.5	38.2	30.89	26.1	16.3	13.7	10.8	8.0	6.0

Table 4. Comparison of a minimum length confidence interval and a traditional interval for Fisher distribution lengths ($v = 0.95$).

An interval length is determined by formula (2).

No.	**Degrees of freedom**		**Length decrease, %**	**No.**	**Degrees of freedom**		**Length decrease, %**
	q_1	q_2			q_1	q_2	
1	15	12	12.1	16	5	10	26.7
2	20	4	15.8	17	5	50	26.1
3	20	12	10	18	8	15	18.5
4	20	15	9.5	19	4	15	31.1
5	30	15	7.3	20	10	30	15.1
6	30	10	8.6	21	12	24	13.1
7	3	50	38.2	22	18	24	9.5
8	20	5	14.2	23	15	4	17.3
9	3	20	38.2	24	15	8	13.4
10	1	50	75	25	10	5	19
11	4	50	30.9	26	10	3	22
12	3	20	38.2	27	24	12	9
13	4	20	31	28	12	24	13.1
14	5	20	26.3	29	20	8	11.5
15	3	10	38.4	30	3	30	38.2

It can be seen from Table 3 that a large enough confidence interval length decrease (about 20%) takes place for $q \leq 6 \ldots 7$. It was also found that a change in the confidence level in the range from 0.9 to 0.99 has little effect on a confidence interval for a variance with a fixed q length decrease.

For Fisher distribution, determined by two parameters, degrees of freedom, finding the area of a confidence interval effective length decrease is therefore much more complicated. The proposed solution is as follows.

According to Table 4 regression parameters were estimated

$$y_t = \alpha_0^0 + \alpha_1^0 (q_{t1})^{-1/2} + \alpha_2^0 (q_{t2})^{-1/2} + \varepsilon_t, \ t = 1,2,\dots \tag{37}$$

where t is the number of degrees of freedom values combination, y_t is a confidence interval length decrease in %, α_i^0, $i = 0, 1, 2$ are unknown regression parameters, ε_t is a random variable.

The standard assumption was made that the sequence $\{\varepsilon_t\}$, $t = 1, 2, \dots$ is a sequence of centered pairwise independent random variables.
The resulting regression

$$\hat{y}_t = \underset{(1.068)}{-14.128} + \underset{(1.468)}{84.003}(q_{t1})^{-1/2} + \underset{(2.325)}{17.892}(q_{t2})^{-1/2}, \ t = 1,2,\dots \tag{38}$$

where $\hat{y}_t$ is y_t estimate, in parentheses under the parameter estimates, their standard errors are indicated.

For the resulting model, the multiple coefficient of determination $R^2 = 0.993$, the significance of F-criterion is $9.33 \cdot 10^{-30}$, the ratio of the standard error of the residuals to the average confidence interval length decrease, equal to $\frac{\sum_{t=1}^{30} y_t}{30}$ was 5.58%. Thus, model (38) is adequate. Using the method described in Cox et al. 1974, based on estimates of the skewness and kurtosis coefficients at the 5% level, the hypothesis of normal noise distribution was adopted in (37).

Let us determine now Fisher distribution degrees of freedom q_1 and q_2 range of values for which length decrease of the confidence interval in percent will be no less than the specified value Y. Let τ denote the number of the combination of q_1 and q_2 for which the decrease in length

$$y_\tau = \min_{y_t \in \Omega} y_t, \quad \Omega = \{y_t : y_t \geq Y, \ t \in \theta\} \tag{39}$$

where θ is a set of combinations of q_1 and q_2.

Let $\tau \neq t$, $t = 1,\dots,T$, where T is the number of combinations q_1 and q_2, according to which the regression parameters (37) were estimated. In the case under consideration $T = 30$. Then, according to the accepted hypotheses about the normal distribution of random variables ε_t, $t = 1, 2,\dots$ and their pairwise uncorrelatedness

$$\frac{y_\tau - \hat{y}_\tau}{S_{y_\tau}} \sim t(q), \tag{40}$$

where $\hat{y}_\tau$ is determined by (38), $t(q)$ is a random variable with Student's distribution with $q = T - 3 = 27$ degrees of freedom.

In (40) $S_{y_\tau} = s(\boldsymbol{x}_\tau' \boldsymbol{K} \boldsymbol{x}_\tau + 1)^{1/2}$, where s is the standard residuals error, $\boldsymbol{K}$ is the estimate of the regression parameter estimates covariance matrix, $x_\tau = [1 \ q_{1\tau} \ q_{2\tau}]'$.

Note that the fulfillment of the condition $\tau \neq t$, $t = 1,\ldots,T$ can always be achieved by changing Y by a small value.
We have from (40)

$$P(y_\tau \geq \hat{y}_\tau - S_{y_\tau} t_p(q)) = 1 - p \tag{41}$$

where p is small, $-t_p(q)$ is a quantile of order p of the distribution $t(q)$.

Let

$$\hat{y}_\tau - S_{y_\tau} t_p(q) \geq Y \tag{42}$$

We get from (41) and (42) $P(y_\tau \geq Y) = P(y_\tau \geq \hat{y}_\tau - S_{y_\tau} t_p(q)) = 1 - p = v$.
Hence according to (39) we obtain $P(y_\tau \geq Y) \geq P(y_\tau \geq Y) = v$, $y_t \in \Omega$.

To obtain values $q_{1\tau}$ and $q_{2\tau}$, satisfying (41), let us solve the problem

$$\left.\begin{array}{c} q_{\tau 2} \rightarrow \max, 3 \leq q_{\tau 2} \leq 100, q_{\tau 2} - \text{integer}, \\ -14.128 + 84.003 q_{\tau 1}^{-1/2} + 17.892 (q_{\tau 2})^{-1/2} \geq t_p(q) s(\boldsymbol{x}_\tau^{'} \boldsymbol{K} \boldsymbol{x}_\tau + 1)^{1/2} + Y \end{array}\right\} \tag{43}$$

where $q_{\tau 1}$ acquires integer values $1,\ldots, \bar{q}_{\tau 1}$, $\bar{q}_{\tau 1}$ is the maximum value for which problem (43) has a solution.

For the probability, $v = 0.95$ the Table 5 shows q_1 values and the corresponding intervals of variation q_2 $[1, q_{2\tau}]$, which provide a confidence interval decreased by no less than $Y = 20\%$ with a probability of no less than 0.95. Wherein $t_{0.05}(2) = 1.703$, $s = 1.261$,

$$\boldsymbol{K} = \begin{bmatrix} 0.154 & -0.284 & -0.77 \\ -0.284 & 1.107 & 0.988 \\ -0.77 & 0.988 & 6.471 \end{bmatrix}.$$

Table 5. The range of Fisher distribution degrees of freedom change, for which confidence interval length decrease will be 20% or more with a probability of at least 0.95.

q_1	1	2	3	4	5	6	7	8	9
q_2	[1; 100]	[1; 100]	[1; 100]	[1; 100]	[1; 100]	[1; 72]	[1; 13]	[1; 5]	[1; 3]

4. Conclusion

In this paper, the necessary and sufficient conditions for a confidence interval of the minimum length l for a certain parameter existence are obtained in the two most frequent cases of its specification: (1) l is proportional to the probability interval of some r. v. length; (2) l is proportional to the difference in reciprocal values of the ends of this interval.

Since in the second case the problem of determining the minimum of l can be multiextremal, it is shown that for χ^2 and Fisher distributions of r. v., there is one minimum, and it is proposed to simplify the original optimization problem by reducing it to one variable optimization.

The issue of use expediency is considered. A procedure is proposed that makes it possible to determine approximately the range of values of the number of Fisher distribution degrees of freedom, for which the confidence interval length will decrease by a given value in comparison with the traditional method.

References

Cox, D.R. and D.V. Hinkley. 1974. Theoretical Statistics. Charman and Hall, London.

Ferentinos, K.K. and K.X. Karakostas. 2006. More on shortest and equal tails. confidence intervals. Communications in Statistics – Theory and Methods, 35(5): 821–829.

Ledermann, W. and L. Emlyn (eds.). 1984. Handbook of Applicable Mathematics. Volume VI: Statistics. Part A. John Wiley & Sons, Chichester-New York-Brisbane-Toronto-Singapore.

Levy, K.J. and S.C. Narula. 1974. Shortest confidence intervals for the ratio of two normal variances. The Canadian Journal of Statistics. 2(1): 83–87.

Minoux, M. 1989. Programmation Mathematique. Bordas et C. N. E. T. – E. N. S. T. Paris.

Poliak, B.T. 1983. Introduction to Optimization. Nauka, Moskow (In Russian).

Tate, R.F. and G.W. Klett. 1959. Optimal confidence intervals for the variance normal distribution. Journal of the American Statistical Association, 54(287): 674–682.

Wilde, D.J. 1964. Optimum Seeking Methods. Prentice-Hall, Inc., Englewood Cliffs, N. J.

Zellner, A. 1971. An Introduction to Bayesian Inference in Econometrics. John Wiley and Sons, Inc., New York-London-Sydney-Toronto.

CHAPTER 6

The Independence Number of the Generalized Wheel Graphs W^{p}_{2k+1}

Oksana S Pichugina,[1,2,*] *Dumitru I Solomon*[3] and *Petro I Stetsyuk*[4]

1. Introduction

Combinatorial Optimization Problems (COPS) are ubiquitous and widespread in modeling of real-world problems of Geometric Design and Operational Research (see, for instance, Pardalos et al. 2013).

A huge contribution to Combinatorial Optimization and Computational Complexity Theory was made by N. Z. Shor. He proposed an original technique for obtaining dual bounds on the optimal value of the objective function in quadratic optimization problems (Shor 2011). This technique includes an algorithm for obtaining the quadratic dual bounds by solving auxiliary non-smooth continuous optimization problems called the r-algorithm (see Shor et al. 1997). Another component of the approach is searching for superflows constraints for the initial problem and then utilizing them to refine the dual bounds. The additional constraints depend on the type of problems and are selected such that the above bounds can be improved.

[1] National Aerospace University "Kharkiv Aviation Institute" (Ukraine).
[2] University of Toronto (Canada).
[3] Academy of Transport, Informatics and Communications (Moldova).
[4] V. M. Glushkov Institute of Cybernetics, National Academy of Sciences of Ukraine.
Emails: atic@mtc.md; stetsyukp@gmail.com
* Corresponding author: o.pichugina@khai.edu

Shor's method of quadratic dual bounds can be used to solve numerous practical problems. One scope of its application is multiobjective optimization problems representable in the form of quadratically constrained quadratic problems (QPs).

Another wide application of Shor's dual bounds technique is Combinatorial Optimization, particularly linear and quadratic binary optimization. Indeed, numerous NP-hard binary optimization problems are formulated as a problem of optimizing linear or quadratic objective functions with additional linear or quadratic constraints (see, for instance, Pardalos et al. 2013, Stetsyuk 2018). The possibility of solving them as a QP consists in replacing the binary condition $x \in B_n = \{0,1\}^n$ with quadratic functional dependencies $x_i^2 - x_i = 0$, $i = \overline{1,n}$. For instance, after the replacement, a linear binary optimization problem turns into a problem of optimizing a linear function with linear and quadratic constraints. Such problems are commonly solved by switching to linear relaxation and solving a new problem on the corresponding combinatorial polytope. An issue in applying this manner to COPs is that most convex hulls of combinatorial sets are described by polytopes with an exponential number of constraints. As a result, linear programming methods become impractical.

At the same time, Shor's method of Lagrange dual bounds allows finding the bounds in a polynomial time on the number of constraints and variables. It utilizes exact nonlinear formulations of these problems, which are often characterized by a relatively small number of constraints. Thereby, as shown by computer experiments, Shor's dual bounds obtained are tighter than linear relaxation ones (see Shor et al. 1997, Shor 2011, Stetsyuk 2008, Stetsyuk 2018). In a monograph by Shor 2011, the efficiency of the dual bounds is illustrated on optimization problems on graphs such as the Maximum Weight Independent Set Problem, the Maximum Cut Problem, the Maximum Graph Bisection Problem, and the Minimal k partition problem, etc. Highly important are the results concerning the Maximum Independent (Stable) Set problem, where the dual bounds obtained by N. Z. Shor are closely related to the well-known Lovasz numbers $\vartheta(G)$ and $\vartheta'(G)$ introduced in Grötschel et al. 2011.

Any Combinatorial Optimization problem allows many different formulations. Depending on the formulation used, a fitting solution method is selected. It turns out that the efficiency of the deriving solutions and their accuracy depends on a model and an optimization method used. Problems of constructing different formulations of COPs are studied in the research domain of continuous functional representations of combinatorial sets (see, for instance, Pichugina et al. 2019, Stoyan et al. 2020), which is a subfield in the Euclidean Combinatorial Optimization Theory (ECO) (Stoyan et al. 2020). For instance, in addition to the above continuous representation of a binary set, it can be represented in the following forms: B_n: $x_1^2 + ... + x_n^2 = n/4$; $0 \le x_i \le 1$, $i = \overline{1,n}$, and B_n: $x_1^2 + ... + x_n^2 = n/4$; $x_1^4 + ... + x_n^4 = n/16$. Two of the three binary set representations are quadratic, thus can be used in Shor's dual bound approach. In ECO, the problem of finding different relaxations of COPs is also investigated. Its solutions can be used for deriving superflows constraints and linear relaxations of linear COPs.

In this chapter, we show the application of Shor's technique of dual bounds for solving the maximum independent set problems (MISPs) on undirected graphs and investigating the complexity of the solutions. This chapter is organized as follows. A combinatorial formulation of MISP is given in Section 2. Section 3 provides the necessary information about the Euclidean combinatorial optimization and functional representations of point sets in Euclidean spaces, including discrete ones. Here, it is also formulated approaches to the relaxation of nonlinear continuous optimization problems. In Section 4, we build several mathematical formulations of MISP in the form of continuous, discrete or partially discrete optimization problems. Also, necessary information about polytope *STAB*(G) and its relaxations is given in this section. Section 5 introduces the general quadratic optimization problem and provides the theoretical foundations of Shor's approach of dual bounds on objective function optimum value, including basic and refined, where the latter utilizes superflows constraints, adding them to the original problem conditions. Section 6 is directly dedicated to obtaining linear and quadratic bounds on the graph independence number $\alpha(G)$. Then polynomial solvability of MISP for t-perfect, h-perfect graphs and W^p_{2k+1}-perfect graphs is justified.

2. Problem statement

Let $G = (V(G), E(G))$ be an undirected graph with vertex set $V(G) = \{1,...,n\}$ and edge set $E(G) \subseteq V(G) \times V(G)$.

An independent set (stable set) in G is a subset of $V(G)$ (further $S \subseteq V(G)$), whose elements are pairwise nonadjacent.

Let $\mathcal{S}(G)$ be a set of all independent sets in G. Then

$$\forall S \in \mathcal{S}(G), \forall i, j \in S, i \neq j : \text{the condition } \{i, j\} \notin E(G) \text{ holds.} \qquad (1)$$

The maximum independent set problem (MISP) is to find an independent set $\mathcal{S}^* \in \mathcal{S}(G)$ of maximum cardinality. A maximal independent set (MIS) or maximal stable set is an independent set that is not a subset of any other independent set, i.e., MISP is a problem of finding an MIS $\mathcal{S}^*$. The size of a maximum independent set is called the stability number of G and is denoted by $\alpha(G)$. Thus the MISP consists in finding the stability number.

In these notations, MISP can be modelled as follows: find

$$\alpha(G) = |S^*| = \max_{S \in \mathcal{S}(G)} |S|. \tag{2}$$

S^* is an MIS in G.

3. Euclidean combinatorial optimization problems and their formalisations

3.1 COPs and ECOPs

In terminologies offered by Stoyan et al. 2020 and Yakovlev et al. 2021, the MISP-model (1), (2) is an example of Combinatorial Optimization Problem's formulations (COPs): find

$$f^* = f(\pi^*) = \underset{\pi \in \Pi' \subseteq \Pi}{extr} f(\pi), \tag{3}$$

where Π is a search domain, which is a combinatorial space, such as permutations, sets of graphs, and Π^* is a feasible set or a feasible domain.

A narrow subclass of COPs allows reformulations in the form of discrete optimization problems over a finite set X' of vectors in Euclidean space. To emphasize that a search domain is a subset of $\mathbb{R}^n$, such formulations are called Euclidean Combinatorial Optimization Problems (ECOPs) (see Stoyan et al. 2020 and Yakovlev et al. 2021): find

$$\varphi^* = \varphi(x^*) = \underset{x \in X' \subseteq X}{\text{extr}} \varphi(x), \tag{4}$$

where X' and X are feasible and search domains, respectively,

$$X' \subseteq X \subseteq \mathbb{R}^n.$$

We will assume that X' and X are images of Π, Π' under a bijective mapping ψ, hence there exists ψ: $\Pi \to X$ such that:

$$X = \psi(\Pi), X' = \psi(\Pi'), \text{ and } \Pi = \psi^{-1}(X), \Pi' = \psi^{-1}(X'). \tag{5}$$

Problems (3) and (4) are equivalent if $f^* = \varphi^*$. Interest in finding a Euclidean reformulation (4) of a COP (3) is in the possibility of solving them as discrete optimization problems with further deriving an optimizer of the initial COP (4) using (5):

$$\pi^* = \psi^{-1}(x^*), \text{where } \pi^* = \underset{x \in X' \subseteq X}{\text{argextr}}\, \varphi(x).$$

Remark 1. If π is a finite set of cardinality n, i.e., $\pi \subseteq M = \{\mu_i\}_{i=\overline{1,n}}$, $|M| = n < \infty$, such as an independent set, a standard way to construct an ECOP associated with such a COP is considering incidence vectors corresponding to a set π:

$$\psi = \chi : \pi \to \chi(\pi) = \{x \in \mathbb{R}^n : x_i = 1, \text{ if } \mu_i \in \pi;\ 0, \text{ otherwise}\}. \tag{6}$$

Clearly, the mapping χ in (6) is bijective, thus satisfying condition (5). Moreover, the cardinality of π is determined as a sum of $\chi(\pi)$-coordinates.

3.2 f-representations with applications

The results of this section are partially based on the work of Pichugina et al. 2020.

Let E be a finite set in $\mathbb{R}^n$ and E' be a superset of E and

$$\mathcal{F} = \{f_j(x)\}_{j=\overline{1,M}}, \tag{7}$$

where f_i: $E \rightarrow \mathbb{R}^n$ are continuous functions for $j = \overline{1,m}$.

Definition 1. A representation of E by functional dependencies:

$$f_j(x) = 0,\ j = \overline{1,m}, \tag{8}$$

$$f_j(x) \leq 0,\ j = \overline{m+1,M} \tag{9}$$

is called a continuous functional representation (an f-representation) of E on E'.

Further, we restrict our consideration to the case where $E' = \mathbb{R}^n$ is simply calling the collection of constraints (8), (9) by an f-representation of E. Thus, the system of constraints (8), (9) is an f-representation of E if and only if it coincides with a solution set of the system.

In the f-representation of E: (a) (8) is a strict part; (b) (9) is a non-strict part; (c) m is an order, particularly, m' and $m'' = m - m'$ are orders of its strict and non-strict parts, respectively.

Typology of f-representations can be given in different ways, such as according to: (a) the type of functions (7) including linear, nonlinear, differentiable, smooth, convex, polynomial, trigonometrical, etc.; (b) the combination of parameters m, m', m''.

Definition 2. System (8), (9) is called:

- a strict f-representation E (further referred to as E.SR) if it contains only a strict part meaning that $m' = m$, $m'' = 0$;
- a non-strict (further referred to as E.NR) if an f-representation contains only a non-strict part hence $m' = 0$, $m'' = m$;
- a mixed (further referred to as E.MR), if it contains both strict and non-strict parts, i.e., $m'(m - m') > 0$.

System of constraints (8), (9) is called an irredundant f-representation of a set E (E.IR) if the exclusion of any of its constraints leads to a formation of a proper superset of E. Otherwise, the representation is a redundant f-representation of E.

Example 1. Let E be a finite set of numbers $E = \{e_1,...,e_k\} \subset \mathbb{R}^1$. The following polynomial constraint can be used:

$$(x - e_1) \cdot ... \cdot (x - e_k) = 0 \tag{10}$$

as an f-representation of E.

The f-representation (10) is polynomial, strict, having the order 1.

Most common application of the expression (10) is a continuous reformulation of binary constraint $x \in \{e_1, e_2\} = \{0,1\}$ by $(x - e_1)\cdot(x - e_2) = x(x - 1)$ resulting in equation

$$x^2 - x = 0. \tag{11}$$

The main application of f-representations of combinatorial sets is a reformulation of discrete optimization problems as continuous with further application of nonlinear continuous optimization methods or their combination with discrete optimization techniques for solving these problems.

For instance, the Quadratic Unconstrained Binary Problem (QUBO) (see, for instance, Kochenberger et al. 2014):

$$f(x) = x^{\mathrm{T}}Qx \to \max, \tag{12}$$

where $x \in B_n = \{x \in \mathbb{R}^n : x_i \in \{0,1\}, i = \overline{1,n}\}$, $Q \in \mathbb{R}^{n\times n}$ is the symmetric matrix of the dimension n. It can be rewritten as a continuous quadratic optimization problem (12) with equality constraints

$$x_i^2 - x_i = 0, i = \overline{1,n}. \tag{13}$$

Evidently, there exists an infinite number of f-representations of any finite point set such as E. Finding new such representations results in deriving new formulations of the original COP. In turn, it opens up prospects for the application of existing and the development of new methods for solving these problems.

Also, f-representations are a powerful source for designing different relaxations of the original combinatorial optimization problem (**COP**): find

$$f^* = f(x^*) = \max f(x), \tag{14}$$

$$x \in E. \tag{15}$$

Relaxations themselves can be used for approximate solutions of COPs and getting bounds on optimal value f^*. Also, combining relaxations of different types in some cases allows the guarantee of obtaining a global solution of a COP (for details, see Pardalos et al. 2013, Shor et al. 1997, Shor 2011, Stoyan et al. 2020).

For COPs, the most common is a polyhedral relaxation (**PR**), where a condition a condition

$$x \in P = \text{conv}E \tag{16}$$

replaces the combinatorial constraint (15).

For instance, a PR of QUBO is (12), $x \in conv\ B_n = [0,1]^n$. Another relaxation called spherical (**SR**) was proposed in by Yakovlev et al. 2021 for sets inscribed in a hypersphere $S_r(a)$ with a radius r centered at a. Particularly, B_n is inscribed in a hypersphere having parameters $a = (0.5,...,0.5) \in \mathbb{R}^n$, $r = \sqrt{n}/2$. In general, an SR has a form of (14), $x \in S_r(a)$.

The relaxations PR and SR are closely connected with f-representations of such a set E, which analytically expresses the fact that $E = P \cap S_r(a)$. Respectively, condition (15) is equivalent to two – (16) and

$$x \in S_r(a). \tag{17}$$

Eliminating either (16) or (17), we get an SR or a PR. Iterative combining PRs and SRs of different dimensions are the basis of the Polyhedral-Spherical method of nonlinear optimization on combinatorial sets inscribing in a hypersphere introduced in Pichugina et al. 2021.

This way of constructing different relaxations of COPs can be generalized as the following: (a) firstly, deriving an irredundant f-representation of E; (b) secondly, eliminating a non-empty subset of constraints from further consideration (let a new feasible domain be $E'' \supset E$); (c) lastly, solving a problem (14) on E''.

Replacing some equality constraints in an irredundant f-representation by the corresponding inequalities also results in a relaxation of the original COP.

When direct nonlinear optimization methods are applied to a problem with the objective function (14) and constraints (8), (9) (to the model, we will further refer to as a continuous reformulation model, **CRM**), it is important to utilize irredundant f-representations in order to deal with as few constraints as possible.

At the same time, if dual methods are used, such as the Lagrange Multipliers Method, utilizing redundant f-representations has shown high efficiency in practice in some cases (see, for example, Shor et al. 1997, Shor 2011, Stetsyuk et al. 2007, Stetsyuk 2018). Therefore, some ways to construct redundant f-representations are now outlined:

- multiplying or adding two or more constraints (8), while saving the equality sign, such as $\forall i,\ j = \overline{1,m}: i \neq j$

$$f_i(x) \cdot f_j(x) = 0; \tag{18}$$

$$f_i(x) + f_j(x) = 0. \tag{19}$$

- multiplying or adding two or more constraints out of (8), (9), while for multiplication, choosing a sign depending on a number of involved constraints of type (9), for addition, taking a sign ≤. Namely,

$$\forall i,j : i \le m < j, i = \overline{1,m}, j = \overline{m+1,M} \quad f_i(x)\cdot f_j(x) = 0; \tag{20}$$

$$\forall i,j : i \ne j, i,j = \overline{m+1,M} \quad f_i(x)\cdot f_j(x) \ge 0; \tag{21}$$

$$\forall i,j : i \le m < j, i,j = \overline{1,M} \quad f_i(x) + f_j(x) \le 0. \tag{22}$$

Remark 2. An example of a relaxation of a combinatorial constraint $x \in \{e_1, ..., e_k\} \subset \mathbb{R}^1$, where $e_1 < ... < e_k$, is $x \in [e_1, e_k]$ or $e_1 \le x \le e_k$.

4. MISP-related graphs and polytopes

4.1 MISP formalizations

Taking into account Remark 1, the following MISP-ECOP formulation can be found.

Let $x = (x_1,...,x_n)$ be a vector of variables. MISP aims to find:

$$\alpha(G) = \mathbf{e}^{\mathrm{T}}x^* = \max \sum_{i\in V(G)} x_i = \max \mathbf{e}^{\mathrm{T}}x \tag{23}$$

subject to constraints

x is a preimage of a stable set $S \in \mathcal{S}(G)$ under mapping ψ given by (6),

i.e.,

$$x = \chi(S), \forall S \in \mathcal{S}(G), \tag{24}$$

Respectively, $x^* = \chi(S^*)$.

Here, $\mathbf{e}$ is an n-dimensional vector of ones.

The MISP model (23), (24) is still a COP due to the form (24) of the stable set constraint.

Let us find different Euclidean forms of this constraint, thus formulating various ECOPs of MISP.

A combinatorial vertex constraint

$$x_i \in \{0,1\}, i = \overline{1,n}, \tag{25}$$

evidently holds.

(25) implies that any variable x_i can be in two states: $x_i = 1$ when $i \in A$ or $x_i = 0$ when $i \notin A$ for $A \subseteq V(G)$, while $x = \chi(A)$.

By definition of an independent set $S \in \mathcal{S}$ (see (1)), adjacent vertices of the corresponding induced subgraph cannot be simultaneously in S, thus there are only three possible combinations of pairs of variables incident to edges of G:

$$\forall (i, j) \in E(G) \quad (x_i, x_j) \in B = \{(0,0), (0,1), (1,0)\}. \tag{26}$$

This means that the sum of variables x_i, x_j takes values 0, 1 only, i.e.,

$$x_i + x_j \in \{0,1\}, (i, j) \in E(G). \tag{27}$$

To the combinatorial MISP-model (23), (25), (27), we will refer to as (edge-constraint MISPModel 1, **MISP.EM1**).

The combinatorial edge constraint (27) can be rewritten by applying the f-representation (11) to $x = x_i + x_j$. This substitution yields

$$(x_i + x_j)^2 - (x_i + x_j) = 0, (i, j) \in E(G).$$

Simplifying the expression with further applying (25) results in

$$(x_i^2 - x_i) + (x_j^2 - x_i) + 2x_i x_j = 2x_i x_j = 0, (i, j) \in E(G),$$

wherefrom an equivalent combinatorial formulation of MISP (further **MISP.EM2**) is obtained having the form of (23) subject to the combinatorial vertex constraint (25) and nonlinear conditions

$$\text{(quadratic edge constraints)}: \quad x_i x_j = 0, (i, j) \in E(G). \tag{28}$$

One more combinatorial MISP model (further **MISP.EM3**) has the form of (23), (25),

$$\text{(linear edge constraints)}: \quad x_i + x_j \leq 1, (i, j) \in E(G). \tag{29}$$

Indeed, by Remark 2, the constraint (27) can be relaxed to two-sided inequalities $0 \leq x_i + x_j \leq 1$, $(i, j) \in E(G)$. The left one $0 \leq x_i + x_j$ is redundant because of presence of (25). Moreover, a solution set of x_i, $x_j \in \{0,1\}$, $x_i + x_j \leq 1$ is exactly set B (see (26)). (29) implies that, among adjacent vertices, at most one can be selected as a member of a stable set.

MISP.EM3 is a famous formulation (see, for example, Grötschel et al. 2011, Shor 2011, Schrijver et al. 2002) of the problem MISP, which relaxation (23) subject to the continuous edge constraints (29) and continuous vertex constraints

$$\text{(vertex constraints)}: \quad 0 \leq x_i \leq 1, i = \overline{1, n} \tag{30}$$

is a linear program (further referred to as **MISP.LR1**) in which a solution yields an upper bound on $\alpha(G)$.

Finally, in MISP.EM3, by replacing the binary constraint (25) with its continuous representation (13) we come to a well-known continuous MISP-formulation presented in Shor 2011 (further referred to as **MISP.EM4**). It has the form of (13), (23), (28).

Remark 3. In addition to MISP.EM1-MISP.EM4, two more edge-constraint MISP-models (further MISP.EM5-MISP.EM6) can be formed if all six possible combinations of vertex constraints (combinatorial and continuous) and edge constraints (combinatorial, linear and quadratic) are selected.

Another source of MISP-reformulations lies in applying properties of cliques in G. Let $\mathcal{Q}(G)$ be a clique set in G. Then the following constraint is satisfied for any graph G:

$$\text{(linear clique constraints)}: \quad \sum_{i \in V(Q)} x_i \leq 1 \quad \forall Q \in \mathcal{Q}(G), \tag{31}$$

where $V(Q)$ is a vertex set of a clique Q.

(31) is a generalization of the edge constraints (29). It means that, since a clique is a complete subgraph in G, in its stable set, there can be present at most one vertex of the clique. Respectively, it is degenerated into (29) if only cliques of order 2 are considered.

This discussion leads to one more famous combinatorial MISP-formulation (further a clique-constraint MISP-Model 1, **MISP. CM1**) having the form of (23), (25), (31), which linear relaxation involving constraints (30), (31) (further referred to as a **MISP.LR2**) is, generally, tighter than MISP.LR1. An issue with utilizing this relaxation is that the number of clique constraints can be exponential, making it impractical to get an upper bound on $\alpha(G)$ with the help of this relaxation.

Now, we introduce a way to produce quadratic equality constraints from linear inequality constraints (31) by conducting in the opposite direction of the above process of producing the linear edge constraints (29) from the corresponding combinatorial ones. For that, we take into account that our variables are binary, and we replace (31) in MISP.CM1 by a combinatorial clique constraint:

$$\text{(combinatorial clique constraint)} \quad \sum_{i \in V(Q)} x_i \in \{0,1\} \quad \forall Q \in \mathcal{Q}(G). \tag{32}$$

Applying now the f-representation (11) to $x = \sum_{i \in V(Q)} x_i$, we obtain:

$$\forall Q \in \mathcal{Q}(G) \quad (\sum_{i \in V(Q)} x_i)^2 - \sum_{i \in V(Q)} x_i = \sum_{i,j \in V(Q)} x_i x_j - \sum_{i \in V(Q)} x_i =$$
$$= 2 \cdot \sum_{i<j, i,j \in V(Q)} x_i x_j + \sum_{i \in V(Q)} (x_i^2 - x_i) = 0.$$

Finally, applying the f-representation (11) to the expression, we come to an equivalent form of (31) in the form of quadratic equations:

$$\text{(quadratic clique constraint)} \quad \sum_{i<j, i,j \in V(Q)} x_i x_j = 0 \quad \forall Q \in \mathcal{Q}(G). \tag{33}$$

In such a way, we have derived another quadratic MISP-formulation utilizing clique constraints having the form of (23), (25, (33) (further referred to as **MISP.CM2**).

Remark 4. In the same manner as in Remark 4, six clique-constraints' MISP-models MISP.CM1MISP.CM6 can be derived by combining vertex constraints (combinatorial or continuous) with involved clique constraints (combinatorial (32), linear (31) or quadratic (33)).

Among MISP.EM1-MISP.EM6 and MISP.CM1-MISP.CM6, there are only two continuous models. These are MISP.EM2 and (13), (23), (33) (further referred to as **MISP.CM3**). These two can be solved by nonlinear optimization methods listed in Section 4 and by general nonlinear solvers such as IPopt or APopt or quadratic solvers supported non-convex constraints and objective functions such as CPLEX. All others contain one or two binary constraints. They can be solved by integer optimization solvers, linear or quadratic dependent on either quadratic constraints are involved, or nonlinear solvers supporting integer variables such as CPLEX and Gurobi.

4.2 The polytope STAB(G) and its relaxations

A feasible set X' in MISP is a set of all incident vectors of stable sets in G. So,

$$X' = \{x \in X : \pi = \chi^{-1}(x) \in \mathcal{S}(G)\}, \tag{34}$$

where the mapping χ is given by (6), while a search domain is $X = \{0,1\}^n$.

Respectively, the polyhedral relaxation of MISP (see (16)) is to optimize (23) subject to a constraint

$$x \in P' = conv(X'), \tag{35}$$

where X' is given by (34).

The polytope P' conventionally is denoted as $STAB(G)$, thus

$$STAB(G) = conv\{\chi(S) : S \in \mathcal{S}(G)\}.$$

Let a polyhedral outer approximation of $STAB(G)$ given by linear constraints $\mathcal{L}$ be denoted as $\mathcal{L}STAB(G)$. Thus,

$$\mathcal{L}STAB(G) = \{x \in \mathbb{R}^n : x \text{ satisfies linear constraints } \mathcal{L}\}, \tag{36}$$

such that

$$STAB(G) \subseteq \mathcal{L}STAB(G). \tag{37}$$

STAB(G) is bounded, i.e., it is a polytope.

Definition 3. If for a graph G

$$\mathcal{L}STAB(G) = STAB(G), \tag{38}$$

then the graph G is called $\mathcal{G}(\mathcal{L})$ graph, where function $\mathcal{G}$ is given in two-line notation in Table 1.

Table 1. Correspondence of $\mathcal{L}$-polytopes to $\mathcal{G}(\mathcal{L})$-types of graphs.

$\mathcal{L}$	F	Q	C	H	W^1_{2k+1}	W^p_{2k+1}
$\mathcal{G}(\mathcal{L})$	bipartite	perfect	t-perfect	h-perfect	W^1_{2k+1}-perfect	W^p_{2k+1}-perfect

Above, some valid linear inequalities for $STAB(G)$ were already listed. These are vertex constraints (30), linear edge constraints (29), and linear clique constraints (31). Without loss of generality, suppose that the vertex constraints (30) are satisfied. Since we can restrict our search by a unit hypercube $[0,1]^n$, it allows refining (36) as the following:

$$\mathcal{L}STAB(G) = \{x \in [0,1]^n : x \text{ satisfies constraints } \mathcal{L}'\}. \tag{39}$$

Here, $\mathcal{L}'$ is constraint $\mathcal{L}'$ without (30).

Let

$$F' = \mathcal{L}' = \{x \text{ satisfies (29)}\},$$
$$Q' = \mathcal{L}' = \{x \text{ satisfies (31)}\}.$$

According to Table 1, for $\mathcal{L} = F$, we have the relation

$$FSTAB(G) = STAB(G), \tag{40}$$

which is a polytope of independent sets bipartite graphs (see, for instance, Asratian et al. 2008). The polytope $FSTAB(G)$ is called a fractional polytope (fractional matching polytope, FMP) (Shor et al. 1997, Shor et al. 1997). A bipartite graph is normally defined as a graph that does not contain any odd-length cycles. (40) leads to another definition of a bipartite graph as a graph satisfying this condition.

Likewise, for $\mathcal{L} = Q$, the same relation holds for a polytope of independent sets in a perfect graph G (see Trotignon 2015 for details). It has the form of $QSTAB(G) = STAB(G)$, while the polytope $QSTAB(G)$ is called a clique polytope in Shor 2011. Usually, a perfect graph is defined equivalently as the one in which the chromatic number of every induced subgraph equals the clique number.

From Table 1, we see that there exist four more families of different graphs satisfies the relation (38). They are related to utilization either (29) or (31) along with singling outing and applying other valid inequalities of $STAB(G)$.

Let us list some of them required for the construction of the table.

$$\text{(odd − cycle constraints)}: \quad \sum_{i \in V(C_{2k+1})} x_i \le k \quad \forall C_{2k+1} \in \mathcal{C}(G), \tag{41}$$

where C_{2k+1} is an odd cycle in graph G, $k \in \mathbb{N} \backslash \{1\}$, $\mathcal{C}(G)$ is a set of cycles in G.

Similarly,

$$\text{(even − cycle constraints)}: \quad \sum_{i \in V(C_{2k})} x_i \le k \quad \forall C_{2k} \in \mathcal{C}(G), \tag{42}$$

where $C_{2k} \in \mathcal{C}(G)$ is an even cycle in graph G, $k \in \mathbb{N} \backslash \{1\}$.

Let a vertex set and an edge set of an even cycle C_{2k} be $V(C_{2k}) = \overline{1,2k}$, $E(C_{2k}) = \{\{1,2\}, \{2,3\},..., \{2k-1, 2k\}, \{1, 2k\}\}$. For the cycle, the inequality (42) holds since the maximum independent set in C_{2k} consists of k vertices with either all even or odd labels. If $k = 1$, the cycle degenerates into an edge, whose independence set consists of at most $k = 1$ vertices. Thus (42) is valid for any $k \in \mathbb{N}$. Adding one more vertex into the circle C_{2k}, such that an odd cycle C_{2k+1} is formed with

$$\begin{aligned} &V(C_{2k+1}) = \overline{1, 2k+1}, \\ &E(C_{2k+1}) = \{\{1,2\},\{2,3\},...,\{2k,2k+1\},\{1,2k+1\}\}, \end{aligned} \tag{43}$$

does not increase the C_{2k}-independence number. It is clear from the above discussion that the value of k is a tight upper bound on a sum of incidence vector components associated with cycles C_{2k} and C_{2k+1}. Thus (41), (42) are always satisfied.

Note that the even-cycle constraints (42) are redundant and can be obtained by a summation of the edge constraints (29) over the edge set of C_{2k}. In contrast, the odd-cycle constraints (41) are candidates for participating in an irredundant H-representations of *STAB*(G) for certain graphs. In this case, these are face-defining constraints.

Next family of valid constraints of *STAB*(G) concerns odd wheels in G. First one is (see Stetsyuk 2018, Stetsyuk et al. 2019):

$$\text{(odd − wheel constraints)} \quad \sum_{i \in V(C_{2k+1})} x_i + kx_{i_{2k+2}} \le k, \forall W^1_{2k+1} \in \mathcal{W}^1(G), \tag{44}$$

where W^1_{2k+1} is an odd-wheel graph on $2k + 1$ vertices in a graph G, which is a join of an odd cycle $C_{2k+1} \in \mathcal{C}(G)$ with a vertex $2k + 2$ connected to each of the cycle vertices, $k \in \mathbb{N}$, $\mathcal{W}^1(G)$ is a set of odd-wheel graphs with a single central vertex.

Suppose the cycle C_{2k+1} satisfies (43), while $V(W^1_{2k+1}) = \overline{1,2k+2}$, hence

$$E(W^1_{2k+1}) = E(C_{2k+1}) \cup \{\{1,2k+2\},\{2,2k+2\},...,\{2k+1,2k+2\}\}.$$

(44) is valid for the wheel due to only two options exist for a stable set in W^1_{2k+1}: the central vertex $2k + 2$ of the wheel is included in its independent set $S(W^1_{2k+1}) \in \mathcal{S}(G)$, hence $x_{2k+2} = 1$. Since this vertex is adjacent to all other vertices of the wheel, the other vertices cannot be included in $S(W^1_{2k+1})$. Respectively, other variables are null. So, the first term in (44) takes zero value, and the second takes the value of k. Thus the whole inequality turns into an identity $k \equiv k$ in this case.

If the central vertex $2k + 2$ is not in the independent set $S(W^1_{2k+1})$, then $x_{i_{2k+2}} = 0$, and the independent set involves only vertices of the odd cycle C_{2k+1}. In this case, the second term on the left-hand side (44) is cancelled, turning the odd-wheel inequality into the odd-cycle inequality (41), which is valid.

This means that the inequality (44) holds for all odd wheels of type W^1_{2k+1}.

Let us generalize (44) from an odd-wheel graph W^1_{2k+1} into an odd-wheel graph W^p_{2k+1}, where $p \in \mathbb{N}$ is a parameter. The graph W^p_{2k+1} is a join of an odd circle $C_{2k+1} \in \mathcal{C}(G)$ and a p-clique Q_p, where $k, p \in \mathbb{N}$. Let $\mathcal{W}(G)$ be a set of W^p_{2k+1}-type graphs, i.e., $\mathcal{W}(G) = \{W^p_{2k+1}\}_{k,\, p\in\mathbb{N}}$. The generalized inequality for (44) looks like

$$\text{(generalized odd – wheel constraints)} \quad \sum_{i\in V(C_{2k+1})} x_i + k \sum_{j\in V(Q_p)} x_j \le k, \ \forall W^p_{2k+1} \in \mathcal{W}(G), \tag{45}$$

where $k, p \in \mathbb{N}$.

In order to show that (45) holds for any $W^p_{2k+1} \in \mathcal{W}(G)$, the same arguments as for the odd-wheel constraints can be applied. Namely, if vertices of the clique Q_p are involved in a stable set of W^p_{2k+1}, then $\alpha(W^p_{2k+1}) = 1$ (**Case 1**); otherwise, only C_{2k+1}-vertices can participate in the set, while their sum does not exceed k according to (44) (**Case 2**).

Other valid constraints of *STAB*(G) can be found in the literature, particularly, in Shor 2011, Stetsyuk et al. 2007, Stetsyuk 2018, such as anti-hole constraints.

Different combinations of (41) and (45) together with either linear edge or clique constraints, taken as $\mathcal{L}$, induce various classes of $\mathcal{G}(\mathcal{L})$-type graphs.

Remark 5. For bipartite graphs, the relation (38) follows from the following theorem.

Theorem 1. [Grötschel et al. 2011] The inequalities (29), (30) give a full description of *STAB*(G) if and only if G is bipartite.

For perfect graphs, we rely on the following statement.

Theorem 2. [Lovász 1979] Graph G is perfect if and only if

$$STAB(G) = QSTAB(G). \tag{46}$$

Theorem 2 can be interpreted as any perfect graph G can be defined as a graph satisfying the relation (46).

It is known from Shor et al. 1997 that if $\exists k \in \mathbb{Z}$ such that $C_{2k+1} \in \mathcal{C}(G)$, then the polytope $FSTAB(G)$ is not integral, respectively, $FSTAB(G) \neq STAB(G)$. Adding the odd-cycle constraints, in some cases, makes the polytope $\mathcal{L}STAB(G)$ integral. The same is true for $QSTAB(G)$, implying that it occurs that $FSTAB(G) \neq STAB(G)$, but adding the odd-cycle constraints can make the obtained polytope integral.

In this regard, let

$$T' = \mathcal{L}' = \{x \text{ satisfies } (29), (41)\};$$
$$H' = \mathcal{L}' = \{x \text{ satisfies } (31), (41)\}.$$

From Table 1, it is seen that $TSTAB(G)$ is a polytope of independent sets of a t-perfect graph, while $HSTAB(G)$ is the one for an h-perfect graph G (Sbihi et al. 1984). The family of t-perfect graphs including bipartite, almost bipartite, series-parallel graphs, and nearly bipartite planar graphs as subclasses, was introduced in Chvátal 1975. The class of h-perfect graphs was introduced in Sbihi et al. 1984. It is a superclass of perfect and t-perfect graphs. These two are interconnected as follows $\mathcal{T}STAB(G) \supseteq \mathcal{H}STAB(G)$ because (31) is stronger than (29). Note that, in contrast to bipartite and perfect graphs, t-perfect and h-perfect graphs are defined directly through the relation (38).

Finally, Table 1 indicates one more class of graphs and polytopes with a full description that involves the generalized odd-wheel constraint (45). Let us introduce a polytope $\mathcal{L}STAB(G)$ such that

$$W_{2k+1}^{p'} = \mathcal{L}' = \{x \text{ satisfies } (29), (41), (45)\}. \tag{47}$$

It induces $W_{2k+1}^{p}\, STAB(G) = STAB(G)$ as a polytope of an independent sets of a w_{2k+1}^{p}-perfect graph G for $k, p \in \mathbb{N}$.

Respectively, w_{2k+1}^{1}-perfect graphs (odd-wheel-perfect graphs) form a subclass of G is a subclass of w_{2k+1}^{p}-perfect graphs corresponding to $p = 1$.

Replacing in (47) the edge constraint (29) by the clique one (31) results in forming another family of graphs further referred to as a-perfect graphs, in which polytope $ASTAB(G)$ of independent sets is defined as

$$A' = \mathcal{L}' = \{x \text{ satisfies } (31), (41), (45)\}. \tag{48}$$

Finally, eliminating odd-cycle constraints from (47), (48) leads to introducing two more types of graphs and polytopes. Namely, let

$$B' = \mathcal{L}' = \{x \text{ satisfies, (29), (45)}\},$$
$$C' = \mathcal{L}' = \{x \text{ satisfies, (31), (45)}\}.$$

The corresponding classes of graphs are b-perfect and c-perfect.

In the next section, it will be shown that MISP is polynomially solvable for all listed families of graphs.

5. Shor's dual bounds on Q_0^*

Naum Shor introduced a new approach to finding upper bounds on the optimal value of the objective function in quadratic problems based on applying the Lagrange multiplies method. The main advantage of the bounds is that they can be found in polynomial time on a number of variables and constraints. The complexity of the approach is $O(n^3) + O(M^2)$, where n is the problem dimension, and M is the number of constraints. A disadvantage is a necessity, in some cases, to solve a non-smooth optimization problem in their evaluation process.

5.1 The simple bound ψ^*

Let us consider a quadratic optimization problem in the following formulation: it is required to find

$$Q_0^* = Q_0(x^*) = \sup_{x \in \mathbb{R}^n} Q_0(x) \tag{49}$$

subject to constraints

$$\begin{aligned} Q_i(x) &\le 0, \quad i = 1, \ldots, m, \\ Q_i(x) &= 0, \quad i = m+1, \ldots, M, \end{aligned} \tag{50}$$

where $M \ge 0$.

Here, for each $i = 0,...,M$, function $Q_i(x)$ is representable as $Q_i(x) = (K_i x, x) + (b_i, x) + c_i$, where K_i is a symmetric matrix of the dimension n, $b_i \in \mathbb{R}^n$, c_i is a scalar. This means that we restrict our consideration with quadratic (if $K_i \neq \mathbf{0}$) and linear (if $K_i = \mathbf{0}$) functions (further, $\mathbf{0}$ is a null matrix of the dimension n).

If problem (49), and (50) are incompatible, we assume that $Q_0^* = -\infty$. Generally, this problem is multi-extremal and belongs to the class of NP-hard problems. An upper bound on Q_0^* can be obtained as follows. Let $u = (u', u'')$ such that $u' = (u_1,...,u_m) \in \mathbb{R}^m_+$, $u'' = (u_{m+1},...,u_M) \in \mathbb{R}^{M-m}$ be a vector of Lagrange multipliers corresponding to the constraints (50). Namely, the multipliers in u' correspond to inequality constraints while u'' is associated with equality-type constraints.

In these notations, the following Lagrangian is associated with the problem (49), (50):

$$L(x,u) = Q_0(x) + \sum_{i=1}^{M} u_i Q_i(x) = (K(u)x, x) + (b(u), x) + c(u), \tag{51}$$

where

$$K(u) = K_0(x) + \sum_{i=1}^{M} u_i K_i(x), \quad b(u) = b_0 + \sum_{i=1}^{M} u_i b_i, \quad c(u) = c_0 + \sum_{i=1}^{M} u_i c_i.$$

Let X be a feasible domain of the problem (49), (50) and $\mathcal{U}^+ = \mathbb{R}_+^m \times \mathbb{R}^{M-m}$. Then $X \times \mathcal{U}^+$ is a feasible domain of the unconstrained optimization problem:

$$L(x^*, u^*) = Q_0^* = \sup_{x \in \mathbb{R}^n, u \in \mathcal{U}^+} L(x,u) \tag{52}$$

associated with the problem (49), (50). Clearly, for any $(x', u') \in X \times \mathcal{U}^+$, there holds $L(x', u') \geq Q_0^*$. This means that a function

$$\psi(u) = \sup_{x \in \mathbb{R}^n} L(x,u) = \sup_{x \in \mathbb{R}^n} [(K(u)x, x) + (b(u), x) + c(u)] \tag{53}$$

serves as an upper bound on Q_0^* for each $u \in \mathcal{U}^+$.
The best one from all the bounds (53) depending on u is

$$\psi^* = \inf_{u \in \mathcal{U}^+} \psi(u). \tag{54}$$

The domain $\mathcal{U}^+$ can be partitioned as follows: $\mathcal{U}^+ = \Omega^- \cup \Omega^0 \cup \Omega^+$, where $\Omega^- = \{u : K(u) \prec \mathbf{0}\}$, $\Omega^+ = \{u : K(u) \succ \mathbf{0}\}$, $\Omega^0 = \{u : K(u) \preceq \mathbf{0}\} \backslash \Omega^-$. Let λ_{max} $(K(u))$ be the maximum eigenvalue of the matrix $K(u)$. Then these three sub-domains can be represented in the following way: $\Omega^- = \{u \in \mathcal{U}^+ : \lambda_{max} [K(u)] < 0\}$, $\Omega^0 = \{u \in \mathcal{U}^+ : \lambda_{max} [K(u)] = 0\}$, $\Omega^+ = \{u \in \mathcal{U}^+ : \lambda_{max} [K(u)] > 0\}$.

If $u \in \Omega^+$, then $\psi(u) = +\infty$, i.e., this function is unbounded from above, hence,

$$Q_0^* = \psi^* = +\infty. \tag{55}$$

For $u \in \overline{\Omega}^-$, where $\overline{\Omega}^- = \Omega^- \cup \Omega^0$, then function $L(x, u)$ is convex. In this case, the problem of finding $\psi(u)$ is a convex optimization one. Moreover, for $u \in \Omega^-$, the matrix $K(u)$ is non-singular hence the value of $\psi(u)$ can be found by solving the linear algebraic system

$$\frac{\partial L(x,u)}{\partial u} = 2K(u)x + b(u) = 0 \tag{56}$$

with consecutive substitution of the solution into $\psi(u)$. From (56), it follows that $x = x(u) = -\frac{1}{2}K^{-1}(u)b(u)$ and $\psi(u) = -\frac{1}{4}(K^{-1}(u)b(u), b(u)) + c(u)$.

If $u \in \Omega^0$, then the matrix $K(u)$ is singular. If the system (56) is compatible, it can have multiple solutions, while function $\psi(u)$ is non-smooth. Generally, if $\mathcal{U}^+ \neq \emptyset$, then there is a nontrivial upper bound on Q_0^* given by (54).

With any predetermined accuracy, the bound ψ^* can be found in polynomial time by methods for minimizing convex non-differentiable functions, such as the ellipsoid method or specific versions of r-algorithms (see Shor 2011). For instance, an outline of the r-algorithm with an adaptive step control is given in detail in Shor et al. 1997.

If ψ^* is attained at $u^* \in \Omega^-$, then

$$\psi^* = \psi(u^*) = Q_0^* = Q_0(x(u^*)),$$

where $x(u^*)$ is a solution of the system (56) for $u = u^*$. Otherwise, ψ^* is attained at the boundary of the domain Ω^-. In this case, there may exist a so-called "duality gap"

$$\Delta^* = \psi^* - Q_0^* > 0.$$

5.2 The improved bound ψ_1^*

A way to reduce the gap Δ^*, thus improving the bound ψ^* was proposed by N. Z. Shor in Shor 2011. It is associated with introducing and utilizing superflows constraints in the problem (49), (50). To the new optimization problem, we will refer to it as Problem 1. Superflows constraints leave a feasible domain of the problem unchangeable. Hence an optimizer or a set of optimizers remains the same. At the same time, a problem of finding a new upper bound ψ_1^* on Q_0^* similar to finding ψ^*, where the superflows constraints are also utilized, has another form, namely, it is of higher dimension with another Lagrangian $L_1(x, U)$ different from $L(x, u)$, i.e., $L_1(x, U) \neq L(x, u)$. Here, U is a vector of Lagrangian multipliers of Problem 1. It turns out that ψ_1^* is at least as tight as ψ^*. Sometimes, in this case, it is more accurate than ψ^*. Thus, utilization of superflows constraints is capable od reducing the duality gap Δ^* resulting in achieving the duality gap Δ_1^* associated with Problem 1, which is lower than Δ^*, hence $\Delta_1^* \leq \Delta^*$.

Suppose that all superflows constraints added to the problem (49), (50) than turn it into Problem 1 are quadratic, i.e., Problem 1 has the form of (49) subject to constraints (50),

$$Q_i(x) \leq 0, \quad i = M+1, \ldots, M+l;$$
$$Q_i(x) = 0, \quad i = M+l, \ldots, M+L,$$

where $0 \leq l \leq L$, each function $Q_i(x) = (K_i x, x) + (b_i, x) + c_i$ is either linear or quadratic, i.e., K_i is a symmetric matrix of the dimension n, $b_i \in \mathbb{R}^n$, $c_i \in \mathbb{R}^1$ for $i = \overline{M+1, M+L}$.

The Lagrangian of Problem 1 is

$$L_1(x,U) = Q_0(x) + \sum_{i=1}^{M+L} u_i Q_i(x) = (K_1(U)x, x) + (b_1(U), x) + c_1(U), \quad (57)$$

where

$$U = \{\{u\}, u_{M+1}, \ldots, u_{M+L}\}, u_{M+1} \geq 0, \ldots, u_{M+l} \geq 0,$$

$$K_1(U) = K_0(x) + \sum_{i=1}^{M+L} u_i K_i(x), \quad b(U) = b_0 + \sum_{i=1}^{M+L} u_i b_i,$$

$$c(U) = c_0 + \sum_{i=1}^{M+L} u_i c_i.$$

The same arguments as for the problem (49), (50) are valid for Problem 1, hence

$$\psi_1(U) = \sup_{x \in \mathbb{R}^n} L_1(x,U),$$

where $U \in \mathcal{U}_1^+ = \mathcal{U}^+ \times \mathbb{R}_+^l \times \mathbb{R}^{L-l}$, is an upper bound on Q_0^* for any $U \in \mathcal{U}_1^+$.

In the family of the bounds $\{\psi_1(U)\}_{U \in \mathcal{U}_1^+}$, the most accurate is

$$\psi_1^* = \inf_{U \in \mathcal{U}_1^+} \psi_1(U). \quad (58)$$

Lemma 1. The bounds (54) and (58) a related as the following:

$$\psi_1^* \leq \psi^*. \quad (59)$$

Proof. Take an arbitrary $u \in \mathcal{U}^+$ an extend it by L zeros. The obtained vector $U = (u, 0, \ldots, 0) \in \mathcal{U}_1^+$. Moreover, $L_1(x, U) = L(x, u)$, wherefrom $\psi_1(U) = \psi(u)$.

At the whole,

$$\psi_1^* = \inf_{U \in \mathcal{U}_1^+} \psi_1(U) \leq \inf_{u \in \mathcal{U}^+} \psi(u) = \psi^*.$$

Lemma 1 states that the upper bound ψ_1^* is at least as accurate as ψ^*, i.e., it makes sense to consider Problem 1 and solve the problem of finding ψ_1^* instead of ψ^*, because involving superflows constraints capable of improving an upper bound in ψ^* and reduce the duality gap.

Now the question arises of which redundant constraints can improve the upper bound ψ^* on Q_0^*. There are several ways to construct them listed in Sec. 3.2 that are based on the utilization of frepresentations of combinatorial sets. Two equality types can be singled out from them: Type 1 – linear combinations of constraints' components such as (19) and (22); Type 2 – utilizing multiplication of constraints' components, e.g., (18), (20), (21).

The next lemma establishes that adding constraints of Type 1 to the original problem (49), (50) does not improve the bound ψ^*.

Lemma 2. If, for Problem 1, $l = 0$ and

$$\exists \lambda_{ik} \in \mathbb{R}^1, i = \overline{m+1, M}, k = \overline{l+1, L},$$
$$\text{such that } Q_{M+k} = \sum_{i=m+1}^{M} \lambda_{ik} Q_i \text{ for } k = \overline{l+1, L}, \tag{60}$$

then

$$\psi_1^* = \psi^*. \tag{61}$$

Proof. If (60) holds then the Lagrangian (57) becomes

$$L_1(x,U) = Q_0(x) + \sum_{i=1}^{M} u_i Q_i(x) + \sum_{k=1}^{L} u_{M+k} Q_{M+k}(x) =$$
$$= Q_0(x) + \sum_{i=1}^{M} u_i Q_i(x) + \sum_{k=l+1}^{L} u_{M+k} \sum_{i=m+1}^{M} \lambda_{ik} Q_i =$$
$$= Q_0(x) + \sum_{i=1}^{m} u_i Q_i(x) + \sum_{i=m+1}^{M} Q_i(x)(u_i + \sum_{k=l+1}^{L} \lambda_{ik}) = L(x,u'),$$

where $u' \in \mathbb{R}^M$, $u_i' = u_i$, $u_i' \geq 0$, $i = \overline{1,m}$; $u_i' = u_i + \sum_{k=l+1}^{L} \lambda_{ik} \in \mathbb{R}^1$, $i = \overline{m+1, M}$, i.e., it has a form of the Lagrangian (51). Hence problems of maximization of $L(x, u)$ and $L_1(x, U)$ are equivalent. The same concerns (58), wherefrom (61) directly follows.

Lemma 2 says that the equality part of the original model (49), (50) used as a base for superflows equality constraints obtained by linear combinations is not capable of improving the bound ψ^* induced by Problem 1 with better bound ψ_1^* by solving the correspondent Problem 1.

Remark 6. Taking into account the number of constraints in the original problem and Problem 1, the computational complexity of finding ψ^* by the ellipsoid method and r-algorithm is $O(n^3) + O(M^2)$, while ψ_1^* can be found in time $O(n^3) + O((M+L)^2)$. This means that the polynomial solvability of getting the Shor's bounds ψ^*, ψ_1^* is ensured only if the number of constraints is fixed or polynomially depends on n.

6. Upper bounds on $\alpha(G)$

This section discusses two types of upper bounds on $\alpha(G)$ – linear approximation implying solution of polyhedral relaxations and dual bounds consisting of solution of MISP in continuous quadratic formulation by Shor's r-algorithm. It continues the research started in Stetsyuk et al. 2007, Stetsyuk 2018, Stetsyuk et al. 2019.

6.1 Linear approximation bound $\alpha_{\mathcal{L}}(G)$

Since objective function in MISP is linear, a polyhedral relaxation of MISP is

$$\alpha(G) = \max \sum_{i \in V(G)} x_i, \quad x \in STAB(G),$$

i.e., it is equivalent to the binary linear problem MISP. An issue is that an H-representation of $STAB(G)$ is generally unknown and may consist an exponential number of constraints. That is why LP-programs are solved for various outer approximations of $STAB(G)$.

Introduce the following linear program (LP-program):

$$\alpha^*_{\mathcal{L}}(G) = \max \sum_{i \in V(G)} x_i, \quad x \in \mathcal{L}STAB(G), \tag{62}$$

where $\mathcal{L}STAB(G)$ is a polytope satisfying (37).

Clearly,

$$\alpha^*_{\mathcal{L}}(G) \geq \alpha(G), \tag{63}$$

i.e., $\alpha^*_{\mathcal{L}}(G)$ is an upper bound on $\alpha(G)$. The bound is exact if (38) holds. By definition, it is true for polytopes with $\mathcal{L}$ listed in Table 1. For them,

$$\alpha^*_{\mathcal{L}}(G) = \alpha(G), \tag{64}$$

i.e., a MISP solution is reduced to solving its linear relaxation over $\alpha^*_{\mathcal{L}}(G)$.

If $\mathcal{L} = F$, we deal with the polytope $FSTAB(G)$ described by n double-vertex inequalities (30) and $|V(G)| = m \leq \frac{n(n-1)}{2}$ edge constraints (29).

All other polytope corresponding to $\mathcal{L} \in \mathbf{L} = \{a,b,c,Q,C,H,W^1_{2k+1}, W^p_{2k+1}\}$ includes at least one family of constraints (31), (41), (44), (45). Each of them, generally, involves an exponential number of constraints.

This means that, among $\alpha^*_{\mathcal{L}}(G)$, $\mathcal{L} \in \mathbf{L}$, the only estimate $\alpha^*_F(G)$ is guaranteed to be found in polynomial time on n and m by linear programming techniques. Hence, a MISP is polynomially solvable for any bipartite graph with the help of LP-methods. Regarding other classes of MISP-related polytopes listed in Table 1, there exist nonlinear programming methods, such as ellipsoid methods and r-algorithms (see, for instance, Shor et al. 1997, Shor 2011) enabling to find an upper bound on $\alpha(G)$ which is at least as precise as $\alpha^*_{\mathcal{L}}(G)$.

6.2 Dual bounds $\psi^*(G)$ and $\psi^*_1(G)$

In this section, we introduce dual bounds $\psi^*(G)$ and $\psi^*_1(G)$ on $\alpha(G)$ corresponding to ψ^* and ψ^*_1 for MISP on a graph G, respectively. They are associated with construction quadratic formulations of MISP with polynomial number of constraints and applying results of Section 5.

6.2.1 MISP "superflows" constraints with applications

The first step will be the construction of superflows constraints that can be found by multiplying functions of MISP-formulations presented in Section 4.1. Therefore, some redundant constraints will be constructed following the approaches to construct relaxations described above. Then, connections with some MISP valid constraints will be established.

Let us introduce denotations for (29) and components of the 2-sided constraints (30):

$$f_{ij} = x_i + x_j - 1 \le 0, (i,j) \in E(G);$$
$$f'_k = -x_k \le 0, k = \overline{1,n};$$
$$f''_k = x_k - 1 \le 0, k = \overline{1,n}.$$

By (21),

1. $$f'_k \cdot f'_{k'} = (-x_k)(-x_{k'}) \ge 0, k, k' = \overline{1,n}$$

or

$$x_k x_{k'} \ge 0, k, k' = \overline{1,n}. \tag{65}$$

2. $$f''_k \cdot f''_{k'} = (x_k - 1)(x_{k'} - 1) \ge 0, k, k' = \overline{1,n}$$

or

$$x_k x_{k'} - x_k - x_{k'} + 1 \ge 0, k, k' = \overline{1,n}. \tag{66}$$

3. $$f_{ij} \cdot f'_k = (x_i + x_j - 1)(-x_k) \ge 0$$

after simplification becoming

$$x_i x_k + x_j x_k \le x_k, (i,j) \in E(G), k = \overline{1,n}.$$

$$x_k(x_i + x_j) \le x_k, (i,j) \in E(G), k = \overline{1,n}. \tag{67}$$

4. $$f_{ij} \cdot f''_k = (x_i + x_j - 1)(x_k - 1) \ge 0$$

resulting in

$$x_i x_k + x_j x_k \ge x_k + x_i + x_j - 1, (i,j) \in E(G), k = \overline{1,n}.$$

or

$$x_k + x_i + x_j - x_i x_k - x_j x_k \le 1, (i,j) \in E(G), k = \overline{1,n}. \tag{68}$$

5. $$f'_{k'} \cdot f''_{k''} = (-x_{k'})(x_{k''} - 1) \geq 0, k', k'' = \overline{1,n}$$

resulting in

6. $$x_{k'}x_{k''} \leq x_{k'}, k', k'' = \overline{1,n}.$$

$$f'_{k'} \cdot f'_{k''} = x_{k'}x_{k''} \geq 0, k', k'' = \overline{1,n}. \tag{69}$$

7. $$f''_{k'} \cdot f''_{k''} = (x_{k'} - 1)(x_{k''} - 1) \geq 0, k', k'' = \overline{1,n}$$

or

$$x_{k'} + x_{k''} - x_{k'}x_{k''} \leq 1, k', k'' = \overline{1,n}. \tag{70}$$

Further in this section, we will show how adding the "superflows" constraints (67) can improve an upper bound on $\alpha(G)$. All the derived "superflows" constraints (65)–(70) can be used in the same manner separately or in a combination.

6.2.2 The bounds $\psi^*(G)$

For MISP, as a basic model (49), (50), the model MISP.EM4 is chosen. It consists of the objective function (23), vertex quadratic constraints (13), and edge quadratic constraints (28). To specify that a MISP is solved for a particular graph G, $\psi^*(G)$ denotes the upper bound ψ^* in this case, while $Q_0^* = \alpha(G)$. It follows that $\psi^*(G)$ is an upper bound on $\alpha(G)$, i.e., $\psi^*(G) \geq \alpha(G)$.

The following theorem establishes a connection of $\psi^*(G)$ with $\alpha_{\mathcal{L}}(G)$.

Let us assume that a feasible domain $\mathcal{X}$ of the problem (49), (50) is a finite set and a polytope $\mathcal{P} = conv\mathcal{X}$.

Theorem 3. If $Q_0(x)$ is a linear function and linear constraints of $\mathcal{P}$ follow from constraints (49), (50), then

$$Q_0^{**} = \max_{x \in \mathcal{P}} Q_0(x) \geq Q_0^*,$$

i.e., Q_0^{**} is an upper bound on Q_0^*.

Corollary 1. If constraints of $\mathcal{L}STAB(G)$-polytope follow from constraints of a Euclidean formulation of MISP, then

$$\psi^*(G) \geq \alpha_{\mathcal{L}}(G). \tag{71}$$

Theorem 4. For a clique $Q_p \in \mathcal{Q}(G)$, the linear clique constraint (31) holds for any incidence vector $x \in X$ satisfying constraints of the continuous model MISP.EM4.

Proof. Take an arbitrary $Q_p \in \mathcal{Q}(G)$. Let its vertices be denoted $\{j_1,...,j_p\} \subseteq \{1,...,n\}$. By assumption, the constraints (13) and (28) are fulfilled at x. Particularly,

$$x_{j_i}^2 - x_{j_i} = 0, i = \overline{1,p}; \tag{72}$$

$$x_{j_s} x_{j_t} = 0, \quad s,t = \overline{1,p},\ s \neq t. \tag{73}$$

Consecutively applying (72), (73), we get:

$$x_{j_1} \sum_{i \in V(Q_p)} x_i = x_{j_1} \sum_{i=1}^{p} x_{j_i} = x_{j_1}^2 + \sum_{i=2, i\neq 1}^{p} x_{j_1} x_{j_i} = x_{j_1}^2 = x_{j_1}.$$

Similarly, for a fixed $i' \in \{1,...,p\}$,

$$x_{j_{i'}} \sum_{i \in V(Q_p)} x_i = x_{j_{i'}} \sum_{i=1}^{p} x_{j_i} = x_{j_{i'}}^2 + \sum_{i=1, i\neq i'}^{p} x_{j_{i'}} x_{j_i} = x_{j_{i'}}^2 = x_{j_{i'}}. \tag{74}$$

Due to $x_{j_{i'}} \in \{0,1\}$, consider two situations. The first one is if there exists $j' = \overline{1,p}$ such that $x_{j_{i'}} = 1$. Choosing such $x_{j_{i'}}$ and substituting it into (74) we get:

$$\sum_{i=1}^{p} x_{j_i} = \sum_{i \in V(Q_p)} x_i = 1. \tag{75}$$

Another situation occurs if there in no $i' = \overline{1,p}$ such that $x_{j_{i'}} = 1$. In this case, $x_{j_{i'}} = 0$, $i = \overline{1,p}$. Respectively,

$$\sum_{i \in V(Q_p)} x_i = 0. \tag{76}$$

Uniting (75) with (76) yields exactly the linear clique inequality (31) for the clique Q_p. Due to the random choice of Q_p from $\mathcal{Q}(G)$, the inequality holds $\forall Q_p \in \mathcal{Q}(G)$ at x, hence (31) is true.

Corollary 2. The polytope $CSTAB(G)$ satisfies conditions of Cor. 1. That is why, relations (64) and (71) can be combined for $\mathcal{L} = C$ and yield

$$\alpha_C^*(G) = \psi^*(G) = \alpha(G).$$

Remark 7. This corollary establishes the polynomial solvability of MISP on *t*-perfect graphs by solving it in the continuous formulation MISP.EM4 by the ellipsoid method or r-algorithms.

Corollary 3. The linear edge constraint (29) holds for any incidence vector $x \in X$ satisfying constraints of the continuous model MISP.EM4.

From this corollary, it follows that Cor. 2 works for $\mathcal{L} = C$ too. Thus relations

$$\alpha_Q^*(G) = \psi^*(G) = \alpha(G)$$

are satisfied. Respectively, on perfect graphs, MISP is polynomially solvable by the ellipsoid method or r-algorithms in the form of MISP.EM4.

6.2.3 The bound $\psi_1^*(G)$

For h-perfect, wheel-perfect and generalized wheel-perfect graphs, we will show a polynomial solvability of Problem 1 for MISP. Problem 1 will be constructed from the basic model MISP.EM4 by adding superflows constraints (67) (further referred to as MISP.EM7).

First, we extend the results of Cor. 1 to this case.

Corollary 4. If constraints of $\mathcal{L}STAB(G)$-polytope follow from constraints of Problem 1, which is a Euclidean formulation of MISP with superflows constraints, then

$$\psi_1^*(G) \geq \alpha_{\mathcal{L}}(G). \tag{77}$$

Theorem 5. For any odd cycle $C_{2k+1} \in \mathcal{C}(G)$, the odd-cycle constraint (41) is satisfied for any incidence vector x associated with its independent set in G and satisfied constraints of the continuous model MISP.EM7.

Proof. Let vertices of the cyrcle C_{2k+1} be denoted as $\{i_1, i_2, ..., i_{2k+1}\} \subseteq \{1, ..., n\}$, then $(x_{i_1}, ..., x_{i_{2k+1}})$ be an incidence vector associated with the corresponding independent set in C_{2k+1}. By assumption, x satisfies the constraints (13), (28), and (67), particularly,

$$x_{i_j}^2 - x_{i_j} = 0, j = \overline{1, 2k+1}; \tag{78}$$

$$x_{i_1} x_{i_{2k+1}} = 0, \quad x_{i_r} x_{i_{r+1}} = 0, \ r = \overline{1, 2k} \tag{79}$$

$$x_{i_r}(x_{i_{2s-1}} + x_{i_{2s}}) \leq x_{i_r}, \quad r = \overline{2, 2k}, s = \overline{1, k}. \tag{80}$$

Let us fix $j = 1$. Likewise proof of Theorem 4, we will apply (78)–(80) to the following relation:

$$x_{i_1} \sum_{i \in V(C_{2k+1})} x_i = x_{i_1} \sum_{r=1}^{2k+1} x_{i_r} = x_{i_1}^2 + x_{i_1} x_{i_2} + x_{i_1} \sum_{s=2}^{k} (x_{i_{2s-1}} + x_{i_{2s}}) + x_{i_1} x_{i_{2k+1}} \leq x_{i_1} + 0 + (k-1) x_{i_1} = k x_{i_1}.$$

Generalizing this reasoning into $j \in \{1,\dots,2k+1\}$, we get two situations depending on parity of j. Situation 1 when j is odd, namely, $j = 2k' + 1$ for some $k' \in \mathbb{N}_+$. Then

$$\begin{aligned} x_{i_j} \sum_{i \in V(C_{2k+1})} x_i &= x_{i_j} \sum_{r=1}^{2k+1} x_{i_r} = x_{i_j} x_{i_{j_1}} + x_{i_j} x_{i_{j_2}} + x_{i_j} \sum_{s=1}^{k'} (x_{i_{2s-1}} + x_{i_{2s}}) + \\ &+ x_{i_j}^2 + x_{i_j} \sum_{s=k'+1}^{k} (x_{i_{2s}} + x_{i_j} x_{i_{2s+1}}) \le 0 + 0 + (k'-1) x_{i_j} + x_{i_j} + (k-k') x_{i_j} = \\ &= k x_{i_j}. \end{aligned} \tag{81}$$

If j is even, namely, $j = 2k'$ for some $k' \in \mathbb{N}$. Then we face the situation 2 when

$$\begin{aligned} x_{i_j} \sum_{i \in V(C_{2k+1})} x_i &= x_{i_j} \sum_{r=1}^{2k+1} x_{i_r} = x_{i_j} \sum_{s=1}^{k'-1} (x_{i_{2s-1}} + x_{i_{2s}}) + x_{i_{j-1}} x_{i_j} + x_{i_j}^2 + \\ &+ x_{i_j} \sum_{s=k'+1}^{k} (x_{i_{2s-1}} + x_{i_{2s}}) + x_{i_j} x_{i_{2s+1}} \le \\ &\le 0 + 0 + (k'-1) x_{i_j} + x_{i_j} + (k-k') x_{i_j} = k x_{i_j}. \end{aligned} \tag{82}$$

Together, (81), (82) imply that

$$x_{i_r} \sum_{i \in V(C_{2k+1})} x_i \le k x_{i_r}, \quad \forall r = \overline{1, 2k+1},$$

i.e., for the cycle C_{2k+1}, the odd-cycle constraint (41) holds. Due to its random choice, the constraint is valid for any odd cycle in G.

Theorem 6. For an odd-wheel $W^{p}_{2k+1} \in \mathcal{W}(G)$, the generalized odd-wheel constraint (45) is satisfied for any incidence vector x associated with its independent set in G and satisfied constraints of the continuous model MISP. EM7.

Proof. Vertices of an odd cyrcle C_{2k+1} will be denoted by $I = \{i_1, i_2, \dots, i_{2k+1}\}$, and the vertices of the clique Q_p we will be denoted $J = \{j_1, \dots, j_p\}$, where $I \cap J = \emptyset$. For them, in addition to (72), (73) and (78)–(80), the following quadratic constrains are valid, where vertices from and J are involved:

$$x_{i_r} x_{j_s} = 0, \quad r = \overline{1, 2k}, \; s = \overline{1, p}, \tag{83}$$

Given that for the vertex $i_1 \in V(C_{2k+1})$ the equation $x^2_{i_1} = x_{i_1}$ is valid by (78), and from the equations (79) we get $x_{i_1} x_{i_2} = 0$ and $x_{i_1} x_{i_{2k+1}} = 0$; from the constraint (83) we obtain $x_{i_1}x_{j_s} = 0$, $s = 1,...,p$; from the inequalities (80) we have $x_{i_1}(x_{i_r} + x_{i_{r+1}}) \le x_{i_1}$, $r = 3,5,...,2k-1$. From here, it follows that

$$x_{i_1}(\sum_{i\in V(C_{2k+1})} x_i + k\sum_{j\in V(Q_p)} x_j) = x_{i_1}(\sum_{r=1}^{2k+1} x_{i_r} + k\sum_{s=1}^{p} x_{j_s}) =$$

$$= x_{i_1}^2 + x_{i_1}x_{i_2} + x_{i_1}\sum_{r=3}^{2k} x_{i_r} + x_{i_1}x_{i_{2k+1}} + k\sum_{s=1}^{p} x_{i_1}x_{j_s} =$$

$$= x_{i_1} + x_{i_1}(x_{i_3} + x_{i_4}) + \ldots + x_{i_1}(x_{i_{2k-1}} + x_{i_{2k}}) \le kx_{i_1}.$$

Similarly, for all $r \in \{1,...,n\}$, we have:

$$x_{i_r}(\sum_{i\in V(C_{2k+1})} x_i + k\sum_{j\in V(Q_p)} x_j) \le kx_{i_r}, \quad r = \overline{1, 2k+1}. \tag{84}$$

We are given that, for a vertex $j_1 \in V(Q_p)$, the equation $x_{j_1}^2 = x_{j_1}$ is true by (72); by (73), we have $x_{j_1}x_{j_s} = 0$, $s = \overline{2,p}$; with respect to (83) we have $x_{i_r}x_{j_1} = 0$, $r = \overline{1,2k+1}$. Wherefrom we can derive the following relation:

$$x_{j_1}(\sum_{i\in V(C_{2k+1})} x_i + k\sum_{j\in V(Q_p)} x_j) = x_{j_1}(\sum_{r=1}^{2k+1} x_{i_r} + k\sum_{s=1}^{p} x_{j_s}) =$$

$$= \sum_{r=1}^{2k+1} x_{i_r}x_{j_1} + kx_{j_1j_1}^2 + k\sum_{s=2}^{p} x_{j_1}x_{j_s} = kx_{j_1}.$$

Similarly, we get

$$x_{j_s}(\sum_{i\in V(C_{2k+1})} x_i + k\sum_{j\in V(Q_p)} x_j) = kx_{j_s}, \quad s = \overline{1, p}. \tag{85}$$

If we add inequality (84) and inequality (85) multiplied by the value of k, we obtain another inequality:

$$(\sum_{r=1}^{2k+1} x_{i_r} + k\sum_{s=1}^{p} x_{j_s})(\sum_{i\in V(C_{2k+1})} x_i + k\sum_{j\in V(Q_p)} x_j) \le k(\sum_{r=1}^{2k+1} x_{i_r} + k\sum_{s=1}^{p} x_{j_s}).$$

It results in another inequality:

$$(\sum_{i\in V(C_{2k+1})} x_i + k\sum_{j\in V(Q_p)} x_j)^2 - k(\sum_{i\in V(C_{2k+1})} x_i + k\sum_{j\in V(Q_p)} x_j) \le 0,$$

wherefrom we get

$$\sum_{i\in V(C_{2k+1})} x_i + k\sum_{j\in V(Q_p)} x_j \le k \quad \forall\, C_{2k+1} \subset \mathcal{C}(G),\ Q_p \subset \mathcal{Q},$$

which is exactly the generalized odd-wheel constraint (45).

Theorem 7. MISP in a formulation MISP.EM7 by the ellipsoid method and r-algorithms for h-perfect and W^p_{2k+1}-perfect graphs.

Proof. From Theorems 5 and 6, it follows that $\mathcal{L}STAB(G)$-polytopes for $\mathcal{L} \in \{H, W^p_{2k+1}\}$ satisfy conditions of Cor. 4, thus (77) holds and becomes

$$\begin{aligned} &\psi_1^*(G) \geq \alpha_H(G), \\ &\psi_1^*(G) \geq \alpha_{W^p_{2k+1}}(G). \end{aligned} \tag{86}$$

For these two cases, (64) takes the form of

$$\begin{aligned} &\alpha_H^*(G) = \alpha(G), \\ &\alpha^*_{W^p_{2k+1}}(G) = \alpha(G). \end{aligned} \tag{87}$$

Together, (86) with (87) result in

$$\begin{aligned} &\alpha_H(G) = \psi_1^*(G) = \alpha_H(G), \\ &\alpha_{W^p_{2k+1}}(G) = \psi_1^*(G) = \alpha_{W^p_{2k+1}}(G). \end{aligned} \tag{88}$$

Hence, the statement of Theorem 7 is true.

7. Conclusion

This work is dedicated to applying Shor's technique of quadratic dual bounds to MIS problems and deriving classes of polynomially solvable MISPs. In particular, the paper substantiates the polynomial solvability of MISPs on wheel-perfect graphs W_p and their generalizations, such as generalized wheel-perfect graphs W^p_{2k+1}. The theory of continuous functional representations of discrete sets was applied in designing a variety of different formulations of MISP. These results underly the theoretical justification of the polynomial solvability of MISP for some classes of graphs. In such a way, we provide a new proof of MISP polynomial solvability for perfect, t-perfect, h-perfect graphs and substantiate this for the first time for a family of W^p_{2k+1}-perfect graphs. In particular, with the help of functional representations techniques, all known so far continuous, binary and partially binary mathematical models were constructed, and several new models were obtained. Based on this, new families of superflows constraints of the polytope $STAB(G)$ of incidence vectors of independent sets in a graph G were also derived. As expected, these families of constraints will allow proving a polynomial solvability of other classes of graphs, in particular, superclasses of W^p_{2k+1}-perfect graphs. The presented technique of obtaining new mathematical formulations of COPs and the models themselves has theoretical value. In the future, the results can be applied in computation experiments to identify more efficient models in certain cases of input data.

References

Asratian, A.S., T.M.J. Denley and R. Häggkvist. 2008. Bipartite Graphs and their Applications. Cambridge University Press, Cambridge.

Cheng, E. and W.H. Cunningham. 1997. Wheel inequalities for stable set polytopes. Mathematical Programming, 77(2): 389–421.

Chvátal, V. 1975. On certain polytopes associated with graphs. Journal of Combinatorial Theory, Series B, 18(2): 138–154.

Grötschel, M., L. Lovász and A. Schrijver. 2011. Geometric Algorithms and Combinatorial Optimization. Springer, Berlin, Heidelberg.

Kochenberger, G. et al. 2014. The unconstrained binary quadratic programming problem: a survey. Journal of Combinatorial Optimization, 28(1): 58–81.

Lovász, L. 1979. On the Shannon capacity of a graph. IEEE Transactions on Information Theory, 25(1): 1–7.

Lovász, L. and M.D. Plummer. 2009. Matching Theory. Chelsea Pub Co, Providence, R.I.

Pardalos, P.M., D.-Z. Du and R.L. Graham (eds.). 2013. Handbook of Combinatorial Optimization. Springer, New York.

Pichugina, O. and O. Kartashov. 2019. Signed permutation polytope packing in VLSI design. pp. 1–6. *In*: 2019 IEEE 15th International Conference on the Experience of Designing and Application of CAD Systems (CADSM).

Pichugina, O. and S. Yakovlev. 2020. Euclidean combinatorial configurations: continuous representations and convex extensions. pp. 65–80. *In*: Lytvynenko, V. et al. (eds.). Lecture Notes in Computational Intelligence and Decision Making. Springer International Publishing, Cham.

Pichugina, O. and L. Koliechkina. 2021. The constrained knapsack problem: models and the polyhedral-ellipsoid method. pp. 233–247. *In*: Strekalovsky, A. et al. (eds.). Mathematical Optimization Theory and Operations Research: Recent Trends. Springer International Publishing, Cham.

Sbihi, N. and J.P. Uhry. 1984. A class of h-perfect graphs. Discrete Mathematics, 51(2): 191–205.

Schrijver, A. 2002. Combinatorial Optimization: Polyhedra and Efficiency. Springer, Berlin; New York.

Shor, N.Z. and P.I. Stetsyuk. 1997. Modified r-algorithm to find the global minimum of polynomial functions. Cybernetics and System Analysis, 33(4): 482–497.

Shor, N.Z. 2011. Nondifferentiable Optimization and Polynomial Problems. Springer, New York; London.

Stetsyuk, P.I. and B.M. Chumakov. 2007. On properties of one upper bound of N. Z. Shor for the weighted the stable set number. pp. 271–272. In Computing Optimization Issues. Kyiv.

Stetsyuk, P.I. 2008. On new properties of Shor's bounds for the weighted independence number of a graph. pp. 164–173. In Proceedings of the International Conference "50 years to the V. M. Glushkov Institute of Cybernetics of the NAS of Ukraine". Kyiv.

Stetsyuk, P.I. 2018. Dual Bounds in Quadratic Extremal Problems. Chisinau, Eureka.

Stetsyuk, P.I. and O.S. Pichugina. 2019. Shor's bounds for the weighted independence number. Scientific Bulletin of Uzhhorod University. Series of Mathematics and Informatics, 35(2): 71–81.

Stoyan, Y.G. and S.V. Yakovlev. 2020. Theory and methods of euclidian combinatorial optimization: current status and prospects. Cybernetics and System Analysis, 56(3): 366–379.

Trotignon, N. 2015. Perfect graphs: a survey. arXiv: 1–53. https://doi.org/10.48550/arXiv.1301.5149.

Yakovlev, S., O. Pichugina and L. Koliechkina. 2021. A lower bound for optimization of arbitrary function on permutations. pp. 195–212. *In*: Babichev, S. et al. (eds.). Lecture Notes in Computational Intelligence and Decision Making. Springer International Publishing, Cham.

CHAPTER 7

Approximations for Estimating Some Options Using the Inverse of the Laplace Transform

Robert Mnatsakanov[1] and *Omar Purtukhia*[2,*]

1. Introduction

In this work, we propose several approximations for the evaluation of some of the option prices based on the inversion of the scaled version of the Laplace transform which was suggested by Mnatsakanov and Sarkisian 2013. The proposed method is also applied to the Black-Scholes model for the estimation of option prices. In addition, we precisely solve the problem of pricing Asian options in the Bachelier model.

Financial markets have become increasingly more sophisticated in recent years and therefore, have the offered products. More complex financial options and derivatives have replaced the simple buy/sell trade deals of earlier years (see Lee and Sheen 2009). The determination of the actual value of financial options is a major concern for market participants.

[1] West Virginia University, Department of Mathematics, Morgantown, WV 26506, USA.
[2] Ivane Javakhishvili Tbilisi State University: Department of Mathematics; A. Razmadze Mathematical Institute, 13 University st., Tbilisi, Georgia.
* Corresponding author: o.purtukhia@gmail.com

While the writer of an option contract might be largely concerned by how much to charge for the contract and what his profit margin would be, the holder of the contract wants to be certain that he is paying a fair price and that he stands to make gains from the exercise of the contract. These and many other concerns have motivated a large volume of research, too numerous to mention, in the area of option pricing.

Financial derivatives such as American and European call-and-put options are referred to as plain vanilla products. Derivative securities which have certain features that make them more complex than the commonly traded plain vanilla products are called exotic options or simply, exotics. In other words, the payoff functions of derivative securities with more complicated forms than standard European or American call-and-put options are known as Exotic Options.

One such kind of Exotic Option is the so-called Binary Option. It is an option with discontinuous payoff function. The simplest examples of Binary Options are call and put options "cash or nothing". The payoff function of the call option has the form $BC_T = QI_{\{S_T>K\}}$, and for the put option --$BP_T = QI_{\{S_T<K\}}$, where K is the strike price at the time of execution T. It is also common the Binary Option "an asset or nothing".

These are the same conditions as in the "cash or nothing" option, but the difference is that owner of the call option receives the price of the asset S_T instead of amount Q. The Standard European Call Option (i.e., the option with the payoff function: $(S_T - K)^+$) is equivalent to a long position (the bought asset) in the "an asset or nothing" option and short position (the sold asset) in the "cash or nothing" option when $Q = K$.

These products are traded in the over-the-counter (OTC) derivative market. Exotic options are important aspects of the portfolio of an investment bank because they are usually more profitable than plain vanilla products. Examples of exotic options are the compound option, chooser option, barrier option, binary/digital option, lookback option, constant proportion portfolio insurance (CPPI), cliquet or ratchet option, variance swap, rainbow option, and Bermudan option (see James 2003, Hull 2006).

Let $V(S, t)$ be the pay-off function. When $t = T$, for a European call option, the pay-off is defined by

$$V(S, T) = max(S - K,0) = (S - K)^+.$$

Similarly, for a put option,

$$V(S, T) = (K - S)^+.$$

From Wilmott et al. 1995, it has been shown that $V(S, t)$ is a unique solution to the partial differential equation

$$\begin{cases} \frac{1}{2}\sigma^2 S^2 \frac{\partial^2 V(S,t)}{\partial S^2} + rS\frac{\partial V(S,t)}{\partial S} - \frac{\partial V(S,t)}{\partial t} - rV(S,t) = 0, \\ V(S,t) = h(S), S > 0, \\ \lim_{S\to 0} V(S,t) = r_1(t), \\ \lim_{S\to\infty} V(S,t) = r_2(t),\ t \in [0,T], \end{cases}$$

where $r_1(t)$, $r_2(t)$ are chosen appropriately to match $h(S)$.

The widely-used formula by Black and Scholes 1973, Merton 1973, provides an exact pricing formula for the simplest model in the case of constant coefficients. However, such a technique is not applicable in the general case with time and space-dependent coefficients. Hence, numerical methods are required for the evaluation of non-standard options.

To evaluate special options such as the American put and call options, Asian options, discretely monitored barrier options, and options with non-standard pay-offs that are characterized by discontinuities which arise at each monitoring date, carefully chosen numerical methods are required to avoid spurious oscillations when low volatility is assumed.

Several numerical methods have been used to solve the Black-Scholes equation. In Seydel 2017, Glasserman 2003, Tavella and Randall 2000, Wilmott et al. 1995 and the references there in, one can find well-known numerical methods for option pricing. Since financial markets are prone to stochastic fluctuations, stochastic approaches like the Monte Carlo methods, which are based on formulating and simulating stochastic differential equations, provide natural tools for simulating asset prices. Time marching methods like the Crank-Nicholson and Explicit and Implicit finite-difference methods are used with suitable spatial discretization schemes.

A major drawback of these time-marching schemes is that they usually require as many time steps as spatial meshes to balance errors arising from discretization. Lee and Sheen 2009 claimed that in particular, for the estimation of basket options of reasonable size, the usual time marching schemes seem to be too slow in practice since the cost of solving an elliptic system to advance to the next time step is usually computationally expensive.

Some related works in which the Laplace transform method was applied include Fu et al. 1998, Geman and Yor 1996, Lee and Sheen 2009, Mallier and Alobaidi 2000, Pelsser 2000, Tagliani and Milev 2013. Most of the earlier works in which the Laplace transform method was applied to solve the Black-Scholes equation were used to obtain the analytic solutions of the various options they studied rather than develop an efficient numerical scheme.

In Geman and Yor 1996, and Pelsser 2000, the Laplace transform is applied for the pricing of a double barrier option and in Mallier and Alobaidi 2000, the pricing of the American call option is considered. The Mellin transform method which is similar to the Laplace transform was used to obtain the analytic solution of an option price by Cruz-Baes and Gonzaelez-Rodriguez 2005, Jodar et al. 2005, Panini and Srivastava 2005.

Tagliani and Milev 2013 suggested a method they called the *mixed* method for a discretely monitored barrier option. The method involves solving the resulting ordinary differential equation (ODE) by computing the Laplace transform of the Black-Scholes equation by a finite-difference scheme and then transforming the solution of the ODE with the well-known Post-Widder inversion formula, see Cohen 2007. The authors showed that the *mixed* method is positivity-preserving, satisfies the discrete maximum principle, is spurious oscillations free, and is convergent to the exact solution.

Goffard et al. 2017, applied a Laplace transform inversion method involving an orthogonal projection of the probability density function with respect to a probability measure that belongs to the Natural Exponential Family of Quadratic Variance Function (NEF-QVF) to compute bivariate probability distributions from their Laplace transforms.

In this work, we present applications of the moment recovered approximation method in the estimation of the prices of some well-known financial options and derivatives. We apply the scaled Laplace transform method to solve the Black-Scholes equation. Our approach does not require either change of variables or solving diffusion equations. The resulting ordinary differential equation (ODE) is a Euler equation which has a closed-form solution. Moreover, based on the Glonti-Purtukhia generalization (see Glonti and Purtukhia 2017) of Clark-Ocone formula, we precisely solve the problem of pricing of Asian type option in the Bachelier model, which is impossible in the Black-Scholes model. Finally, we derive a constructive integral representation for one class of path-dependent Brownian functionals.

2. Auxiliary concepts and results

We apply the scaled Laplace transform method to solve the Black-Scholes equation which depends on one stock asset. Applying the moment recovered (MR) inversion method, we obtain approximations of the European-style put and call options. For the rate of approximation of the inversion method (see Fadahunsi and Mnatsakanov 2018).

Put Option. Let $P(S, \tau)$ be the value of a European put option, where S is the current value of the underlying asset and $\tau = T - t$ is the time left till maturity. Assume that the volatility σ and the risk-free interest rate r depend only on

S (i.e., $\sigma = \sigma(S)$ and $r = r(S)$). The price $P(S, \tau)$ satisfies the Black-Scholes equation

$$\frac{1}{2}\sigma^2 S^2 \frac{\partial^2 P(S,\tau)}{\partial S^2} + rS\frac{\partial P(S,\tau)}{\partial S} - \frac{\partial P(S,\tau)}{\partial \tau} - rP(S,\tau) = 0, \tag{1}$$

with boundary conditions

$$P(S, 0) = max\{K - S, 0\},$$

$$P(S, \tau) \to Ke^{-r\tau} \text{ as } S \to 0,\ P(S, \tau) \to 0 \text{ as } S \to \infty.$$

To solve (1), we take the scaled Laplace transform of $P(S, \tau)$ and its partial derivatives as follows:

$$\mu_\lambda(P(S,\cdot)) = L_{P(S,\cdot),b}(\lambda) = \int_0^\infty e^{-\lambda \, ln(b)\tau} P(S,\tau)d\tau, \tag{2}$$

and

$$\mu_\lambda\left(\frac{\partial P(S,\cdot)}{\partial \tau}\right) = L_{\frac{\partial P(S,\cdot)}{\partial \tau},b}(\lambda) = \int_0^\infty e^{-\lambda \ln(b)\tau} \frac{\partial P(S,\cdot)}{\partial \tau} d\tau. \tag{3}$$

Applying integration by parts, we get

$$\mu_\lambda\left(\frac{\partial P(S,\cdot)}{\partial \tau}\right) = -P(S,0) + \lambda \ln(b)\mu_\lambda(P(S,\cdot)), \tag{4}$$

$$\begin{aligned}\mu_\lambda\left(\frac{\partial P(S,\cdot)}{\partial \tau}\right) &= L_{\frac{\partial P(S,\cdot)}{\partial \tau},b}(\lambda) = \int_0^\infty e^{-\lambda \ln(b)\tau} \frac{\partial P(S,\cdot)}{\partial \tau} d\tau = \\ \frac{\partial}{\partial S}\int_0^\infty e^{-\lambda \ln(b)\tau} P(S,\tau)d\tau &= \frac{\partial}{\partial S}\mu_\lambda(P(S,\cdot)),\end{aligned} \tag{5}$$

and

$$\mu_\lambda\left(\frac{\partial^2 P(S,\cdot)}{\partial S^2}\right) = L_{\frac{\partial^2 P(S,\cdot)}{\partial S^2},b}(\lambda) = \int_0^\infty e^{-\lambda \ln(b)\tau} \frac{\partial^2 P(S,\tau)}{\partial S^2} d\tau = \frac{\partial^2}{\partial S^2}\mu_\lambda(P(S,\cdot)). \tag{6}$$

Transforming the boundary conditions, we get

$$\begin{aligned}&\lim_{S\to 0}\mu_\lambda(P(S,\cdot)) = \lim_{S\to 0}\int_0^\infty e^{-\lambda \ln(b)\tau} P(S,\tau)d\tau = \\ &\int_0^\infty e^{-\lambda \ln(b)\tau} \lim_{S\to 0} P(S,\tau)d\tau = \int_0^\infty e^{-\lambda \ln(b)\tau} Ke^{-r\tau} d\tau = \frac{K}{r + \lambda \ln(b)}.\end{aligned} \tag{7}$$

Applying (2)–(7), (1) becomes,

$$\frac{1}{2}\sigma^2 S^2 \frac{\partial^2 \mu_\lambda P(S,\cdot)}{\partial S^2} + rS\frac{\partial \mu_\lambda P(S,\cdot)}{\partial S} - (r + c\lambda)\mu_\lambda P(S,\cdot) = -P(S,0), \tag{8}$$

with boundary conditions

$$\mu_\lambda(P(S,\cdot)) \to \frac{K}{r+c\lambda} \text{ as } S \to 0 \text{ and } \mu_\lambda P(S,\cdot) \to 0 \text{ as } S \to \infty, \tag{9}$$

Solving the homogeneous part of (8), we get

$$\frac{1}{2}\sigma^2 S^2 \mu_\lambda'' + rS\mu_\lambda' - (r+\lambda\ln(b))\mu_\lambda = 0, \tag{10}$$

$$S^2\mu_\lambda'' + \frac{2r}{\sigma^2}S\mu_\lambda' - \frac{2(r+\lambda\ln(b))}{\sigma^2}\mu_\lambda = 0, \tag{11}$$

and observe that (10) is an Euler equation. By setting $\mu_\lambda = CS^\gamma$, $\mu_\lambda' = C\gamma S^{\gamma-1}$ and $\mu_\lambda'' = C\gamma(\gamma-1)S^{\gamma-2}$, for some γ, $C \in R$. Substituting this into (11), we get

$$S^2(CS^\gamma)'' + \frac{2r}{\sigma^2}S(CS^\gamma)' - \frac{2(r+\lambda\ln(b))}{\sigma^2}CS^\gamma = 0,$$

from here

$$CS^\gamma\left[\gamma(\gamma-1) + \frac{2r}{\sigma^2}\gamma - \frac{2(r+\lambda\ln(b))}{\sigma^2}\right] = 0.$$

Solving the equation

$$\gamma^2 + \left(\frac{2r}{\sigma^2} - 1\right)\gamma - \frac{2(r+\lambda\ln(b))}{\sigma^2} = 0,$$

we get

$$\gamma_1 = \frac{-\left(r - \frac{1}{2}\sigma^2\right) + \sqrt{\left(r - \frac{1}{2}\sigma^2\right)^2 + 2\sigma^2(r+\lambda\ln(b))}}{\sigma^2}, \tag{12}$$

and

$$\gamma_2 = \frac{-\left(r - \frac{1}{2}\sigma^2\right) - \sqrt{\left(r - \frac{1}{2}\sigma^2\right)^2 + 2\sigma^2(r+\lambda\ln(b))}}{\sigma^2}. \tag{13}$$

Hence, the solution to (11) is

$$\mu_\lambda^{(c)}(P(S,\cdot)) = C_1 S^{\gamma 1} + C_2 S^{\gamma 2}, \tag{14}$$

where the superscript (c) indicates that the solution is the complementary solution of (8), which is also the solution in cases when $S \geq K$.

When $K \geq S$ at maturity T, i.e., $\tau = 0$, $P(S, 0) = K - S$. So, the non-homogeneous part of (8) corresponds to cases when the strike K is greater than the underlying stock price. Hence, the put option will be exercised.

Let $\mu_\lambda^{(p)}(P(S,\cdot)) = AS + B$ be a particular solution of (10), for some $A, B \in R$. Then, $\mu_\lambda^{(p)'} = A$ and $\mu_\lambda^{(p)''} = 0$. Substituting these into (8) we get

$$rSA - (r + \lambda\, ln(b))(AS + B) = -(K - S),$$

from here

$$-(r + \lambda\, ln(b))B - \lambda\, ln(b)AS = -(K - S). \tag{15}$$

Comparing the coefficients of the left and right sides (15), we obtain

$$A = -\frac{1}{\lambda \ln(b)} \text{ and } B = \frac{K}{r + \lambda \ln(b)}.$$

Hence,

$$\mu_\lambda^{(p)}(P(S,\cdot)) = -\frac{1}{\lambda \ln(b)} S + \frac{K}{r + \lambda \ln(b)}. \tag{16}$$

Combining (14) and (16), we get the general solution of (8)

$$\mu_\lambda(P(S,\cdot)) = L_{P(S,\cdot),b}(\lambda) = C_1 S^{\gamma 1} + C_2 S^{\gamma 2} - \frac{1}{\lambda \ln(b)} S + \frac{K}{r + \lambda \ln(b)}.$$

Thus,

$$\mu_\lambda(P(S,\cdot)) = \begin{cases} C_1 S^{\gamma 1} + C_2 S^{\gamma 2} - \frac{1}{\lambda \ln(b)} S + \frac{K}{r + \lambda \ln(b)}, & S < K; \\ C_1 S^{\gamma 1} + C_2 S^{\gamma 2}, & S \geq K. \end{cases}$$

We have that $\gamma_1 \geq 0 \geq \gamma_2$. Therefore in the case when $S < K$, $C_2 = 0$ ensures the boundedness of the derivative $\mu_\lambda'(P(S,\cdot))$. In the case when $S \geq K$, $C_1 = 0$ ensures that the value of the option goes to zero when the stock price goes to infinity. Thus, the general solution to (8) reduces to

$$\mu_\lambda(P(S,\cdot)) = \begin{cases} C_1 S^{\gamma 1} - \frac{1}{\lambda \ln(b)} S + \frac{K}{r + \lambda \ln(b)}, & S < K; \\ C_2 S^{\gamma 2}, & S \geq K, \end{cases} \tag{17}$$

where (17) satisfies the boundary conditions in (9). Next, we solve for C_1 and C_2:

$$\mu_\lambda(P(S,\cdot))_{S=K} = C_1 K^{\gamma 1} - \frac{K}{\lambda \ln(b)} + \frac{K}{r + \lambda \ln(b)}, \quad S < K, \tag{18}$$

$$\mu_\lambda(P(S,\cdot))|_{S=K} = C_2 K^{\gamma 2}, \quad S \geq K, \tag{19}$$

$$\mu_{\lambda}'(P(S,\cdot))_{S=K} = \gamma_1 C_1 K^{\gamma_1 - 1} - \frac{1}{\lambda \ln(b)}, \quad S < K, \tag{20}$$

$$\mu_{\lambda}'(P(S,\cdot))|_{S=K} = \gamma_2 C_2 K^{\gamma_2 - 1}, \quad S \geq K. \tag{21}$$

Setting (18) = (19) and (20) = (21), we get

$$C_1 K^{\gamma 1} - \frac{K}{\lambda \ln(b)} + \frac{K}{r + \lambda \ln(b)} = C_2 K^{\gamma 2}, \tag{22}$$

$$\gamma_1 C_1 K^{\gamma_1 - 1} - \frac{1}{\lambda \ln(b)} = \gamma_2 C_2 K^{\gamma_2 - 1}. \tag{23}$$

Multiply (22) by γ_2/K and subtract (23) to get

$$\gamma_2 C_1 K^{\gamma_1 - 1} - \gamma_1 C_1 K^{\gamma_1 - 1} + \frac{\gamma_2}{r + \lambda \ln(b)} - \frac{\gamma_2}{\lambda \ln(b)} + \frac{1}{\lambda \ln(b)} = 0.$$

Solving for C_1, we get

$$C_1 = \left[\frac{\gamma_2}{r + \lambda \ln(b)} - \frac{\gamma_2 - 1}{\lambda \ln(b)} \right] \cdot \frac{K^{1-\gamma_1}}{\gamma_1 - \gamma_2}. \tag{24}$$

Substituting in (23), we get

$$C_2 = \left[\frac{\gamma_1}{r + \lambda \ln(b)} - \frac{\gamma_1 - 1}{\lambda \ln(b)} \right] \cdot \frac{K^{1-\gamma_2}}{\gamma_1 - \gamma_2}. \tag{25}$$

Finally, to estimate the value of the put option, we apply the moment recovered (MR) inversion method to (17). Thus, the value of the European put option is approximated by

$$P_{\alpha,b}(S,\tau) = \frac{[\alpha b^{-\tau}] \ln(b) \Gamma(\alpha + 2)}{\alpha \Gamma([\alpha b^{-\tau}] + 1)} \sum_{m=0}^{\alpha - [\alpha b^{-\tau}]} \frac{(-1)^m \mu_{m + [\alpha b^{-\tau}]}(P(S,\cdot))}{m!(\alpha - [\alpha b^{-\tau}] - m)!},$$

where

$$\Gamma(\alpha) = \int_0^{\infty} x^{\alpha - 1} e^{-x} \, dx$$

is the gamma function and

$$\mu_{m+[\alpha b^{-\tau}]}(P(S,\cdot)) = \begin{cases} C_1 S^{\gamma 1} - \dfrac{1}{\ln(b)(m + [\alpha b^{-\tau}])} S + \dfrac{K}{r + \lambda \ln(b)(m + [\alpha b^{-\tau}])}, & S < K; \\ C_2 S^{\gamma 2}, & S \geq K. \end{cases}$$

In C_1 and C_2, $\lambda \equiv m + [\alpha b^{-\tau}]$.

Call Option. Recall, the Black-Scholes equation for a European style call option is

$$\frac{1}{2}\sigma^2 S^2 \frac{\partial^2 C(S,\tau)}{\partial S^2} + rS\frac{\partial C(S,\tau)}{\partial S} - \frac{\partial C(S,\tau)}{\partial \tau} - rC(S,\tau) = 0, \qquad (26)$$

with boundary conditions

$$C(S,0) = max\{S - K, 0\},$$

$$C(S,\tau) \to 0 \text{ as } S \to 0,\ C(S,\tau) \to S \text{ as } S \to \infty.$$

We take the scaled Laplace transform of (26) and solve the resulting ordinary differential equation with boundary conditions $\mu_\lambda(C(S,\cdot)) \to \frac{S}{c\lambda}$ as $S \to \infty$ and $\mu_\lambda(C(S,\cdot)) \to 0$ as $S \to 0$. We get the general solution

$$\mu_\lambda(C(S,\cdot)) = \begin{cases} C_1 S^{\gamma 1} + C_2 S^{\gamma 2} + \dfrac{1}{\lambda \ln(b)} S - \dfrac{K}{r + \lambda \ln(b)}, & S > K; \\ C_1 S^{\gamma 1} + C_2 S^{\gamma 2}, & S \le K, \end{cases}$$

where γ_1, γ_2, C_1 and C_2 are as defined in (12), (13), (24), and (25), respectively.

We observe that in the case when $S > K$, $C_1 = 0$ ensures the boundedness of the derivative $\mu'_\lambda(C(S,\cdot))$ and in the case when $S \le K$, $C_2 = 0$ ensures that the option value goes to zero as the stock price goes to zero (see Kumar 2008). Thus, the general solution reduces to

$$\mu_\lambda(C(S,\cdot)) = \begin{cases} C_2 S^{\gamma 2} + \dfrac{S}{\lambda \ln(b)} - \dfrac{K}{r + \lambda \ln(b)}, & S > K; \\ C_1 S^{\gamma 1}, & S \le K, \end{cases} \qquad (27)$$

where (27) satisfies the boundary conditions.

Applying the MR inversion method, the call option can be approximated by

$$C_{\alpha,b}(S,\tau) = \frac{[\alpha b^{-\tau}]\ln(b)\Gamma(\alpha+2)}{\alpha\Gamma([\alpha b^{-\tau}]+1)} \sum_{m=0}^{\alpha-[\alpha b^{-\tau}]} \frac{(-1)^m \mu_{m+[\alpha b^{-\tau}]}(C(S,\cdot))}{m!(\alpha - [\alpha b^{-\tau}] - m)!},$$

where $\lambda \equiv m + [\alpha b^{-\tau}]$ in γ_1, γ_2, C_1 and C_2.

3. Double-barrier options

A double-barrier option is a type of financial option where the option to exercise depends on the price of the underlying asset crossing or reaching two barriers namely: L (lower barrier) and U (upper barrier). The payoff depends on whether the underlying asset price reaches L or U during the transaction period.

The option knocks out (becomes worthless) if either barrier is reached during its lifetime. If neither barrier is reached by maturity T, the option pays the standard Black-Scholes pay-off $max\{0.S(T) - K\}$, where K, the strike price of the option, satisfies $L < K < U$.
A double-barrier knock-out call option satisfies the following equation:

$$\frac{1}{2}\sigma^2 S^2 \frac{\partial^2 B(S,\tau)}{\partial S^2} + rS\frac{\partial B(S,\tau)}{\partial S} - \frac{\partial B(S,\tau)}{\partial \tau} - rB(S,\tau) = 0, \qquad (28)$$

with boundary conditions

$$B(S, 0) = max\{S - K, 0\} I_{[L,U]}(S),$$

$$B(S, \tau) \to 0 \text{ as } S \to 0 \text{ or } S \to \infty,$$

where $\tau = T - t$ and $B(S, \tau)$ is the price of a Barrier call option and

$$B(S,0) = \max\{S - K, 0\} I_{[L,U]}(S) = \begin{cases} S - K, & S \in [L,U] \text{ and } S > K; \\ 0, & S \notin [L,U] \text{ and } S \leq K. \end{cases}$$

Taking the scaled Laplace transform of (28) and solving the resulting ordinary differential equation, we get the general solution

$$\mu_\lambda(B(S,\cdot)) = \begin{cases} C_2 S^{\gamma 2} + \dfrac{S}{\lambda \ln(b)} - \dfrac{K}{r + \lambda \ln(b)}, & S > K \text{ and } S \in [L,U]; \\ C_1 S^{\gamma 1}, & S \leq K \text{ and } S \notin [L,U]. \end{cases}$$

Applying the MR inversion method, a barrier call option can be approximated by

$$B_{\alpha,b}(S,\tau) = \frac{[\alpha b^{-\tau}] \ln(b) \Gamma(\alpha + 2)}{\alpha \Gamma([\alpha b^{-\tau}] + 1)} \sum_{m=0}^{\alpha - [\alpha b^{-\tau}]} \frac{(-1)^m \mu_{m+[\alpha b^{-\tau}]}(B(S,\cdot))}{m!(\alpha - [\alpha b^{-\tau}] - m)!},$$

where $\lambda \equiv m + [\alpha b^{-\tau}]$ in γ_1, γ_2, C_1 and C_2.

Hedging of the Knock-Out Barrier Option

For one functional of Brownian motion, which is the payoff function of Knock-Out Barrier Option in the case of Bachelier's model of financial market, the Clark's integral representation with an explicit form of integrand was obtained by Glonti and Purtukhia 2014a. This functional is a multiplication of the payoff function of European Call Option and an indicator of some event. Hence, it is impossible to use Clark-Ocone's formula (Ocone 1984) because the indicator of the event is not Malliavin differentiable if the probability of this event is not equal to zero or one. There, according to the generalized

Clark-Ocone formula obtained by Glonti and Purtukhia 2017, is derived the explicit form of the integrand. This integrand is the optimal hedging strategy replicating the Knock-Out Barrier Option in the case of Bachelier's model.

Let $B = (B_t)$, $t \in [0,T]$ be the Brownian motion on the probability space $(\Omega, \Im, P)$. Let $(\Im_t^B)$, $t \in [0,T]$ be the natural filtration generated by the Brownian motion. Assume that the $\Im_t^B$-measurable F_T functional has the following form:

$$F_T = (B_T - K)^+ I_{\{B_T^* \leq L\}}, \tag{29}$$

where $B_T^* = \max_{0 \leq t \leq T} B_t$, K and L are positive constants such that $L \geq K$ and I_A is the indicator of event A.

The following stochastic integral representation is fulfilled (see, Theorem 2 in Glonti and Purtukhia 2014a)

$$F_T = EF_T - \int_0^T \frac{2(L-K)}{\sqrt{T-t}} \varphi\left(\frac{L-B_t}{\sqrt{T-t}}\right) dB_t + \int_0^T \left\{ \Phi\left(\frac{B_t - K}{\sqrt{T-t}}\right) - \Phi\left[\frac{B_t - 2(L-K)}{\sqrt{T-t}}\right] \right\} dB_t, \tag{30}$$

where

$$\varphi(x) = \frac{1}{\sqrt{2\pi}} exp\left\{-\frac{x^2}{2}\right\} \text{ and } \Phi(x) = \frac{1}{\sqrt{2\pi}} \int_{-\infty}^{x} exp\left\{-\frac{z^2}{2}\right\} dz.$$

Consider now Knock-Out Barrier Option in the financial market, represented by Bachelier 's model. In this case the stock price S is described by

$$S_t = S_0 + \mu t + \sigma B_t.$$

The payoff of Up-and-Out Call Barrier Option is (see Musela and Rutkowski 1999) the functional F_T from (29).

Consider the unique martingale (risk neutral) measure $\bar{P} \sim P$ such that (see Shyriaev 1999) $d\bar{P} = Z_T dP$ with

$$Z_T = exp\left\{-\frac{\mu}{\sigma} B_t - \frac{1}{2}\left(\frac{\mu}{\sigma}\right)^2 t\right\}.$$

It is well-known that under this measure the following relation is fulfilled:

$$Law\{S_0 + \mu t + \sigma B_t,\ t \leq T | \bar{P}\} = Law\{S_0 + \sigma B_t,\ t \leq T | P\}$$

and if $\sigma = 1$

$$E^{\bar{P}}\,[F_T | \Im_t^B] = E[\bar{F}_T | \Im_t^B],$$

where $\bar{F}_T$ is given in (29) with $\bar{K} = K - S_0$ and $\bar{L} = L - S_0$.

Now, using representation (30), it is easy to find the optimal hedging strategy.

4. Asian options

Asian options belong to the so-called path-dependent derivatives. They are among the most difficult to price and hedge both analytically and numerically. Basket options are even harder to price and hedge because of the large number of state variables. Several approaches have been proposed in the literature, including Monte Carlo simulations, tree-based methods, partial differential equations, and analytical approximations among others. The last category is the most appealing because most of the other methods are very complex and slow.

In the 80th of the past century, it turned out (see, Harison and Pliska 1981) that the martingale representation theorems (along with the Girsanov's measure change theorem) play an important role in modern financial mathematics. The well-known Clark (see, Clark 1970) formula says nothing on finding explicitly the integrand which takes part in the integral representation. In the case when the Wiener functional has the stochastic derivative, the Ocone-Clark's (see, Ocone 1984) formula is fined explicitly this integrand.

Our approach (see, Jaoshvili and Purtukhia 2005), within the classical Ito's calculus, allows one to construct explicitly integrand by using both the standard L_2 theory and the theories of weighted Sobolev spaces, allows one to construct explicitly the integrand if the Brownian functional has no stochastic derivative. Later, we proposed an approach that instead of the smoothness of the functional required only the smoothness of the Levy martingale associated with it and generalized the Clark-Ocone formula for stochastically non-smooth functionals (see, Glonti and Purtukhia 2017).

Asian Options also are a type of Exotic Option. The payoff function of this option depends on the average value of the price of an asset during a certain period of the option lifetime. The payoff function of the Asian Option by definition has the following representation: $C_T^A = [AS(T_0, T) - K]^+$, where

$$A_S(T_0,T) = \frac{1}{T-T_0}\int_{T_0}^{T} S_t dt$$

is the arithmetic mean of the prices of an asset at time interval $[T_0, T]$, K is a strike and $S = (S_t)$ $(0 \le t \le T)$ is geometrical Brownian motion.

The main difficulty in pricing and hedging of the Asian Option is that the random variable $A_S(T_0, T)$ is not lognormal distributed and therefore, it is rather difficult to obtain explicit formulas of pricing of this option.

Let us now recall some concepts and facts from Nualart 2006.

Let B_t be a Brownian motion on a standard filtered probability space $(\Omega, \Im, \Im_t, P)$ and let $\Im_t = \Im_t^B$ be the augmentation of the filtration generated by B.

Definition 1. The class of smooth Brownian functionals $\wp$ is the class of a random variable which has the form $F = f(B_{t_1},...,B_{t_n})$, $f \in C_p^{\infty}(R^n)$, $t_i \in [0,T]$, $n \geq 1$, where $C_p^{\infty}(R^n)$ is the set of all infinitely continuously differentiable functions $f: R^n \to R$ such that f and all of its partial derivatives have polynomial growth.

Definition 2. The stochastic (Malliavin) derivative of a smooth random variable $F \in \wp$ is the stochastic process $D_t^B F$ given by

$$D_t^B F = \sum_{i=1}^{n} \frac{\partial f}{\partial x_i}(B_{t_1},..., B_{t_n}) I_{[0,t_i]}(t).$$

Definition 3. D^B is closable as an operator from $L_2(\Omega)$ to $L_2(\Omega; L_2[0,T])$. We will denote its domain by $D_{2,1}^B$. That means, $D_{2,1}^B$ is equal to the adherence of the class of smooth random variables with respect to the norm

$$||F||_{2,1} = \{E[|F|^2 + (||D^B_{\cdot} F||^2_{L_2([0,T])})]\}^{1/2}.$$

In general, as is familiar in Malliavin calculus, we introduce the norm

$$||F||_{p,1} = \{E[|F|^p + (||D^B_{\cdot} F||^2_{L_2([0,T])})^{p/2}]\}^{1/p},$$

where $D^B_{\cdot}$ is the Malliavin derivative operator and $D_{p,1}^B$ denotes the Banach space which is the closure of the class of smooth Brownian functionals $\wp$ with respect to the norm $||\cdot||_{p,1}$ $(p \geq 1)$.

When the random variable F belongs to the Hilbert space $D_{2,1}^B$, it turns out that the integrand in the Clark representation can be identified as the optional projection of the Malliavin derivative of F.

Theorem 1. (see, Ocone 1984). If F is differentiable in Malliavin sense, $F \in D_{2,1}^B$, then the stochastic integral representation is fulfilled

$$F = EF + \int_0^T E[D_t^B | \Im_t^B] dB_t \quad \text{(P -a.s.)}.$$

Theorem 2. (see, Glonti and Purtukhia 2017). Suppose that $G_t = E[F|\Im_t^B]$ is Malliavin differentiable $G_t \in D_{2,1}^B$ for almost all $t \in [0,T)$. Then we have the stochastic integral representation

$$G_t = F = EF + \int_0^T v_u \, dB_u \quad (P\text{ -a.s.}),$$

where

$$v_u := \lim_{t \uparrow T} E[D_u^B G_t | \Im_u^B] \text{ in the } L_2([0,T] \times \Omega).$$

It is clear that there are also such functionals which don't satisfy even the weakened conditions, i.e., the non-smooth functionals whose conditional mathematical expectations is not stochastically differentiable too. In particular, to such functional belongs the integral type functional $\int_0^T u_s \, ds$ with non-smooth integrand $u_s(\omega)$.

It is well-known that if $u_s(\omega) \in D^B_{2,1}$ for all s, then $\int_0^T u_s(\omega)\, ds \in D^B_{2,1}$ and

$$D^B_t \{\int_0^T u_s(\omega)\, ds\} = \int_0^T D^B_t\, [u_s(\omega)]ds.$$

But if $u_s(\omega)$ is not differentiable in Malliavin sense, then the Lebesgue average (with respect to ds) also is not differentiable in Malliavin sense (see, for example, Glonti and Purtukhia 2014b).

Asian options give the holder of the contract the right to buy or sell an asset for its average price over a pre-set transaction period. Consider a time interval $[0,T]$, and let $\{A(x),\ x \geq 0\}$ be the average underlying price. Then

$$A(x) = \frac{1}{x}\int_0^x S_u du.$$

Hence pay-off at maturity T is

$$[A(T) - K]^+ = max\{A(T) - K,\ 0\}.$$

The price V of an Asian option is a function of the underlying asset price S, the average price A, and time to maturity T.

Bachelier Model

Let on the probability space $(\Omega, \Im, P)$ is given the Brownian motion $B = (B_t)$, $t \in [0,T]$ and $(\Im^B_t)$, $t \in [0,T]$ is the natural filtration generated by the Brownian motion B. Consider the Bachelier market model with risk-free asset price evolution described by

$$dM_t = rM_t\, dt,\ \ M_0 = 1.$$

where $r \geq 0$ is an interest rate and risky asset price evolution

$$dS_t = \mu dt + \sigma dB_t,\ \ S_0 = 1,$$

where $\mu \in R$ is appreciation rate and $\sigma > 0$ is volatility coefficient.

Let

$$Z_T = exp\left\{-\frac{\mu - r}{\sigma}B_T - \frac{1}{2}\left(\frac{\mu - r}{\sigma}\right)^2 T\right\}$$

and $\widetilde{P}$ is the measure on $(\Omega, \Im^B_T)$ such that

$$d\widetilde{P} = Z_T dP.$$

From Girsanov's Theorem it follows that under this measure (martingale measure)

$$\tilde{B}_t = B_t + \frac{\mu - r}{\sigma}t$$

is the standard Brownian motion and

$$dS_t = rdt + \sigma d\widetilde{B}_t,\ S_0 = 1$$

or

$$S_t = 1 + rt + \sigma \widetilde{B}_t.$$

Consider the problem of "replication" the European Option of Exotic Type with the pay-off of integral type $G = C_T^A$, i.e., one needs to find a trading strategy (β_t, γ_t), $t \in [0,T]$ such that the capital process

$$X_t = \beta_t M_t + \gamma_t S_t,\ X_T = G$$

under the self-financing condition

$$dX_t = \beta_t dM_t + \gamma_t dS_t.$$

Let for simplicity $r = 0$. According to the well-known properties of the Malliavin derivative, Brownian motion and conditional mathematical expectation, using the fact that $\int_0^t B_s ds \sim N(0, \frac{t^3}{3})$ (see, (35) in Appendix), it is not difficult to verify that

$$D_t G = I_{\left\{1+\frac{\sigma}{T}\int_0^T B_s ds - K > 0\right\}} \frac{\sigma}{t}\int_0^T D_t^B B_s ds = I_{\left\{1+\frac{\sigma}{T}\int_0^T B_s ds - K > 0\right\}} \frac{\sigma(T-t)}{T}$$

and

$$E[D_t G \mid \Im_t^B] = \frac{\sigma(T-t)}{T} P\{1 + \frac{\sigma}{t}\int_0^T B_s ds - K > 0 \mid \Im_t^B\} =$$

$$= \frac{\sigma(T-t)}{T} P\{\int_0^T (T-s) dB_s > \frac{T}{\sigma}(K - \frac{rT}{2} - 1) \mid \Im_t^B\} =$$

$$= \frac{\sigma(T-t)}{T} P\{\int_0^T (T-s) dB_s + \int_0^T (T-s) dB_s > \frac{T}{\sigma}(K-1) \mid \Im_t^B\} =$$

$$= \frac{\sigma(T-t)}{T} P\{x + \int_t^T (T-s) dB_s > \frac{T}{\sigma}(K-1)\}\Big|_{x=\int_0^t (T-s)dB_s} =$$

$$= \frac{\sigma(T-t)}{T} P\{N(0,1) > \frac{\sqrt{3}}{\sqrt{(T-t)^3}}[\frac{T}{\sigma}(K-1) - x]\}\Big|_{x=\int_0^t (T-s)dB_s} =$$

$$= \frac{\sigma(T-t)}{T}\left[1 - \Phi\left(\frac{\sqrt{3}}{\sqrt{(T-t)^3}}[\frac{T}{\sigma}(K-1) - x]\right)\right]\Big|_{x=\int_0^t (T-s)dB_s}.$$

Hence, we obtain the following integral representation

$$G = EG + \int_0^T \frac{\sigma(T-t)}{T}\left[1-\Phi\left(\frac{\sqrt{3}}{\sqrt{(T-t)^3}}[\frac{T}{\sigma}(K-1)-\int_0^t (T-s)dB_s]\right)\right]dB_t.$$

Therefore, we can find the component γ_t of the hedging strategy $\pi = (\beta_t, \gamma_t)$, $t \in [0,T]$, which is defined by integrand of last representation and is equal

$$\gamma_t = \frac{T-t}{T}\left[1-\Phi\left(\frac{\sqrt{3}}{\sqrt{(T-t)^3}}[\frac{T}{\sigma}(K-1)-\int_0^t (T-s)dB_s]\right)\right].$$

Moreover, we have

$$EG = \frac{1}{\sqrt{2\pi T^3/3}}\int_{-\infty}^{+\infty}(1+\frac{\sigma}{T}x-K)^+ e^{-\frac{x^2}{2T^3/3}}dx =$$

$$= \frac{1}{\sqrt{2\pi T^3/3}}\int_{T(K-1)/\sigma}^{+\infty}(1+\frac{\sigma}{T}x-K)e^{-\frac{x^2}{2T^3/3}}dx := J_1 + J_2,$$

where

$$J_1 := \frac{(1-K)}{\sqrt{2\pi T^3/3}}\int_{T(K-1)/\sigma}^{+\infty} e^{-\frac{x^2}{2T^3/3}}dx = \frac{(1-K)}{\sqrt{2\pi}}\int_{T(K-1)/\sigma}^{+\infty} e^{-\frac{x^2}{2T^3/3}}d\left(\frac{x}{\sqrt{T^3/3}}\right) =$$

$$= \frac{(1-K)}{\sqrt{2\pi}}\int_{\frac{T(K-1)/\sigma}{\sqrt{T^3/3}}}^{+\infty} e^{-\frac{x^2}{2}}dx = (1-K)\left[1-\Phi\left(\frac{K-1}{\sigma}\sqrt{\frac{3}{T}}\right)\right]$$

and

$$J_2 := \frac{1}{\sqrt{2\pi T^3/3}}\frac{\sigma}{T}\int_{T(K-1)/\sigma}^{+\infty} x e^{-\frac{x^2}{2T^3/3}}dx = -\frac{1}{\sqrt{2\pi T^3/3}}\frac{\sigma}{T}\frac{T^3}{3}\int_{T(K-1)/\sigma}^{+\infty} d\left(e^{-\frac{x^2}{2T^3/3}}\right) =$$

$$= \frac{1}{\sqrt{2\pi T^3/3}}\frac{\sigma T^2}{3}e^{-\frac{x^2}{2T^3/3}}\Big|_{T(K-1)/\sigma}^{+\infty} = -\sigma\sqrt{\frac{T}{3}}\varphi\left(\frac{x}{\sqrt{T^3/3}}\right)\Big|_{T(K-1)/\sigma}^{+\infty} =$$

$$= \sigma\sqrt{\frac{T}{3}}\varphi\left(\frac{T(K-1)/\sigma}{\sqrt{T^3/3}}\right) = \sigma\sqrt{\frac{T}{3}}\varphi\left(\frac{K-1}{\sigma}\sqrt{\frac{3}{T}}\right).$$

Further, we can find the capital process

$$X_t = \widetilde{E}\,[C_T^A|\Im_t^B] = \widetilde{E}[C_T^A] + \int_0^t \gamma_s\, d\widetilde{B}_s;$$

the second component β_t of hedging strategy π:

$$\beta_t = X_t - \gamma_t S_t$$

and the price C of this option:

$$C = \tilde{E}[C_T^A] = (1-K)\left[1-\Phi\left(\frac{K-1}{\sigma}\sqrt{\frac{3}{T}}\right)\right] + \sigma\sqrt{\frac{T}{3}}\varphi\left(\frac{K-1}{\sigma}\sqrt{\frac{3}{T}}\right).$$

Black-Scholes Model

Consider now the Black-Scholes market model with risk-free asset price evolution described by

$$dM_t = rM_t\,dt,\ M_0 = 1.$$

where $r \geq 0$ is an interest rate and for risky asset price evolution we have

$$dS_t = \mu S_t\,dt + \sigma S_t\,dB_t,\ S_0 = 1,$$

where $\mu \in R$ is appreciation rate, and $\sigma > 0$ is volatility coefficient.

Unlike the Bachelier model, it is not possible to determine the distribution of pay-off function of Asian option here and it is impossible precisely solve the problem of pricing.

Geman and Yor 1996 established that the Asian option can be valued by the Laplace transform method. They showed that the price $V(t)$ is given as

$$V(t) = \frac{4S_t}{\sigma^2 T} exp\{-r(T-t)\}C^{(v)}(h,q),$$

where r is the constant interest rate,

$$V = \frac{2r}{\sigma^2} - 1; h = \frac{\sigma^2}{4}(T-t); q = \frac{\sigma^2}{4S_t}\left\{KT - \int_0^t S_u du\right\},$$

and the Laplace transform of $C^{(v)}(h, q)$ with respect to h is given as

$$\int_0^\infty e^{-\lambda h}C^{(v)}(h,q)dh = \frac{\int_0^{1/2q} e^{-x}x^{\frac{\mu-v}{2}-2}(1-2qx)^{\frac{\mu+v}{2}+1}dx}{\lambda(\lambda-2-2v)\Gamma(\frac{\mu-v}{2}-2)}.$$

Applying the MR-inversion method, we approximate $V(t)$ by

$$V_{\{\alpha,b\}}(t) = \frac{4S_t}{\sigma^2 T}e^{-r(T-t)}\frac{[\alpha b^{-h}]\ln(b)\Gamma(\alpha+2)}{\alpha\Gamma([\alpha b^{-h}]+1)} \times$$

$$\times \sum_{m=0}^{\alpha-[\alpha b^{-h}]} \frac{(-1)^m \mu_{m+[\alpha b^{-h}]}(C^{(v)}(\cdot,q))}{m!(\alpha-[\alpha b^{-h}]-m)!}$$

with values of LHS of previous equation evaluated at $h = \dfrac{\sigma^2}{4}(T-t)$.

5. Appendix

Constructive integral representation

Here we will consider one class of Brownian functionals, and for them an integrand will be found for the Clarke stochastic integral representation in an explicit form. More precisely, we will study path-dependent Brownian functionals of the following type

$$G(n, K) = [(\int_0^T B_s\, ds)^n - K]^+,$$

which in a particular case will be the pay-off function of an Asian option in the Bachelier model. Indeed, we have

$$[A(T) - K]^+ = \left[\frac{1}{T}\int_0^T B_s ds - K\right]^+ = \frac{1}{T}\left[\int_0^T B_s ds - KT\right]^+ = \frac{1}{T}G(1, KT).$$

Remark. In addition, note that if a functional F has a Clarke representation with an integrand φ, then its linear transformation $aF + b$ also has a Clarke representation with an integrand $a\varphi$.

Let us introduce the following notation:

$$erf(x) = \frac{2}{\sqrt{\pi}}\int_0^x e^{-t^2} dt = \frac{2}{\sqrt{\pi}}\sum_{r=0}^{\infty}\frac{(-1)^r x^{1+2r}}{(1+2r)r!};$$

$$\mu = \sqrt{T^3/3}, \sigma = \sqrt{(T-t)^3/3};$$

$$\alpha(2i-1) = 0,\ \alpha(2i) = 1;$$

$$\beta(2i-1, x) = 0, \beta(2i, x) = (2i-1)!!\sqrt{\frac{\pi}{2}}\left[1 - erf\left(\frac{x}{\sqrt{2}}\right)\right];$$

$$\gamma(i, x) = e^{-x^2/2}\cdot\sum_{r=1}^{[i/2]-\alpha(i)+1}\frac{(i-1)!!}{(i-2r+1)!!}x^{i-(2r-1+\alpha(i))},$$

where $[i/2]$ denotes the integer part of $i/2$.

Lemma 1. For any natural number $n \geq 1$ and a real number y the following relation holds:

$$\int_y^{+\infty} x^{2n-1}e^{-x^2/2}dx = e^{-y^2/2}\sum_{r=1}^{n}\frac{(2n-2)!!}{(2n-2r)!!}y^{2n-2r}. \tag{31}$$

Proof. For $n = 1$, due to the standard integration technique, we have

$$\int_y^{+\infty} xe^{-x^2/2}dx = -\int_y^{+\infty} d(e^{-x^2/2}) = e^{-y^2/2}.$$

Suppose now that (31) is true for n, and prove its validity for $n+1$. Using the partial integration formula, we easily obtain that

$$\int_y^{+\infty} x^{2n-1}e^{-x^2/2}dx = -\int_y^{+\infty} x^{2n}d(e^{-x^2/2}) = -x^{2n}e^{-x^2/2}\Big|_y^{+\infty} +$$

$$+2n\int_y^{+\infty} x^{2n-1}e^{-x^2/2}dx = y^{2n}e^{-y^2/2} + e^{-y^2/2}\sum_{r=1}^{n}\frac{2n(2n-2)!!}{(2n-2r)!!}y^{2n-2r} =$$

$$= e^{-y^2/2}\sum_{r=1}^{n+1}\frac{(2(n+1)-2)!!}{(2(n+1)-2r)!!}y^{2(n+1)-2r}.$$

Hence, the method of mathematical induction completes the proof of the lemma.

Analogously one can verify the validity of the following result.

Lemma 2. For any natural number $n \geq 1$ and a real number y the following relation holds:

$$\int_y^{+\infty} x^{2n}e^{-x^2/2}dx = (2n-1)!!\sqrt{\frac{\pi}{2}}\left[1-erf\left(\frac{y}{\sqrt{2}}\right)\right]+$$
$$+ye^{-y^2/2}\sum_{r=1}^{n}\frac{(2n-1)!!}{(2n-2r+1)!!}y^{2n-2r}. \tag{32}$$

Lemma 3. Combining relations (31) and (32) from the previous lemmas, we conclude that

$$\int_y^{+\infty} x^n e^{-x^2/2}dx = \beta(n,y) + y^{\alpha(n)}e^{-y^2/2}\sum_{r=1}^{[n/2]-\alpha(n)+1}\frac{(n-1)!!}{(n-2r+1)!!}y^{n-(2r-1+\alpha(n))}. \tag{33}$$

Theorem 3. For any odd natural number, the functional $G := G(n, K)$ admits the following stochastic integral representation

$$G = EG + \frac{n}{\sqrt{2\pi}}\sum_{i=0}^{n-1}3^{-i/2}C_{n-1}^{i}\int_0^T (T-t)^{1+3i/2}\delta(i,t)dB_t,$$

where

$$EG = \mu^n\frac{1}{\sqrt{2\pi}}\left\{(n-1)!!\sqrt{\frac{\pi}{2}}\left[1-erf\left(\frac{x}{\sqrt{2}}\right)\right]+\frac{K^{1/n}}{\mu}\gamma(n,\frac{K^{1/n}}{\mu})\right\}-K\left[1-\Phi\left(\frac{K^{1/n}}{\mu}\right)\right]$$

and

$$\delta(i,t) = \left\{x^{n-1-i}\left[\beta\left(i,\frac{K^{1/n}-x}{\sigma}\right)+\left(\frac{K^{1/n}-x}{\sigma}\right)^{\alpha(i)}\gamma\left(i,\frac{K^{1/n}-x}{\sigma}\right)\right]\right\}\Bigg|_{x=\int_0^t (T-S)dB_s}.$$

Proof. It is not difficult to see that the random variable $\int_0^T B_s\,ds$ has a normal distribution with parameters zero and $T^3/3$. Indeed, due to the stochastic version of integration by parts, we have

$$\int_0^T B_s\,ds = sB_s|_0^T - \int_0^T s dB_s = \int_0^T (T-s)dB_s. \tag{34}$$

Hence, we easily ascertain that

$$\int_0^T B_s\,ds \cong N(0,E\int_0^T (B_s)^2\,ds) = N(0,\ T^3/3):= N(0,\mu^2) \tag{35}$$

and

$$\int_t^T (T-s)dB_s \cong N(0,\ (T-t)^3/3):= N(0,\sigma^2). \tag{36}$$

Using the relation (35) and Lemma 1, we can write

$$EG = \frac{1}{\sqrt{2\pi}\mu}\int_{-\infty}^{+\infty}(x^n - K)^+ e^{-x^2/(2\mu^2)}dx = \frac{1}{\sqrt{2\pi}\mu}\int_{-\infty}^{+\infty}(x^n-K)I_{\{x>K^{1/n}\}}e^{-x^2/(2\mu^2)}dx =$$

$$= \frac{1}{\sqrt{2\pi}}\int_{-\infty}^{+\infty}(\mu^n x^n - K)I_{\{x>\mu^{-1}K^{1/n}\}}e^{-x^2/2}dx = \frac{\mu^n}{\sqrt{2\pi}}\int_{\mu^{-1}K^{1/n}}^{+\infty}x^n e^{-x^2/2}dx -$$

$$-\frac{\mu^n K}{\sqrt{2\pi}}\int_{\mu^{-1}K^{1/n}}^{+\infty}e^{-x^2/2}dx = \mu^n\frac{1}{\sqrt{2\pi}}\left\{(n-1)!!\sqrt{\frac{\pi}{2}}\left[1-erf\left(\frac{x}{\sqrt{2}}\right)\right]+\frac{K^{1/n}}{\mu}\gamma(n,\frac{K^{1/n}}{\mu})\right\}-$$

$$-K\left[1-\Phi\left(\frac{K^{1/n}}{\mu}\right)\right].$$

According to the rule of stochastic differentiation of a composite function (see, Proposition 1.2.4 in Nualart 2006), using the well-known properties of the Malliavin derivatives and the relation (34), we have

$$D_t^B I_{\left\{\left(\int_0^T B_s ds\right)^n - K>0\right\}} n\left(\int_0^T B_s ds\right)^{n-1}\int_0^T I_{[0,s]}(t)ds = n(T-t)\left(\int_0^T B_s ds\right)^{n-1} I_{\left\{\int_0^T (T-s)dB_s > K^{1/n}\right\}}.$$

Further, due to the well-known properties of conditional mathematical expectation and Brownian motion, based on the relation (36), we can write that

$$E[D_t^B G\,|\,\Im_t^B] = n(T-t)\times$$

$$\times E\left[\left(x+\int_t^T(T-s)dB_s\right)^{n-1} I_{\left\{x+\int_t^T (T-s)dB_s>K^{1/n}\right\}}\right]\Bigg|_{x=\int_0^t (T-s)dB_s} =$$

$$= n(T-t)\frac{1}{\sqrt{2\pi}\sigma}\left[\int_{-\infty}^{+\infty}(x+y)^{n-1}I_{\{x+y>K^{1/n}\}}e^{-y^2/(2\sigma^2)}dy\right]\Bigg|_{x=\int_0^t (T-s)dB_s} =$$

$$= n(T-t)\sum_{i=0}^{n-1} C_{n-1}^{i} x^{n-1-i} \frac{1}{\sqrt{2\pi}\sigma}\left[\int_{K^{1/n}-x}^{+\infty} y^{i} e^{-y^{2}/(2\sigma^{2})} dy\right]\Bigg|_{x=\int_0^t (T-s)dB_s} =$$

$$= n(T-t)\sum_{i=0}^{n-1} C_{n-1}^{i} x^{n-1-i} \frac{\sigma^{i}}{\sqrt{2\pi}}\left[\int_{(K^{1/n}-x)/\sigma}^{+\infty} y^{i} e^{-y^{2}/2} dy\right]\Bigg|_{x=\int_0^t (T-s)dB_s}.$$

Based on the relations obtained above by the Clarke-Ocone formula (see, Theorem 1) and using relation (33), the proof of the theorem is easily completed.

Acknowledgement

This work is supported by the project CPEA-LT-2016/10003 "Advanced Collaborative Program for Research Based Education on Risk Management in Industry and Services under Global Economic, Technological and Environmental Changes: Enhanced Edition".

References

Black, F. and M. Scholes. 1973. The pricing of options and corporate liabilities. The Journal of Political Economy, 81(3): 637–654.

Clark, M.C. 1970. The representation of functionals of Brownian motion by stochastic integrals. J. Ann. Math. Stat., 41: 1282–1295.

Cohen, A.M. 2007. Numerical Methods for Laplace Transform Inversion. Springer, New York.

Cruz-Bàez, D.I. and J.M. González-Rodriguez. 2005. A different approach for pricing European options. pp. 373–378. *In*: Isabel, M. and G. Planas (eds.). The Proceedings of the 8th WSEAS International Conference on Applied Mathematics (MATH`05), Tenerife, Canary Islands, Spain.

Fadahunsi, A.I. and R.M. Mnatsakanov. 2018. Recovery of ruin probability and value at risk from the scaled Laplace transform inversion. Journal of Computational and Applied Mathematics, 342: 249–262.

Fu, M.C., D.B. Madan and T. Wang. 1998. Pricing of continuous Asian options: a comparison of Monte Carlo and Laplace transform inversion methods. Journal of Computational Finance, 2: 49–74.

Geman, H. and M. Yor. 1996. Pricing and hedging double barrier options: a probabilistic approach. Mathematical Finance, 6(4): 365–378.

Glasserman, P. 2003. Monte Carlo Methods in Financial Engineering. Springer, New York.

Glonti, O. and O. Purtukhia. 2014a. Clark's representation of Wiener functionals and hedging of the Barrier Option. Bulletin of the Georgian National Academy of Sciences, 8(1): 32–39.

Glonti, O. and O. Purtukhia. 2014b. Hedging of European option of integral type. Bulletin of the Georgian National Academy of Sciences, 8(3): 4–13.

Glonti, O. and O. Purtukhia. 2017. On one integral representation of functionals of Brownian motion. Theory of Probability & Its Applications, 61(1): 133–139.

Goffard, P.O., S. Loisel and D. Pommeret. 2017. Polynomial approximations for bivariate aggregate claims amount probability distributions. Methodology and Computing in Applied Probability, 19: 151–174.

Harrison, J.M. and S.R. Pliska. 1981. Martingales and stochastic integrals in the theory of continuous trading. Stochastic Processes and Applications, 11: 215–260.

Hull, J.C. 2006. Options, Futures and Other Derivatives. Prentice Hall, New Jersey.

James, P. 2003. Option Theory. John Wiley & Sons, Ltd., West Sussex.

Jaoshvili, V. and O. Purtukhia. 2005. Stochastic integral representation of functionals of wiener processes. Bulletin of the Georgian National Academy of Sciences, 171(1): 17–20.

Jódar, L., P. Sevilla-Peris, R. Sala and J.C. Cortés. 2005. A new direct method for solving the Black-Scholes equation. Applied Mathematics Letters, 18: 29–32.

Kumar, S.S.S. 2008. Financial Derivatives. Phi Learning Private Limited, New Delhi.

Lee, H. and D. Sheen. 2009. Laplace transformation method for the Black-Scholes equation. International Journal of Numerical Analysis and Modeling, 6(4): 642–658.

Mallier, R. and G. Alobaidi. 2000. Laplace transforms and American options. Applied Mathematical Finance, 7(4): 241–256.

Merton, R.C. 1973. Theory of rational option pricing. The Bell Journal of Economics and Management Science, 4(3): 141–183.

Mnatsakanov, R.M. and K. Sarkisian. 2013. A note on recovering the distributions from exponential moments. Applied Mathematics and Computation, 219: 8730–8737.

Musela, M. and M. Rutkowski. 1997. Martingale Method in Financial Modeling. Springer-Verlag, Berlin.

Nualart, D. 2006. The Malliavin Calculus and Related Topics. Springer-Verlag, Berlin.

Ocone, D. 1984. Malliavin's calculus and stochastic integral representations of functionals of diffusion processes. Stochastics, 12(3-4): 161–185.

Panini, R. and R.P. Srivastav. 2005. Pricing perpetual options using Mellin transforms. Applied Mathematics Letters, 18: 471–474.

Pelsser, A. 2000. Pricing Double Barrier options using Laplace transforms. Finance and Stochastics, 4(1): 95–104.

Shyriaev, A.N. 1999. Essentials of Stochastic Financial Mathematics. Facts, Models, Theory, World Scientific, Singapore.

Seydel, R.U. 2003. Tools for Computational Finance (second edition). Springer-Verlag, London.

Tagliani, A. and M. Milev. 2013. Laplace transform and finite difference methods for the Black-Scholes equation. Applied Mathematics and Computation, 200: 649–658.

Tavella, D. and C. Randall. 2000. Pricing Financial Instruments: The Finite Difference Method. John Wiley & Sons, Inc., New York.

Wilmott, P., S. Howinson and J. Dewynne. 1995. The Mathematics of Financial Derivatives. Cambridge University Press, Cambridge.

CHAPTER 8

A Nash Equilibrium based Model of Agents Coordination through Revenue Sharing and Applications to Telecommunications

*Paolo Pisciella** and *Alexei A Gaivoronski*

1. Introduction

Digitalization is reshaping the way economic sectors conduct business. Moreover, digitalization is the main enabler of the platform economy. Platforms create new market opportunities by facilitating the interaction between different user groups. Platforms offer a wide array of services, such as online marketplaces, social networks, collaborative economy platforms and online gaming. Platform businesses may be defined as businesses creating significant value through the acquisition and/or matching, interaction and connection of two or more customer groups to enable them to transact (Murati (2021), Reillier and Reillier (2017)). In this respect, the platform works as a broker between supply and demand, where the supply side might consist of a single as well as of a group of actors, capturing a part of the value exchanged between these parties. In simple cases the supply side for digital services consists of only two agents: the developer and the platforms themselves. In more complex cases there might be a group of different developers, each contributing with one component to

Norwegian University of Science and Technology, Trondheim, Norway.
Email: Alexei.Gaivoronski@ntnu.no
* Corresponding author: Paolo.Pisciella@ntnu.no

the overall service provision. A platform can therefore be considered an entity coordinating interactions between groups of stakeholders. Such organizations are normally distinct business actors, each interested in the maximization of their own profits. In order to align the efforts of each participant the platform can employ different mechanisms. One of the most used is revenue sharing, widely used in mobile apps development. Nowadays the 30/70 splitting ratio introduced by Apple has *de facto* become a standard in the mobile industry. Nevertheless, a steadily growing share of developers complains about the rigidity of the application of such a business model (Hillegas 2012), with videogames platforms such as Steam or Microsoft trying to define different sharing schemes depending on the importance of the developer.

In this chapter we tackle the problem of aligning incentives between a group of agents engaging in the collaborative provision of service around a platform. We consider the presence of an aggregator and multiple followers, coordinated by the aggregator through a revenue-sharing mechanism. Followers are assumed to consider the choices of their peers when defining their contribution to the platform service. The aggregator uses revenue sharing to incentivize the followers to choose an output level which delivers the highest profit to herself. This will result in providers reaching a Nash Equilibrium (NE) on the level of involvement in the collaborative provision of a given service. Such an equilibrium solution is in line with current industry trends since the delivery of advanced data services is expected to increasingly replace that of basic data services and they require the involvement of multiple contributors. In this respect, our contribution can be placed in the analysis of supply chain coordination.

The novel contribution of this paper consists in the following.

1. We develop a set of bilevel stochastic optimization models for the optimization of aggregator-led platform provision of a bundle of modular services that share some of the same components in different proportions. The problem is viewed as a stochastic Stackelberg game with one leader and multiple followers. The leader (aggregator) decides the revenue distribution, while the followers (service providers) select the level of involvement in the service definition by allocating their provision capacities. This portfolio selection is defined through an appropriate Nash equilibrium. We pursue two complementary approaches to develop solvable optimization models within this framework. The first one extends to the bilevel case of the linear stochastic programming models with a finite number of scenarios of uncertain demand (Sections 3, 4). The second approach considers continuous demand distributions and

involves the analytical study of the set of possible Nash equilibria on the lower level (Sections 5, 6). We obtain an explicit characterization of the set of Nash equilibria that considerably simplifies the resulting integrated bilevel model and facilitates the application of nonlinear programming techniques.

2. We employ a linearization technique for a special class of quadratic complementarity problems in order to obtain the MILP model for the case of demand described by a finite number of scenarios (Section 4 and Appendix).
3. Portfolio theory of investment science (Markowitz (1991)) is extended from the classical single-actor case to the case of multiple competing actors in the single leader/multiple followers setting of our problem. We provide portfolio selection models for risk-averse actors in Section 6 that extend the notion of efficient frontier familiar from investment science. Examples are provided that highlight the new features of multi-actor efficient frontiers not found in the classical case.

Besides, we provide in Section 2 a brief overview of the existing literature on coordination models for supply chains and the related applications in telecommunications, placing this paper in the general research context. Section 5 contains a case study and related numerical results of the application of the models from Sections 3, and 4 to a problem of the design of mobile platform services.

2. Literature review

Literature in incentive definition for supply chain coordination is less extensive than its counterpart on centralized supply chains, where one entity aims to optimize the entire system performance. This is mainly due to the intrinsic difficulties that this approach entails. The policymaker faces the same challenges to optimize system performance but, on top of that, a revenue reallocation scheme is required to maintain the interest and participation of all the agents. Many incentive schemes have been studied assuming deterministic demand. Some research was focused on collaboration patterns stemming from the use of quantity discounts by Crowther (1964), Dolan (1987), Lal and Staelin (1984) and in particular Parlar and Wang (1994) who studied the coordination of production and ordering decisions as a Stackelberg game between a supplier and a retailer considering a quantity discount pricing. However, quantity discounts are usually not able to optimize system performance when there are heterogeneous buyers and/or multiple products.

In addition to quantity discount policies, an alternative approach based on revenue-sharing mechanisms has been proposed. Under this viewpoint, the system is first jointly optimized and after this the revenues are shared between the supplier and the retailers. The effective implementation of such a policy strictly depends on the availability of a revenue-sharing scheme that is acceptable to all parties. Goyal (1976) proposes to distribute the revenues to supply chain members proportionally according to their costs. Jogeklar and Tharthare (1990) allocated the profit by requiring the customers to pay for the processing cost they impose on the producer, while the producer lowers the unit price in return.

The paper closest to ours is that of Gerchak and Wang (2004). They obtain coordination by considering a Stackelberg game between the retailer and a group of suppliers which are supposed to interact among themselves to reach a Nash Equilibrium on quantities delivered to the retailer. In turn, the retailer picks an appropriate vector of revenue shares in order to induce the Nash Equilibrium of the followers which is most profitable to her. The authors find out that the final quantity delivered and assembled, which is determined by the revenue shares set by the assembler to maximize her own profit, is in general not equal to the quantity which would optimize the entire system. The paper of Cachon and Lariviere (2005) presents the opposite viewpoint, letting the suppliers choose the revenue sharing vector, while the retailer selects the quantities to order.

The literature classifies two kinds of business models for the collaborative provision of services around platforms (Ballon and Heesvelde (2011)): a *brokered model* where the provision is hierarchically coordinated by one of the agents who aims at extracting the largest possible share of value out of the provision, and a *distributed model*, where the distribution of bargaining power is balanced among the agents involved. The brokered model can be mathematically described by utilising the concept of Stackelberg Game, as done in Gerchak and Wang (2004), Chernonog et al. (2015), Avinadav and Perlman (2015a,b), Yang and Wang (2017), Govindan and Malomfalean (2019) and solved by means of bilevel coordination problems as in Gaivoronski and Zoric (2008), Pisciella and Zoric (2009), Pisciella (2012), Pisciella and Gaivoronski (2017). Recent research has introduced risk management in the coordination of a single seller and single purchaser (Fan and Shou (2020)).

This paper distinguishes itself from the previous literature by considering the provision of several services that share some of the same components in different proportions in a stochastic setting with both discrete and continuous probability distributions. Besides, we consider more realistic models of service providers who possess different provision capacities and different risk attitudes.

This allows us to develop and solve novel bilevel stochastic optimization models that show how the actors can select the portfolios of services to participate in, something that was not possible with the previous approaches. In addition, due to the complexity of the resulting models, we place the emphasis on the combination of analytical study and numerical solution.

3. Modelling the collaborative service provision

We model the design of a service platform that provides a portfolio of complex modular data services to a population of users. Each service j is defined by a set of modules $i = 1,...,N_j$. For the sake of uniformity we call here "modules" a large set of entities from which a service is actually composed and/or that are necessary for the functioning of the service. For example, it can be a software component or application that makes part of a service. These modules are measured with appropriate module-specific units like man/hours, bandwidth, monetary equivalent, etc. A service j is described by the amount of λ_{ij} of each module i necessary for provision of a service unit

$$\lambda_j = \left(\lambda_{1j},...,\lambda_{ij},...,\lambda_{Nj}\right). \tag{1}$$

The management of the service platform is the objective of an actor described as a Service Aggregator (SA). He/she has to decide which services to include in the platform service portfolio. This actor provides the aggregation module and possibly other modules to the platform services. The most important aspect of his/her role is to offer incentives that will make other actors eager to participate. We shall refer to these other actors uniformly as Service Providers (SP), while their established industrial designation can be various (service provider, developer etc.).

We consider a group I of such service providers. They may decide to contribute their modules to a set of composite platform services aggregated by Service Aggregator, or, alternatively, to engage in service provision outside this service platform. They make their decision by selecting the optimal allocation of their provision capacity among the platform services and off-platform service provision activities. For the sake of clarity, we shall assume here that each service provider supplies only one module i that is different from other modules. Thus, we can consider the provider $i \in I$ as the agent-supplying module i. We denote by $\mathcal{P}_i$ the set of platform services to which the service provider i can contribute, and by ε_i the set of off-platform services to which he/she can supply his/her module. The set of actors who supply modules necessary for the provision of service j is denoted by $\mathcal{I}_j$. We consider the detailed service composition as in (1) only for the platform services.

Service provider i has provision capacity W_i. Denoting by q_{ij} the amount of provision capacity that provider i allocates to service j we obtain the following capacity constraint of the i-th service provider:

$$\sum_{j \in \mathcal{P}_i \cup \mathcal{E}_i} q_{ij} \leq W \tag{2}$$

The Service Aggregator decides the price p_j for a unit of service j offered to the end users. In this paper we take this price as fixed and do not consider the problem of its selection. Since we are focusing on the problem of design of the service platform, it is natural to consider the future demand for different services in the platform service portfolio as uncertain. More specifically, it is assumed that the demand for service $j \in \mathcal{P}_i$ is random and takes the finite number of values d_{js} with probabilities π_s, where s is used to index demand scenarios, $s \in \Omega$. The case when demand is distributed according to a more general and possibly continuous distribution is considered in Section 5.

Taking into account the service composition (1), the demand for module i resulting from demand d_{js} for service j under scenario $s \in \Omega$ will be $\lambda_{ij} d_{js}$. While modeling the design stage, we can assume that the service provider i allocates her capacity q_{ij} to service j before demand is known. Therefore, part of this capacity y_{ijs} may remain spare when the demand scenario s occurs. Thus, the actual volume of the module delivery by provider i to service j under scenario s will be $q_{ij} - y_{ijs}$. It can not exceed the required amount $\lambda_{ij} d_{js}$:

$$q_{ij} - y_{ijs} \leq \lambda_{ij} d_{js}, \qquad j \in \mathcal{P}_i, \quad s \in \Omega. \tag{3}$$

In the case of external opportunities $j \in \mathcal{E}_i$, we denote by d_{js} the direct demand for module i under scenario s. Therefore the demand constraint similar to (3) in the case $j \in \mathcal{E}_i$ takes the form:

$$q_{ij} - y_{ijs} \leq d_{js}, \qquad j \in \mathcal{E}_i, \quad s \in \Omega. \tag{4}$$

If the provider i allocates capacity q_{ij} for supplying component i to service j and amount y_{ijs} goes unused under scenario s then according to (1) the maximal amount of demand for service j that will be served under scenario s is

$$\frac{q_{ij} - y_{ijs}}{\lambda_{ij}},$$

which is true for any $i \in \mathcal{I}_j$. This yields an additional set of constraints on the decisions of the provider i:

$$\frac{q_{ij} - y_{ijs}}{\lambda_{ij}} \leq \frac{q_{kj} - y_{kjs}}{\lambda_{kj}}, k \in \mathcal{I}_j \setminus \{i\} \tag{5}$$

The revenue from the sale of the bundled service is collected by Service Aggregator. In order to incentivize other providers to participate in service provision, he/she should share a part of this revenue with every required service provider. Let us denote by γ_{ij} the share of the revenue from the provision of service j allocated to provider i by the Service Aggregator,

$$\gamma_{ij} \geq 0, \sum_{i \in \mathcal{I}_j} \gamma_{ij} = 1.$$

We have assumed here that Service Aggregator provides the aggregation module to every service j. Then the revenue of provider i from delivering his/her component to service j under scenario s will be

$$\frac{\gamma_{ij}}{\lambda_{ij}} p_j (q_{ij} - y_{ijs})$$

In addition, provider i will get revenue for delivering her module to external opportunities $j \in \mathcal{E}_i$ at price p_{ij} that under scenario s will be

$$p_{ij}(q_{ij} - y_{ijs}).$$

Besides, provider i will bear cost c_i per unit of allocated capacity. We assume that this cost exists irrespective of the volume of actual delivery. Then the expected profit of service provider i is as follows:

$$\sum_{s \in \Omega} \pi_s \sum_{j \in \mathcal{P}_i} \frac{\gamma_{ij}}{\lambda_{ij}} p_j (q_{ij} - y_{ijs}) + \sum_{s \in \Omega} \pi_s \sum_{j \in \mathcal{E}_i} p_{ij}(q_{ij} - y_{ijs}) - c_i \sum_{j \in \mathcal{P}_i \cup \mathcal{E}_i} q_{ij} \tag{6}$$

We shall assume in this section that the service providers are risk neutral and maximize their expected profits as in (6), the case of risk-averse service providers will be considered in Section 6. Therefore, component provider i has the problem of maximizing function (6) subject to constraints (2)–(5). We shall refer to this problem as the *CP* problem.

If not for constraint (5), for fixed γ_{ij} this problem can be recognized as a linear stochastic programming problem with recourse with the first stage variables q_{ij} and recourse variables y_{ijs}, written in the form of deterministic equivalent (see Birge and Louveaux (1997), Kall and Wallace (1994) for further details on such problems). However, constraint (5) makes its solution depend on the decision variables of all providers $i \in I$. Therefore in order to define an appropriate solution for problem *CP* we have to invoke the equilibrium concepts of the game theory, in this case, the notion of Nash equilibrium. Thus, the *CP* problem constitutes a stochastic equilibrium problem.

Before we proceed with the analysis of problem *CP*, let us simplify it dispensing with the secondary features and concentrating on the main issues. First of all, let us assume that there is only one off-platform opportunity for each provider with infinite demand and price $p_i^0 > c_i$. Then from (6) follows that all the capacity that is left after the allocation to platform services will be allocated to this external opportunity. Denoting this capacity by q_i^0 we can reformulate constraint (2) as follows:

$$q_i^0 = W_i - \sum_{j \in \mathcal{P}_i} q_{ij}$$

Substituting this in (6) and remembering that ε_i is a singleton with corresponding q_{ij} and p_{ij} denoted by q_i^0 and p_i^0 we obtain the following expression for profit:

$$\sum_{j \in P_i} \left(\frac{\gamma_{ij} p_j}{\lambda_{ij}} - p_i^0 \right) q_{ij} - \sum_{j \in P_i} \sum_{s \in \Omega} \frac{\gamma_{ij} p_j}{\lambda_{ij}} \pi_s y_{ijs} + W_i (p_i^0 - c_i).$$

Observing that the term $W_i(p_i^0 - c_i)$ does not depend on decision variables, we can reformulate the problem (2)–(6) as

$$SP(\gamma_{ij}) = \max_{q_{ij}, y_{ijs}} \sum_{j \in P_i} \left(\frac{\gamma_{ij} p_j}{\lambda_{ij}} - p_i^0 \right) q_{ij} - \sum_{j \in P_i} \sum_{s \in \Omega} \frac{\gamma_{ij} p_j}{\lambda_{ij}} \pi_s y_{ijs} \tag{7}$$

subject to

$$\sum_{j \in \mathcal{P}_i} q_{ij} \le W_i \tag{8}$$

$$q_{ij} - y_{ijs} \le \lambda_{ij} d_{js}, j \in P_i, s \in \Omega \tag{9}$$

$$\frac{q_{ij} - y_{ijs}}{\lambda_{ij}} \le \frac{q_{kj} - y_{kjs}}{\lambda_{kj}}, \qquad k \in \mathcal{I}_j \setminus \{i\}, j \in \mathcal{P}_i, s \in \Omega \tag{10}$$

$$q_{ij} \ge 0, y_{ijs} \ge 0$$

where we have stressed the dependence of its solution on the revenue shares γ_{ij}. Observe that the objective function in (7) can be equivalently expressed as

$$\sum_{j \in P_i} \sum_{s \in \Omega} \frac{\gamma_{ij} p_j}{\lambda_{ij}} \pi_s \left(q_{ij} - y_{ijs} \right) - p_i^0 \sum_{j \in P_i} q_{ij} \tag{11}$$

Let us consider in more detail the set of constraints (10). Suppose that j is fixed. Since for any $i, k \in \mathcal{I}_j$ the constraint

$$\frac{q_{ij} - y_{ijs}}{\lambda_{ij}} \le \frac{q_{kj} - y_{kjs}}{\lambda_{kj}}$$

belongs to the set of constraints of the problem (7)–(10) of actor i and constraint

$$\frac{q_{kj} - y_{kjs}}{\lambda_{kj}} \leq \frac{q_{ij} - y_{ijs}}{\lambda_{ij}}$$

belongs to the set of constraints of the problem (7)–(10) of actor k then the feasible set of the strategies of both actors will be the same if the inequality signs in these constraints will be changed to equality:

$$\frac{q_{kj} - y_{kjs}}{\lambda_{kj}} = \frac{q_{ij} - y_{ijs}}{\lambda_{ij}}$$

which yields

$$\frac{y_{ijs}}{\lambda_{ij}} = \frac{q_{ij}}{\lambda_{ij}} - \frac{q_{kj}}{\lambda_{kj}} + \frac{y_{kjs}}{\lambda_{kj}}, s \in \Omega.$$

Suppose now that i has the largest ratio q_{ij}/λ_{ij} among all $l \in \mathcal{I}_j$, k has the next largest and

$$\frac{q_{ij}}{\lambda_{ij}} > \frac{q_{kj}}{\lambda_{kj}}.$$

Taking

$$q_{ij}' = \frac{\lambda_{ij}}{\lambda_{kj}} q_{kj}, y_{ijs}' = \frac{\lambda_{ij}}{\lambda_{kj}} y_{kjs}$$

we get

$$\frac{q_{ij}' - y_{ijs}'}{\lambda_{ij}} = \frac{q_{kj} - y_{kjs}}{\lambda_{kj}} = \frac{q_{ij} - y_{ijs}}{\lambda_{ij}}$$

but $q_{ij}' < q_{ij}$. Therefore $(q_{ij}', y_{ijs}', s \in \Omega)$ belong to the feasible set of actor i. Besides, if we substitute decisions $(q_{ij}, y_{ijs}, s \in \Omega)$ of actor i with these decisions then the feasible sets and objective functions of all other actors will remain unchanged and the value of the objective of actor i will increase due to (11). Thus, the new set of strategies composed of strategy $(q_{ij}', y_{ijs}', s \in \Omega)$ of actor i and original strategies of all other actors will dominate the original set of strategies. Therefore, the equilibrium solution of the problems (7)–(10) will remain unchanged if constraint

$$\frac{q_{ij}}{\lambda_{ij}} = \frac{q_{kj}}{\lambda_{kj}}$$

is added to the feasible set of actors i, k. Similarly we get to the same conclusion if several actors have the largest ratio q_{ij}/λ_{ij}. Repeating this argument recursively we obtain that the set of constraints

$$\frac{q_{ij}}{\lambda_{ij}} = \frac{q_{kj}}{\lambda_{kj}}, k \in I_j, j \in \mathcal{P}_i \tag{12}$$

can be added to the problem (7)–(10) of every actor $i \in I$ without changing the equilibrium solution. Constraints (10) in this case can be written equivalently as

$$\frac{y_{ijs}}{\lambda_{ij}} = \frac{y_{kjs}}{\lambda_{kj}}, k \in \mathcal{I}_j \setminus \{i\}, j \in \mathcal{P}_i, s \in \Omega \tag{13}$$

Let us dispense with constraints (13) altogether, considering problems (7)–(9), (12). This will increase the feasible set of strategies for each actor. However, the equilibrium solution will remain unchanged again. To see this, let us consider a fixed j, actors i, $k \in \mathcal{I}_j$ and scenario $s \in \Omega$. Suppose that for some decisions q_{ij}, q_{kj}, y_{ijs}, y_{kjs} feasible for problems (7)–(9), (12) we have

$$\frac{y_{ijs}}{\lambda_{ij}} > \frac{y_{kjs}}{\lambda_{kj}}.$$

Respective constraints (9) for these i, j, k, s take the form

$$\frac{y_{ijs}}{\lambda_{ij}} \geq \frac{q_{ij}}{\lambda_{ij}} - d_{js} \tag{14}$$

$$\frac{y_{kjs}}{\lambda_{kj}} \geq \frac{q_{kj}}{\lambda_{kj}} - d_{js} \tag{15}$$

Let us take

$$y'_{ijs} = \frac{\lambda_{ij}}{\lambda_{kj}} y_{kjs} < y_{ijs}$$

Due to (12) and (15) y'_{ijs} will be feasible for (14). At the same time since $y'_{ijs} < y_{ijs}$ the objective of actor i will increase due to (7). Therefore, the set of strategies where y_{ijs} is substituted by y'_{ijs} will dominate the original set, but for this set the respective constraint (13) is satisfied. Thus, any set of feasible strategies of the problem (7)–(9), (12) is dominated by a feasible set of strategies for which also constraint (10) is satisfied. Therefore the equilibrium solutions of the problems (7)–(9), (12) and (7)–(10), (12) coincide. With this we have proved that the set of equilibrium solutions of the problem (7)–(10) coincides with the set of equilibrium solutions of the following problem:

$$SP(\gamma_{ij}) = \max_{q_{ij}, y_{ijs}} \sum_{j \in P_i} \left(\frac{\gamma_{ij} p_j}{\lambda_{ij}} - p_i^0 \right) q_{ij} - \sum_{j \in P_i} \sum_{s \in \Omega} \frac{\gamma_{ij} p_j}{\lambda_{ij}} \pi_s y_{ijs} \tag{16}$$

subject to

$$\sum_{j \in \mathcal{P}_i} q_{ij} \leq W_i \tag{17}$$

$$q_{ij} - y_{ijs} \leq \lambda_{ij} d_{js}, j \in \mathcal{P}_i, s \in \Omega \tag{18}$$

$$\frac{q_{ij}}{\lambda_{ij}} \leq \frac{q_{kj}}{\lambda_{kj}} \qquad k \in \mathcal{I}_j \setminus \{i\}, j \in \mathcal{P}_i \tag{19}$$

$$q_{ij} \geq 0, y_{ijs} \geq 0$$

This problem has a smaller number of constraints compared to (7)–(10) and for this reason, we shall study its solution.

Let us assign index $i = 1$ to the Service Aggregator. Solving the equilibrium problem (16)–(19) for fixed shares γ_{ij} he will obtain his optimal decisions $q_{1j}(\gamma_j)$ and $y_{1js}(\gamma_j)$ for each service j as the function of shares $\gamma_j = \{\gamma_{ij}, i \in \mathcal{I}_j\}$. Then he selects the shares γ_j, maximizing his equilibrium profit:

$$\max_{\gamma_j} \sum_{j \in P_i} \left(\frac{\gamma_{1j} p_j}{\lambda_{1j}} - p_1^0 \right) q_{1j}(\gamma_j) - \sum_{j \in P_i} \sum_{s \in \Omega} \frac{\gamma_{1j} p_j}{\lambda_{1j}} \pi_s y_{1js}(\gamma_j) \tag{20}$$

subject to

$$\sum_{i \in \mathcal{I}_j} \gamma_{ij} = 1 \tag{21}$$

$$\gamma_{ij} \geq 0$$

This decision process can be described as a Stackelberg game. In this game Service Aggregator is the leader and service providers are the followers. The leader selects revenue shares and communicates them to the followers. The followers select their capacity allocations solving the equilibrium problem (16)–(19). The leader divides the revenue by selecting the shares that maximize his equilibrium profit. In the next section, this model is further reformulated and solved using approaches of bilevel programming (see Bard (1998), Dempe (2002)).

4. MILP reformulation of nonlinear Stackelberg model

In order to solve the bilevel problem with (20)–(21) as the upper level and (16)–(19) as the lower level we formulate the Karush-Kuhn-Tucker conditions for the Nash equilibrium of the lower level and add these conditions to the upper level, obtaining a single-level problem (see Bard (1998)).

The Nash Equilibrium is obtained as the joint solution of the optimality conditions of all the decision problems of all actors (Bazaraa and Shetty

(2006)). Adding these conditions for the problems (16)–(19) for all $i \in I$ to the problem (20)–(21) yields the following single level optimization problem:

$$\max_{q_{ij}, y_{ijs}, \mu_i^c, \mu_{ijs}^d, \mu_{ikj}^e, \gamma_{ij}} \sum_{j \in P_i} \left(\frac{\gamma_{1j} p_j}{\lambda_{1j}} - p_1^0 \right) q_{1j} - \sum_{j \in P_i} \sum_{s \in \Omega} \frac{\gamma_{1j} p_j}{\lambda_{1j}} \pi_s y_{1js} \tag{22}$$

$$\sum_{i \in \mathcal{I}_j} \gamma_{ij} = 1, j \in J$$

$$\mu_i^c + \sum_{s \in \Omega} \mu_{ijs}^d + \sum_{k \in \mathcal{I}_j \setminus \{i\}} \frac{\mu_{ikj}^e}{\lambda_{ij}} \geq \frac{\gamma_{ij} p_j}{\lambda_{ij}} - p_i^0, \quad i \in I, j \in \mathcal{P}_i,$$

$$\mu_{ijs}^d \leq \frac{\gamma_{ij} p_j}{\lambda_{ij}} \pi_s, \quad i \in I, j \in \mathcal{P}_i, \quad s \in \Omega,$$

$$\sum_{j \in \mathcal{P}_i} q_{ij} \leq W_i, i \in I,$$

$$q_{ij} - y_{ijs} \leq \lambda_{ij} d_{js}, \quad i \in I, j \in \mathcal{P}_i \quad s \in \Omega,$$

$$\frac{q_{ij}}{\lambda_{ij}} - \frac{q_{kj}}{\lambda_{kj}} \leq 0, \quad i \in I, j \in \mathcal{P}_i, k \in \mathcal{I}_j \setminus \{i\}, \tag{23}$$

$$\mu_i^c \left(W_i - \sum_{j \in \mathcal{P}_i} q_{ij} \right) = 0, i \in I,$$

$$\mu_{ijs}^d \left(\lambda_{ij} d_{js} - q_{ij} + y_{ijs} \right) = 0, \quad i \in I, j \in \mathcal{P}_i, \quad s \in \Omega,$$

$$\mu_{ikj}^e \left(\frac{q_{kj}}{\lambda_{kj}} - \frac{q_{ij}}{\lambda_{ij}} \right) = 0, \quad i \in I, j \in \mathcal{P}_i, k \in \mathcal{I}_j \setminus \{i\}, \tag{24}$$

$$q_{ij} \left(\mu_i^c + \sum_{s \in \Omega} \mu_{ijs}^d + \sum_{k \in \mathcal{I}_j \setminus \{i\}} \frac{\mu_{ikj}^e}{\lambda_{ij}} - \frac{\gamma_{ij} p_j}{\lambda_{ij}} + p_i^0 \right) = 0, \quad i \in I, j \in \mathcal{P}_i,$$

$$y_{ijs} \left(\frac{\gamma_{ij} p_j}{\lambda_{ij}} \pi_s - \mu_{ijs}^d \right) = 0, \quad i \in I, j \in \mathcal{P}_i, \quad s \in \Omega,$$

$$q_{ij}, y_{ijs}, \mu_i^c, \mu_{ijs}^d, \mu_{ikj}^e, \gamma_{ij} \geq 0$$

This problem can be simplified. Observe that the set of constraints (23) contains the following pairs:

$$\frac{q_{ij}}{\lambda_{ij}} - \frac{q_{kj}}{\lambda_{kj}} \le 0, \frac{q_{kj}}{\lambda_{kj}} - \frac{q_{ij}}{\lambda_{ij}} \le 0, j \in \mathcal{P}, i,k \in \mathcal{I}_j$$

Consequently,

$$\frac{q_{ij}}{\lambda_{ij}} = \frac{q_{kj}}{\lambda_{kj}}, j \in \mathcal{P}, i,k \in \mathcal{I}_j \tag{25}$$

Therefore, complementarity constraints (24) become redundant and can be dropped. Besides, many constraints in (23) become redundant and can be pruned as follows. Let us select for every $j \in \mathcal{P}$ some index $r_j \in \mathcal{I}_j$. Then (23), (25) are equivalent to the following:

$$Q_j = \frac{q_{r_j j}}{\lambda_{r_j j}} = \frac{q_{kj}}{\lambda_{kj}}, j \in \mathcal{P}, i,k \in \mathcal{I}_j \setminus \{r_j\}$$

Therefore, we can make in problem (22) the following substitution

$$q_{ij} = \lambda_{ij} Q_j, i \in I, j \in \mathcal{P}_i$$

substituting maximization with respect to q_{ij} by maximization with respect to Q_j. In addition, let us denote

$$y_{ijs} = \lambda_{ij} z_{ijs}, i \in I, j \in \mathcal{P}_i \quad s \in \Omega,$$

Making all these substitutions we obtain:

$$\max_{\gamma_{ij}, Q_j, z_{ijs}, \mu_i^c, \mu_{ijs}^d, \mu_{ikj}^e} \sum_{j \in \mathcal{P}} \left(\gamma_{1j} p_j - \lambda_{1j} p_1^0\right) Q_j - \sum_{j \in \mathcal{P}} \sum_{s \in \Omega} p_j \pi_s \gamma_{1j} z_{1js} \tag{26}$$

$$\sum_{i \in \mathcal{I}_j} \gamma_{ij} = 1, j \in \mathcal{P} \tag{27}$$

$$\gamma_{ij} p_j - \lambda_{ij} \mu_i^c - \lambda_{ij} \sum_{s \in \Omega} \mu_{ijs}^d - \sum_{k \in \mathcal{I}_j \setminus \{i\}} \mu_{ikj}^e \le \lambda_{ij} p_i^0, \qquad i \in I, j \in \mathcal{P}_i, \tag{28}$$

$$-\gamma_{ij} p_j \pi_s + \lambda_{ij} \mu_{ijs}^d \le 0, \qquad i \in I, j \in \mathcal{P}_i, \quad s \in \Omega, \tag{29}$$

$$\sum_{j \in \mathcal{P}_i} \lambda_{ij} Q_j \le W_i, i \in I, \tag{30}$$

$$Q_j - z_{ijs} \le d_{js}, \qquad i \in I, j \in \mathcal{P}_i \quad s \in \Omega, \tag{31}$$

$$\mu_i^c\left(W_i - \sum_{j\in\mathcal{P}_i}\lambda_{ij}Q_j\right) = 0, i \in I, \tag{32}$$

$$\mu_{ijs}^d\left(d_{js} - Q_j + z_{ijs}\right) = 0, \qquad i \in I, j \in \mathcal{P}_i, \quad s \in \Omega, \tag{33}$$

$$Q_j\left(\lambda_{ij}\mu_i^c + \lambda_{ij}\sum_{s\in\Omega}\mu_{ijs}^d + \sum_{k\in\mathcal{I}_j\setminus\{i\}}\mu_{ikj}^e - \gamma_{ij}p_j + \lambda_{ij}p_i^0\right) = 0, \qquad i \in I, j \in \mathcal{P}_i, \tag{34}$$

$$z_{ijs}\left(\gamma_{ij}p_j\pi_s - \lambda_{ij}\mu_{ijs}^d\right) = 0, \qquad i \in I, j \in \mathcal{P}_i, \quad s \in \Omega, \tag{35}$$

$$\gamma_{ij}, Q_j, z_{ijs}, \mu_i^c, \mu_{ijs}^d, \mu_{ikj}^e \geq 0. \tag{36}$$

This problem has quadratic terms $\gamma_{1j}z_{1js}$ and $\gamma_{1j}Q_j$ in the objective and nonlinear complementary slackness constraints (32)–(35). In order to transform it into a mixed integer linear problem we utilize the following proposition.

Proposition 1 *Consider the problem*

$$\max_{x\geq 0} x^{\mathrm{T}}Rx + c^{\mathrm{T}}x \tag{37}$$

$$A_1x \leq b_1 \tag{38}$$

$$x^{\mathrm{T}}E^{\mathrm{T}}\left(A_1x - b_1\right) = 0 \tag{39}$$

$$A_2x = b_2 \tag{40}$$

where $x \in \mathbb{R}^n$, matrices R, E, A_1, A_2 have dimensions $n \times n$, $m_1 \times n$, $m_1 \times n$, $m_2 \times n$ correspondingly and the following conditions are satisfied

1. Any element of vector Ex coincides with some element of vector x and can not attain zero simultaneously with corresponding element of vector $A_1x - b_1$ at the optimal solution of problem (37)–(40).
2. Feasible set defined by (38)–(40) is bounded.

Let $x = (\hat{x}, \bar{x})$ where $\bar{x}$ contains all elements of x for which exists identical element of Ex and $u = (\hat{u}, \bar{u})$ is a vector with binary components with $\bar{u}$ corresponding to $\bar{x}$. Then the problem (37)–(40) is equivalent to the following mixed integer linear programming (MILP) problem:

$$\max_{x,u,\sigma_1,\sigma_2,\sigma_3} \frac{1}{2}\left(c^{\mathrm{T}}x + b_1^{\mathrm{T}}\sigma_1 + b_2^{\mathrm{T}}\sigma_2\right) \tag{41}$$

$$A_1x \leq b_1 \tag{42}$$

$$A_2 x = b_2 \tag{43}$$

$$\left(R + R^{\mathrm{T}}\right)x - A_1^{\mathrm{T}}\sigma_1 - A_2^{\mathrm{T}}\sigma_2 - E^{\mathrm{T}}\sigma_3 \le -c \tag{44}$$

$$x - Mu \le 0 \tag{45}$$

$$-\left(R + R^{\mathrm{T}}\right)x + A_1^{\mathrm{T}}\sigma_1 + A_2^{\mathrm{T}}\sigma_2 + E^{\mathrm{T}}\sigma_3 + Mu \le M\mathbf{1}_n + c \tag{46}$$

$$-A_1 x + M\bar{u} \le M\mathbf{1}_{m_1} - b_1 \tag{47}$$

$$\sigma_1 - M\bar{u} \le 0 \tag{48}$$

$$-\sigma_1 - M\bar{u} \le 0 \tag{49}$$

$$\sigma_3 + M\bar{u} \le M\mathbf{1}_{m_1} \tag{50}$$

$$x, \sigma_3 \ge 0, u \in \{0,1\}^n$$

where M is a large enough positive number and $\mathbf{1}_n$, $\mathbf{1}_{m_1}$ are column vectors of ones with lengths n and m_1.

The proof of this proposition can be found in the Appendix. In the notations of this proposition (γ_{ij}, Q_j, z_{ijs}, μ_i^c, μ_{ijs}^d, μ_{ikj}^e) corresponds to x, constraints (28)–(31) to constraint (38), constraint (27) to constraint (40), complementarity constraints (32)–(35) to constraint (39). Each row of matrix E in (39) in this case corresponds to one of the variables (Q_j, z_{ijs}, μ_i^c, μ_{ijs}^d). It has one at the place that corresponds to its variable, while all other elements are zeros. Condition 1 is equivalent to the requirement that at the optimal solution of linear programming problem (16)–(19) for any fixed i primal (dual) variables can not be zeros if dual (primal) constraints are satisfied with equality.

Let us introduce binary and dual variables for constraints of problem (26)–(35) that correspond to (σ_1, σ_2, σ_3, u, v) of problem (41)–(50). The following table (Table 1) shows the correspondence between the variables of the two problems.

Let us introduce the constraints of the resulting problem by groups that correspond to (44)–(50) (constraints that correspond to (42), (43) are original constraints (27)–(31)).

1. Constraints that correspond to (44). In this group we place the corresponding variable to the left of the constraint for the purposes of clarity.

Table 1. Correspondence between variables of problem (41)–(50) and problem, resulting from (26)–(35).

Problem (41)–(50)	Problem (26)–(35)
x	$\gamma_{ij}, Q_j, z_{ijs}, \mu_i^c, \mu_{ijs}^d, \mu_{ikj}^e$
σ_1	$\theta_i^c, \theta_{ijs}^d, \theta_{ij}^q, \theta_{ijs}^z$
σ_2	θ_j^γ
σ_3	$\phi_i^c, \phi_{ijs}^d, \phi_{ij}^q, \phi_{ijs}^z$
u	$\delta_{ij}^\gamma, \delta_{ij}^q, \delta_{ijs}^z, \delta_i^c, \delta_{ijs}^d, \delta_{ikj}^e$
$\bar{u}$	$\delta_{ij}^q, \delta_{ijs}^z, \delta_i^c, \delta_{ijs}^d$

$$\gamma_{1j} : p_j Q_j - \sum_{s\in\Omega} p_j \pi_s z_{1js} + \sum_{s\in\Omega} p_j \pi_s \theta_{1js}^z - \theta_j^\gamma \le 0, j \in \mathcal{P}_1 \quad (51)$$

$$\gamma_{ij}, i \ne 1 : \sum_{s\in\Omega} p_j \pi_s \theta_{ijs}^z - \theta_j^\gamma \le 0, i \in I, i \ne 1, j \in \mathcal{P}_i$$

$$Q_j : \gamma_{1j} p_j - \sum_{i\in I_j} \lambda_{ij} \theta_i^c - \sum_{i\in I_j}\sum_{s\in\Omega} \theta_{ijs}^d - \sum_{i\in I_j} \phi_{ij}^q \le \lambda_{1j} p_1^0, j \in \mathcal{P}$$

$$z_{1js} : -p_j \pi_s \gamma_{1j} + \theta_{1js}^d - \phi_{1js}^z \le 0, j \in \mathcal{P}_1 \quad s \in \Omega,$$

$$z_{ijs}, i \ne 1 : \theta_{ijs}^d - \phi_{ijs}^z \le 0, i \in I, i \ne 1, j \in \mathcal{P}_i, s \in \Omega,$$

$$\mu_i^c : \sum_{j\in\mathcal{P}_i} \lambda_{ij} \theta_{ij}^q - \phi_i^c \le 0, i \in I$$

$$\mu_{ijs}^d : \lambda_{ij} \theta_{ij}^q - \lambda_{ij} \theta_{ijs}^z - \phi_{ijs}^d \le 0, i \in I, j \in \mathcal{P}_i, s \in \Omega,$$

$$\mu_{ikj}^e : \theta_{ij}^q \le 0, i \in I, j \in \mathcal{P}_i, k \ne i$$

2. Constraints that correspond to (45):

$$\gamma_{ij} - M\delta_{ij}^\gamma \le 0, i \in I, j \in \mathcal{P}_i \quad (52)$$

$$Q_j - M\delta_{ij}^q \le 0, i \in I, j \in \mathcal{P}_i,$$

$$z_{ijs} - M\delta_{ijs}^z \le 0, i \in I, j \in \mathcal{P}_i, s \in \Omega,$$

$$\mu_i^c - M\delta_i^c \le 0, i \in I,$$

$$\mu_{ijs}^d - M\delta_{ijs}^d \le 0, i \in I, j \in \mathcal{P}_i, s \in \Omega,$$

$$\mu_{ikj}^e - M\delta_{ikj}^e \le 0, i \in I, j \in \mathcal{P}_i, k \ne i,$$

3. Constraints that correspond to (46):

$$-p_jQ_j + \sum_{s\in\Omega} p_j\pi_s z_{1js} - \sum_{s\in\Omega} p_j\pi_s\theta^z_{1js} + \theta^\gamma_j + M\delta^\gamma_{1j} \le M, j \in \mathcal{P} \tag{53}$$

$$-\sum_{s\in\Omega} p_j\pi_s\theta^z_{1js} + \theta^\gamma_j + M\delta^\gamma_{ij} \le M, i \in I, i \ne 1, j \in \mathcal{P}_i$$

$$-\gamma_{1j}p_j + \sum_{i\in I_j} \lambda_{ij}\theta^c_i + \sum_{i\in I_j}\sum_{s\in\Omega} \theta^d_{ijs} + \sum_{i\in I_j} \phi^q_{ij} + M\delta^q_j \le M - \lambda_{1j}p^0_1, j \in \mathcal{P}$$

$$p_j\pi_s\gamma_{1j} - \theta^d_{1js} + \phi^z_{1js} + M\delta^z_{1js} \le M, j \in \mathcal{P} \quad s \in \Omega,$$

$$-\theta^d_{ijs} + \phi^z_{jjs} + M\delta^z_{ijs} \le 0, i \in I, i \ne 1, j \in \mathcal{P}_i, s \in \Omega,$$

$$-\sum_{j\in\mathcal{P}_i} \lambda_{ij}\theta^q_{ij} + \phi^c_i + M\delta^c_i \le M, i \in I$$

$$-\lambda_{ij}\theta^q_{ij} + \lambda_{ij}\theta^z_{ijs} + \phi^d_{ijs} + M\delta^d_{ijs} \le M, i \in I, j \in \mathcal{P}_i, s \in \Omega,$$

$$-\theta^q_{ij} + M\delta^e_{ikj} \le M, i \in I, j \in \mathcal{P}_i, k \ne i$$

4. Constraints that correspond to (47):

$$-\sum_{j\in\mathcal{P}_i} \lambda_{ij}Q_j + M\delta^c_i \le M - W_i, i \in I \tag{54}$$

$$-Q_j + z_{ijs} + M\delta^d_{ijs} \le M - d_{js}, i \in I, j \in \mathcal{P}_i, s \in \Omega,$$

$$-\gamma_{ij}p_j + \lambda_{ij}\mu^c_i + \lambda_{ij}\sum_{s\in\Omega} \mu^d_{ijs} + \sum_{k\in\mathcal{I}j\setminus\{i\}} \mu^e_{ikj} + M\delta^q_{ij} \le M - \lambda_{ij}p^0_i, i \in I, j \in \mathcal{P}_i,$$

$$\gamma_{ij}p_j\pi_s - \lambda_{ij}\mu^d_{ijs} + M\delta^z_{ijs} \le M, i \in I, j \in \mathcal{P}_i, \quad s \in \Omega,$$

5. Constraints that correspond to (48):

$$\theta^c_i - M\delta^c_i \le 0, i \in I \tag{55}$$

$$\theta^d_{ijs} - M\delta^d_{ijs} \le 0, i \in I, j \in \mathcal{P}_i, s \in \Omega,$$

$$\theta^q_{ij} - M\delta^q_{ij} \le 0, i \in I, j \in \mathcal{P}_i,$$

$$\theta^z_{ijs} - M\delta^z_{ijs} \le 0, i \in I, j \in \mathcal{P}_i, s \in \Omega,$$

6. Constraints that correspond to (49):

$$-\theta_i^c - M\delta_i^c \leq 0, i \in I \tag{56}$$

$$-\theta_{ijs}^d - M\delta_{ijs}^d \leq 0, i \in I, j \in \mathcal{P}_i, s \in \Omega,$$

$$-\theta_{ij}^q - M\delta_{ij}^q \leq 0, i \in I, j \in \mathcal{P}_i,$$

$$-\theta_{ijs}^z - M\delta_{ijs}^z \leq 0, i \in I, j \in \mathcal{P}_i, s \in \Omega,$$

7. Constraints that correspond to (50):

$$\phi_i^c + M\delta_i^c \leq M, i \in I \tag{57}$$

$$\phi_{ijs}^d + M\delta_{ijs}^d \leq M, i \in I, j \in \mathcal{P}_i, \quad s \in \Omega$$

$$\phi_{ij}^q + M\delta_{ij}^q \leq M, i \in I, j \in \mathcal{P}_i,$$

$$\phi_{ijs} + M\delta_{ijs}^z \leq M, i \in I, j \in \mathcal{P}_i, \quad s \in \Omega,$$

Finally, we obtain an equivalent expression for the objective function from (41) that yields the MILP equivalent to (26)–(35):

$$\max \frac{1}{2}\left(-p_1^0 \sum_{j \in \mathcal{P}} \lambda_{1j} Q_j + \sum_{i \in I} W_i \theta_i^c + \sum_{i \in I} \sum_{j \in \mathcal{P}_i} \sum_{s \in \Omega} d_{js} \theta_{ijs}^d + \sum_{i \in I} \sum_{j \in \mathcal{P}_i} p_i^0 \lambda_{ij} \theta_{ij}^q + \sum_{j \in \mathcal{P}} \theta_j^\gamma \right) \tag{58}$$

where maximization is performed subject to constraints (27)–(31), (51)–(57) with respect to γ_{ij}, Q_j, z_{ijs}, μ_i^c, μ_{ijs}^d, μ_{ikj}^e, θ_i^c, θ_{ijs}^d, θ_{ij}^q, θ_{ijs}^z, θ_j^γ, ϕ_i^c, ϕ_{ijs}^d, ϕ_{ij}^q, ϕ_{ijs}^z, δ_{ij}^γ, δ_{ij}^q, δ_{ijs}^z, δ_i^c, δ_{ijs}^d, δ_{ik}^e with γ_{ij}, Q_j, z_{ijs}, μ_i^c, μ_{ijs}^d, μ_{ikj}^e, ϕ_i^c, ϕ_{ijs}^d, ϕ_{ij}^q, $\phi_{ijs}^z \geq 0$ and δ_{ij}^γ, δ_{ij}^q, δ_{ijs}^z, δ_i^c, δ_{ijs}^d, δ_{ikj}^e binary. This MILP problem provides the same solution of the original problem (16)–(21) and can be solved by available MILP solvers.

5. Nonlinear models with explicit characterization of Nash equilibria

In this section, we take an alternative route to the analysis of the Stackelberg game with the service aggregator as the leader and the service providers as followers. It begins with an analytical study of the properties of the solution of the profit-maximizing problems of the followers formulated in the most general nonlinear way with possibly continuous distributions. This allows us to find an analytical expression for the Nash equilibrium among the followers and design a numerical algorithm for its computation. After this, we formulate the leader problem which consists of finding the Nash equilibrium that is optimal for the aggregator. The resulting nonlinear programming problem is solvable with available NLP software.

The advantage of this approach consists of the possibility to consider the risk aversion of the agents in a relatively easy way. It is done by adding to the problems of the actors' constraints on risk which are handled similarly to the capacity constraints. Its drawback is that the leader problem is usually nonconvex and additional effort should be spent in order to find the global optimum and not a local one.

We start as in Section 3 by formulating service aggregator/service providers leader/followers' problem. Similarly, to Section 3 we assume that there is infinite external demand for module i and there is the external price $p_i^0 > c_i$ for this module. Then all spare capacity will be utilized for serving the external demand. The service providers' profit similar to (6) becomes

$$\sum_{j\in\mathcal{P}_i} \gamma_{ij} p_j \mathbb{E}_{d_j} \min\left\{d_j, \min_{k\in I_j}\left\{\frac{q_{kj}}{\lambda_{kj}}\right\}\right\} + p_i^0\left(W_i - \sum_{j\in\mathcal{P}_i} q_{ij}\right) - c_i W_i.$$

Here the first term is the sum of expected revenue shares that are due to provider i for his participation in platform services. The served demand is

$$\min\left\{d_j, \min_{k\in I_j}\left\{\frac{q_{kj}}{\lambda_{kj}}\right\}\right\} \tag{59}$$

that is, the smallest value between the actual realization of the demand and the smallest demand serving the capability allocated by the participating providers. The second term is the revenue generated by the satisfaction of the external demand for module i, while the third term is the provision cost which does not depend on the decision variables because the whole capacity is employed. Dropping the terms not depending on the capacity distribution q_{ij} among platform services and adding the capacity constraint we obtain the following problem:

$$\max_{q_{ij}\geq 0} \sum_{j\in\mathcal{P}_i} \gamma_{ij} p_j \mathbb{E}_{d_j} \min\left\{d_j, \min_{k\in I_j}\left\{\frac{q_{kj}}{\lambda_{kj}}\right\}\right\} - p_i^0 \sum_{j\in\mathcal{P}_i} q_{ij} \tag{60}$$

$$\sum_{j\in\mathcal{P}_i} q_{ij} \leq W_i \tag{61}$$

This problem is the exact reformulation of problem (7)–(10) for the case of arbitrary distributions of demand: constraints (9), (10) are incorporated into the nonlinear expression (59) for the satisfaction of demand.

We start by describing the set of Nash equilibria for the game where the payoffs of the agents are defined by (60) and feasible sets of their strategies by (61). This is done by embedding (60)–(61) into the more general problem defined below.

Let us consider $i = 1:N$ actors with actions $x^i \in X^i \subseteq \mathbb{R}^n$. The set X^i of admissible actions is defined as follows:

$$X^i = \left\{x^i \mid x^i \in V^i, x^i \geq 0, F_k^i\left(x^i\right) \leq 0, k = 1:K\right\} \tag{62}$$

The payoff function $F_0^i(x)$, $x = (x^1,...,x^N)$ of agent i has the following form:

$$F_0^i(x) = F_0^i(x^i, z(x)), z(x) = \left(z_1(x),...,z_n(x)\right), z_j(x) = \min_i x_j^i. \tag{63}$$

Let us define the collection $\bar{x} = (\bar{x}^1,...,\bar{x}^N)$ of the agents' strategies as follows

$$\bar{x}^k = \left(\bar{x}_1,...,\bar{x}_n\right),\ \bar{x}_j \leq \min_i \hat{x}_j^i, \forall k \tag{64}$$

where $\hat{x}^i$ is some solution of the following problem

$$\max_{x^i \in X^i} F_0^i(x^i, x^i) \tag{65}$$

The following proposition shows that under certain conditions any $\bar{x}$ that satisfies (64), (65) is a Nash equilibrium and there are no other Nash equilibria in the game of actors with payoff functions (63). The vector inequalities are understood componentwise.

Proposition 2. *Suppose that the following conditions are satisfied.*

1. Functions $F_k^i(x^i, z)$ are nondecreasing for $x^i \in X^i$, namely if y^i, $u^i \in X^i$ and $y^i \geq u^i$ then
$$F_k^i\left(y^i, z\right) \geq F_k^i\left(u^i, z\right)$$
2. Sets $V^i \subseteq \mathbb{R}^+$ are monotone, namely if $u^i \in V^i$ and $0 \leq y^i \leq u^i$, then $y^i \in V^i$.
3. Function $F_0^i(x^i, z)$ is nonincreasing with respect to $x^i \in X^i$, namely if $y^i \geq u^i$ then $(y^i, z) \leq F_0^i(u^i, z)$.
4. Function $F_0^i(x^i, x^i)$ is monotonous to maximum on X^i, namely if y^i, $u^i \in X^i$ are such that all their components coincide except $y_j^i \neq u_j^i$ then $F_0^i(y^i, y^i) \leq F_0^i(u^i, u^i)$ if $y_j^i \leq u_j^i \leq \hat{x}^i$.

Then an arbitrary collection $\bar{x}$ of the agent's strategies is defined in (64)–(65) is the Nash equilibrium for the noncooperative game with the agent's payoffs defined by (63).

Suppose, in addition, that inequality in Condition 3 is strict and for any i, j there exists solution $\hat{x}^i$ of (65) such that for any y^i, u^i defined as in

Condition 4 with $y_j^i = \hat{x}_j^i$ and $y_j^i < u_j^i$ we have $F_0^i(y^i, y^i) > F_0^i(u^i, u^i)$. Then any Nash equilibrium is defined by (64), (65).

In order to apply this proposition to (60)–(61) let us make a substitution

$$q_{ij} = \lambda_{ij} Q_{ij}$$

and take $Q_i = \{Q_{ij}, j \in \mathcal{P}\}$, $Q = \{Q_i, i \in \mathcal{I}\}$. Then (60)–(61) takes the form

$$\max_{Q_i \geq 0} \psi^i \left(Q_i, z(Q)\right) \tag{66}$$

$$\sum_{j \in \mathcal{P}_i} \lambda_{ij} Q_{ij} \leq W_i \tag{67}$$

where,

$$\psi^i \left(Q_i, z\right) = \sum_{j \in \mathcal{P}_i} \gamma_{ij} p_j \mathbb{E}_{d_j} \min\left\{d_j, z_j\right\} - p_i^0 \sum_{j \in \mathcal{P}_i} \lambda_{ij} Q_{ij}, z_j\left(Q\right) = \min_{k \in I_j} Q_{kj} \tag{68}$$

Besides that, we shall consider the problem

$$\max_{Q_{ij} \geq 0} \sum_{j \in \mathcal{P}_i} \gamma_{ij} p_j \mathbb{E}_{d_j} \min\left\{d_j, Q_{ij}\right\} - p_i^0 \sum_{j \in \mathcal{P}_i} \lambda_{ij} Q_{ij} \tag{69}$$

subject to constraint (67). We obtain the following characterization of Nash equilibria in (60)–(61).

Proposition 3 *Suppose that the solution of problem (69) is unique. Then Nash equilibria for* (60)–(61) *exist and for any such equilibrium*

$$q_{ij} = \lambda_{ij} Q_j, i \in I, j \in \mathcal{P}_i$$

$$Q_j \leq Q_{ij}^*, Q_i^* = \arg\max_{Q_{ij} \geq 0} \sum_{j \in \mathcal{P}_i} \gamma_{ij} p_j \mathbb{E}_{d_j} \min\left\{d_j, Q_{ij}\right\} - p_i^0 \sum_{j \in \mathcal{P}_i} \lambda_{ij} Q_{ij}, i \in I \tag{70}$$

where the maximum in (70) is taken subject to (67) and $Q_i = \{Q_{ij}, j \in \mathcal{P}_i\}$.

Proof. The proof consists in checking the conditions of Proposition 2. We take Q_i as x^i in (62), (63), $\psi^i(Q_i, z)$ from (68) as $F_0^i(x^i, z)$, $K = 1$,

$$\sum_{j \in \mathcal{P}_i} \lambda_{ij} Q_{ij} - W_i$$

as $F_1^i(x^i, z)$, $\{Q_i | Q_i \geq 0\}$ as V_i. Then Condition 1 is satisfied due to $\lambda_{ij} \geq 0$. Condition 2 is trivial. Condition 3 is satisfied with strict inequalities because p_i^0, $\lambda_{ij} > 0$ in (68). Condition 4 and the additional conditions are satisfied because $\psi^i(Q_i, Q_i)$ is concave, and has the unique maximum on the feasible set and is separable with respect to components of Q_i. ■

Thus, Proposition 3 shows that the Nash equilibria of (60)–(61) can be described by solving problem (69). It is similar to the multi-product newsvendor problem with a budget constraint (see Hadley and Whitin (1963), Erlebacher (2000), Abdel-Malek and Montanari (2005), Chung and Kirca (2008), Dance and Gaivoronski (2012)). However, the second term in (69) is different from the second term in the classic newsvendor problems and these results can not be utilized here. Therefore, we derive its optimal solution from the following proposition

Proposition 4. *Consider the problem*

$$\max_{x\geq 0}\sum_j\left(a_j\mathbb{E}_{d_j}\min\{d_j,x_j\}-b_jx_j\right) \tag{71}$$

$$\sum_j c_jx_j\leq W \tag{72}$$

with a_j, b_j, $c_j > 0$, denote $H_j(x_j)$ as the cumulative density function of demand d_j, $0 \leq d_j \leq d_j^+$. Suppose that d_j has positive bounded density $h_j(d_j)$ for $d_j \in [0, d_j^+]$. Take

$$x_j(\rho)=H_j^{-1}\left(1-\frac{b_j+\rho c_j}{a_j}\right), \tag{73}$$

where $H_j^{-1}(x) = 0$ for $x \leq 0$ and suppose that ρ^* is a solution of the equation

$$\sum_j c_jx_j(\rho)=W$$

if it exists. Then solution of (71), (72) is unique and equals $x_j(0)$ if

$$\sum_j c_jx_j(0)<W \tag{74}$$

and $x_j(\rho^*)$ otherwise.

The proof of this proposition is given in the Appendix. Applying Proposition 4 to (70) from Proposition 3 we obtain the following characterization of Nash equilibria for (60)–(61).

Proposition 5. *Suppose that demands d_j are distributed on $0 \leq d_j \leq d_j^+$ with cdf $H_j(d_j)$ and positive bounded densities.*

Then Nash equilibria for (60)–(61) exists and for any such equilibrium there exists $Q_j \geq 0$, $j \in \mathcal{P}$ such that

$$q_{ij}=\lambda_{ij}Q_j, i\in I, j\in\mathcal{P}_i \tag{75}$$

$$Q_j \le H_j^{-1}\left(1-\max_{i\in\mathcal{I}_j}\frac{\lambda_{ij}\left(p_i^0+\rho_i^*\right)}{\gamma_{ij}p_j}\right) \tag{76}$$

where,

$$\rho_i^* = \begin{cases} 0 & \text{if} \quad \sum_{j\in\mathcal{P}_i}\lambda_{ij}H_j^{-1}\left(1-\frac{\lambda_{ij}p_i^0}{\gamma_{ij}p_j}\right) < W_i \\ \rho : \sum_{j\in\mathcal{P}_i}H_j^{-1}\left(1-\frac{\lambda_{ij}\left(p_i^0+\rho\right)}{\gamma_{ij}p_j}\right) = W_i & \text{otherwise} \end{cases} \tag{77}$$

Vice versa, any Q_j satisfying (76), (77) generates a Nash equilibrium through (75).

Proof. Assumptions of Proposition 4 are satisfied with $a_j = y_{ij}p_j$, $b_j = p_i^0\lambda_{ij}$, $c_j = \lambda_{ij}$. Applying this proposition we obtain that the solution Q_i^* of problem (69) is unique and

$$Q_{ij}^* = Q_{ij}\left(\rho_i^*\right) = H_j^{-1}\left(1-\frac{\lambda_{ij}\left(p_i^0+\rho_i^*\right)}{\gamma_{ij}p_j}\right). \tag{78}$$

with ρ_i^* from (77). Due to the uniqueness of the solution, we can apply Proposition 3. We substitute (78) into (70) and put maximum inside the argument of H_j^{-1} utilizing its monotonicity, which yields (76). ■

In the most important cases when the service providers do not want to concentrate solely on supplying modules to the aggregator and have enough spare capacity, bound (76) simplifies to

$$Q_j \le H_j^{-1}\left(1-\max_{i\in\mathcal{I}_j}\frac{\lambda_{ij}p_i^0}{\gamma_{ij}p_j}\right).$$

Thus, there is a multitude of Nash equilibria in problem (60)–(61). Being in a privileged position, the aggregator can strive to achieve the equilibrium that maximizes his/her profit. According to (69) and Proposition 5 this equilibrium is the solution of the problem (69) for $i = 1$ subject to additional constraint (76). This is a convex optimization problem. The aggregator can achieve this equilibrium by announcing it to the service providers because they will not have the incentive to deviate from it. In order to compute it the aggregator needs to know only the prices p_i^0 of the external opportunities for service providers, no knowledge of their costs is necessary.

Having described the set of Nash equilibria, we can now formulate the upper-level problem of the aggregator in the framework of the single leader/multiple followers Stackelberg game. As in Sections 3, and 4 we consider the following game structure.

1. The aggregator selects the revenue sharing scheme y_{ij}, $i \in I, j \in \mathcal{P}_i$ and communicates it to service providers.
2. The service providers select the equilibrium capacity allocations q_{ij} maximizing their profits as in (60)–(61).
3. The aggregator selects the revenue shares that maximize his/her profit, taking into account the dependence of the capacity allocations on revenue shares.

Utilizing the characterization of the Nash equilibria from Proposition 5 we obtain the following aggregator's problem

$$\max_{\gamma_{ij}, Q_j \geq 0} \sum_{j \in \mathcal{P}} \gamma_{1j} p_j \mathbb{E}_{d_j} \min\{d_j, Q_j\} - p_1^0 \sum_{j \in \mathcal{P}} \lambda_{1j} Q_j \tag{79}$$

$$\sum_{i \in \mathcal{I}_j} \gamma_{ij} = 1, j \in \mathcal{P} \tag{80}$$

$$\sum_{j \in \mathcal{P}} \lambda_{1j} Q_j \leq W_1 \tag{81}$$

$$Q_j \leq H_j^{-1}\left(1 - \max_{i \in \mathcal{I}_j \setminus \{1\}} \frac{\lambda_{ij}\left(p_i^0 + \rho_i^*\right)}{\gamma_{ij} p_j}\right) \tag{82}$$

where ρ_i^* is obtained from (77). This problem has relatively few variables and constraints, as opposed to the problems derived in Sections 3, and 4. However, we have lost convexity and extra effort should be spent in order to obtain the global and not the local maximum.

6. Extension of portfolio theory for competing risk-averse agents

The approach developed in the previous section allows us to consider risk-averse agents. In our selection of risk measures we follow the traditional choice of variance of profit, mainly because it is easier to treat analytically. Alternative risk measures like CVaR (Rockafellar and Uryasev (2000)) can be also considered. Thus, we assume that agent i wants to limit the variance of profit, which following (60) takes the form

$$R(q) = \mathbb{E}\left(\sum_{j \in \mathcal{P}_i} \gamma_{ij} p_j \left(\min\left\{ d_j, \min_{k \in I_j} \left\{ \frac{q_{kj}}{\lambda_{kj}} \right\} \right\} - \mathbb{E}_{d_j} \min\left\{ d_j, \min_{k \in I_j} \left\{ \frac{q_{kj}}{\lambda_{kj}} \right\} \right\} \right) \right)^2 \quad (83)$$

where $q = \{q_i, i \in I\}$, $q_i = \{q_{ij}, j \in \mathcal{P}_i\}$. The problem of agent i becomes

$$\max_{q_{ij} \geq 0} \sum_{j \in \mathcal{P}_i} \gamma_{ij} p_j \mathbb{E}_{d_j} \min\left\{ d_j, \min_{k \in I_j} \left\{ \frac{q_{kj}}{\lambda_{kj}} \right\} \right\} - p_i^0 \sum_{j \in \mathcal{P}_i} q_{ij} \quad (84)$$

$$\sum_{j \in \mathcal{P}_i} q_{ij} \leq W_i \quad (85)$$

$$R(q) \leq \sigma^2 \quad (86)$$

where σ^2 is some positive number. Making the same substitution $q_{ij} = \lambda_{ij} Q_{ij}$ as in the previous section, we can apply Proposition 2 in order to characterize the Nash equilibria in the game with agents having a payoff (84) and feasible sets (85), (86).

Proposition 6. *Suppose that the solution* $Q_i^* = \{Q_{ij}^*, j \in \mathcal{P}_i\}$ *of problem*

$$\max_{Q_j \geq 0} \sum_{j \in \mathcal{P}} \gamma_{ij} p_j \mathbb{E}_{d_j} \min\{d_j, Q_j\} - p_i^0 \sum_{j \in \mathcal{P}} \lambda_{ij} Q_j \quad (87)$$

$$\sum_{j \in \mathcal{P}_i} \lambda_{ij} Q_j \leq W_i, \quad (88)$$

$$\mathbb{E}\left(\sum_{j \in \mathcal{P}} \gamma_{ij} p_j \left(\min\{d_j, Q_j\} - \mathbb{E}_{d_j} \min\{d_j, Q_j\} \right) \right)^2 \leq \sigma_i^2 \quad (89)$$

is unique for all $i \in I$. *Then Nash equilibria for (84)–(86) exists and any such equilibrium is defined by some* $Q = \{Q_j, j \in \mathcal{P}\}$ *such that*

$$Q_j \leq \min_i Q_{ij}^* \quad (90)$$

taking $q_{ij} = \lambda_{ij} Q_{ij}$. *Vice versa, any* Q *that satisfies (90) defines a Nash equilibrium for (84)–(86).*

Proof. Similar to the proof of Proposition 3 we have to check the satisfaction of the conditions of Proposition 2 in problem (84)–(86). The only novelty in (84)–(86) compared to (60), (61) is the risk constraint (86). Therefore we have to check only Condition 1 that is relevant for this constraint. Taking $R(q)$

from (83) as $F_2^i(x^i, z)$ of Proposition 2 with $x^i = \{q_{ij}, j \in \mathcal{P}_i\}$ we observe that $F_2^i(x^i, z)$ is defined in such a way that does not depend on x^i at all. Therefore Condition 1 of Proposition 2 is satisfied automatically. ■

Similarly to (79), (82) the aggregator computes the optimal profit shares γ_{ij} in the risk-averse case by solving the problem

$$\max_{\gamma_{ij}, Q_j \geq 0} \sum_{j \in \mathcal{P}} \left(\gamma_{1j} p_j \mathbb{E}_{d_j} \min\{d_j, Q_j\} - p_1^0 \lambda_{1j} Q_j \right) \tag{91}$$

$$\sum_{i \in \mathcal{I}_j} \gamma_{ij} = 1, j \in \mathcal{P} \tag{92}$$

$$\sum_{j \in \mathcal{P}} \lambda_{1j} Q_j \leq W_1, \tag{93}$$

$$\mathbb{E}\left(\sum_{j \in \mathcal{P}} \gamma_{1j} p_j \left(\min\{d_j, Q_j\} - \mathbb{E}_{d_j} \min\{d_j, Q_j\} \right) \right)^2 \leq \sigma_1^2 \tag{94}$$

$$0 \leq Q_j \leq \min_{i > 1} Q_{ij}^* (\gamma_i) \tag{95}$$

where $Q_{ij}^*(\gamma_i)$ are solutions of problems (87)–(89).

It is possible to obtain some analytical results that characterize the solutions to these problems in the direction of Proposition 4. However, these characterizations will be considerably more complex. The reasonable alternative is to organize a two-level solution process where the NLP solver of (91)–(95) calls the solver of (87)–(89) when processing the constraints, one example of such computation is presented below. Problem (87)–(89) takes a specific form depending on the considered demand distributions.

Example 7. *Suppose that demands* d_j *are independent and distributed uniformly on* $[0, d_j^+]$. *Then* (87)–(89) *becomes*

$$\max_{Q_j} \sum_{j \in \mathcal{P}} \left(\left(\gamma_{ij} p_j - p_i^0 \lambda_{ij} \right) Q_j - \frac{\gamma_{ij} p_j}{2 d_j^+} Q_j^2 \right)$$

subject to

$$\sum_{j \in \mathcal{P}_i} \frac{\gamma_{ij}^2 p_j^2}{3 d_j^+} Q_j^3 \left(1 - \frac{3}{4 d_j^+} Q_j \right) \leq \sigma_i^2$$

and constraint (88).

The value of σ_1 from (94) and σ_i from (89) are risk tolerances of aggregator and service providers. Solving problems (91)–(95) and (87)–(89) for different values of σ_i one obtains the dependence of actors' profits on their risk tolerances. Having such dependencies, the respective decision makers can select the optimal balance between risk and profit and implement it selecting the optimal distribution of their capacities between the different services that the solutions to these problems provide. This capacity of distribution can be viewed as the *efficient portfolio* of services that each actor decides to contribute to. The dependence of profit on risk is conceptually similar to efficient frontier from finance (see Markowitz (1991)).

Thus, the collection of problems (91)–(95), (87)–(89) can be seen as an extension of the portfolio theory of modern investment science to the multiagent environment of collaborative service provision. The structure of these problems is also similar to portfolio optimization problems from finance. One major novelty here is that, unlike in the traditional portfolio theory, the actors' optimal portfolios depend not only on their own risk tolerances but also on the risk tolerances of other participants. For example, the aggregator, while solving his/her portfolio problem (91)–(95) should take into account also risk tolerances of service providers that enter his/her problem through (95).

Figures 1 and 2 highlight some of the differences between the classical efficient frontiers of portfolio theory and the multiagent efficient frontiers described in this section. They present efficient frontiers (dependence of optimal profit on risk) for the case when demand is distributed as in Example 7 and there are three agents contributing their modules to the composite services: A1, A2 and A3. Agent A1 takes also the role of coordinator. Generally, the figures show a pattern familiar to traditional portfolio theory: the optimal profit increases monotonically while agents become more risk tolerant but the rate of increase diminishes with increasing risk. However, there are also substantial differences that are due to the multiagent nature of our portfolio problem.

Figure 1 shows the effects of nonuniform risk attitudes among the agents. It compares two cases. In the first case, the risk tolerance of all agents in the same and simultaneously increases as shown on the horizontal axis. In the second case the risk tolerance of agent A1 increases as shown on the horizontal axis, but the risk tolerances of remaining agents A2 and A3 are kept constant. One can see that in the first case the profit of A1 and the total profit of all agents increases faster and reaches saturation sooner than in the second case. In other words, when some of the agents exhibit lower levels of risk tolerance, they still can achieve higher levels of profit that are associated with more aggressive risk-taking behavior in the classical case. However, in order for this to happen it is necessary that among them there exists an agent (or agents) who is willing to shoulder more risk than it would be necessary for him/her in the case of other more risk tolerant participants (in this case A1).

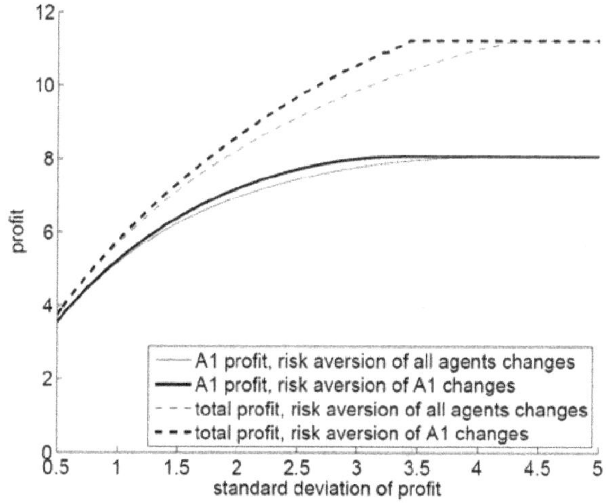

Figure 1. Dependence of profit on admissible risk for uniform and nonuniform risk aversion among agents.

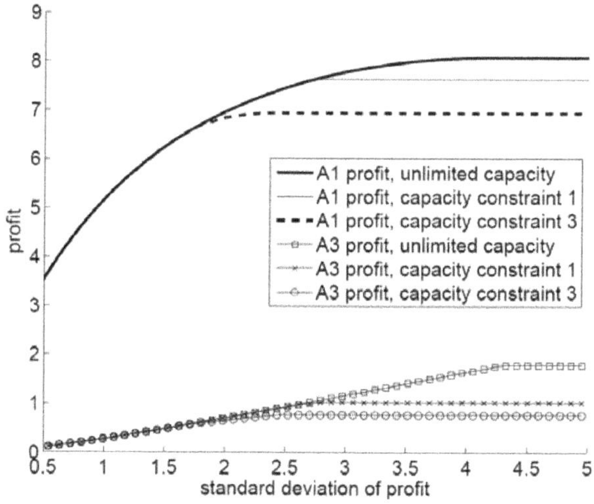

Figure 2. Dependence of profit of agents A1 and A3 on admissible risk for different capacity constraints.

Figure 2 illustrates the effects of capacity constraints and the phenomenon of bottleneck agents, unknown in the classical case. The capacity constraint like (85) or (93) is present in classical portfolio theory in the form of a budget constraint (Zenios (2007)). However, in the classical case it does not affect the trade-off between risk and profit because in the case of single agent the right-hand side of this constraint can be normalized to one and portfolio can be expressed in fractions of the total budget. The multiagent environment is very different in this respect, as Figure 2 shows. Three cases are compared there: no limit on capacity, a specific capacity constraint for each agent (constraint 1) and reduced by 13% constraint for A1, while constraints for A2 and A3 remain unchanged from the previous case (constraint 3). The risk tolerance of all agents is the same and increases as shown on the horizontal axis. One can observe that, in the presence of a capacity constraint, increasing the risk tolerance after some level does not lead to an increase in profit because the risk constraint becomes nonbinding. In addition, a more stringent budget constraint for A1 reduces profit also for A3 and he/she can not compensate for it by increasing his/her risk tolerance. This is because the capacity constraint of A1 becomes a bottleneck for the whole system.

7. Case study in provision of mobile apps

The models described in the previous sections were developed for the business analysis of the delivery of advanced mobile data services. Such services are highly customizable, and are characterized by a high level of modularity and normally offered as *service portfolios*, i.e., packages of services interacting in order to satisfy the end-user's needs. The delivery of these services is normally carried out through a so-called *service platform*, which hosts the services and makes them accessible to the end-users.

Let us assume that two developers engage in the collaborative provision of three different mobile apps to be sold through a platform, such as the Apple Store or the Google Play store and let us assume that one of the developers (A3) can charge an ever increasing price to sell its service off platform (or targeting other platforms with a large enough customer base). This will gradually incentivize such developer to exit the collaborative pattern and sell his/her service using a different channel. Figure 3 shows how the capacities of each developer are allocated for different levels of market power of the aforementioned developer. When the price that the developer can charge to sell his/her service off platform is low, every provider will invest entirely within the platform. As the price for such service increases, the Content Provider will move his/her investment outside the platform. To maintain a fixed balance over the different service amounts within the platform service portfolios the aggregator (platform operator, in this case) will have to grant an increasingly higher share of the revenues to the developer, which in turn will earn higher profits out of the collaborative provision within the platform. Conversely, the platform operator, which was earlier gaining extra profits, will have to give up on an increasing part of his/her revenue share to counterbalance the increase in market power for the developer. This is shown in Figures 5 and 6. The bias of the revenue share in favour of the Content Provider will eventually reduce the profits for the other developer (A2) and the platform operator (A1), which will therefore opt for reducing the amount of their service provision to the platform as well. Figure 3 shows this effect of reduction of the total capacity dedicated to the platform for each of the providers. This reallocation of capacity for all the Providers will rebalance the revenue sharing scheme in favour of the platform operator, as shown in Figure 6, and a consequent slowdown of his/her decrease in profits. When the market power for developer A3 keeps increasing, the profit for this latter will increase further, while profits for the remaining providers will keep decreasing. It interesting to notice how the composition of the the offered mobile apps by the platform changes as a consequence of the market power increase for developer (A3). Figure 4 shows

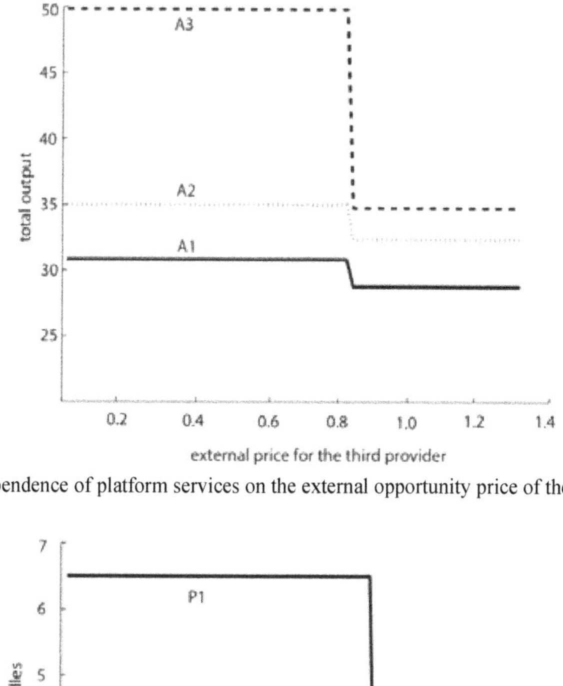

Figure 3. Dependence of platform services on the external opportunity price of the third provider.

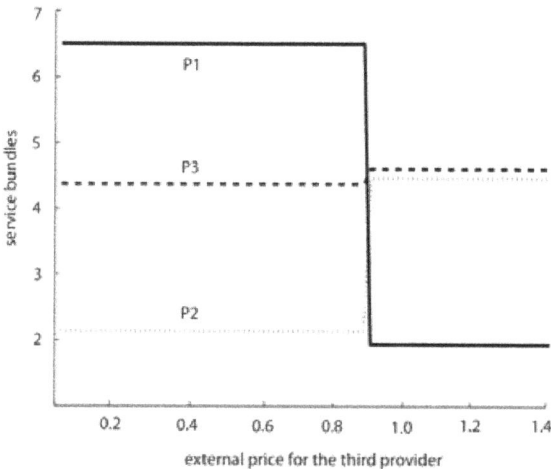

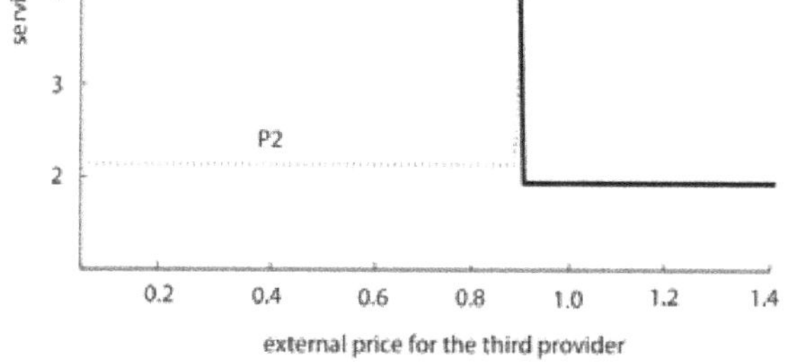

Figure 4. Dependence of the output of services on the external opportunity price of the third provider.

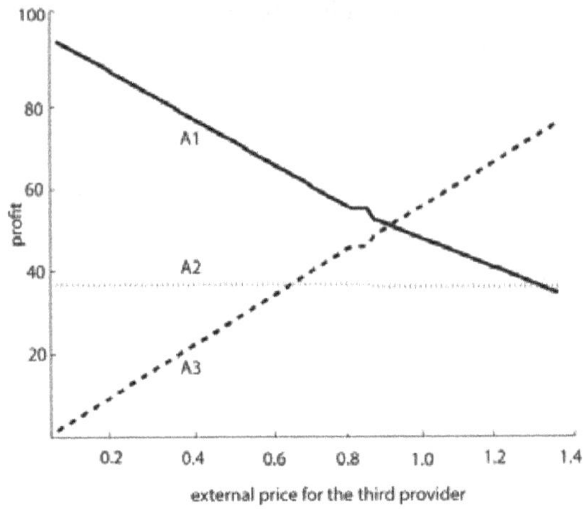

Figure 5. Dependence of profits on the external opportunity price of the third provider.

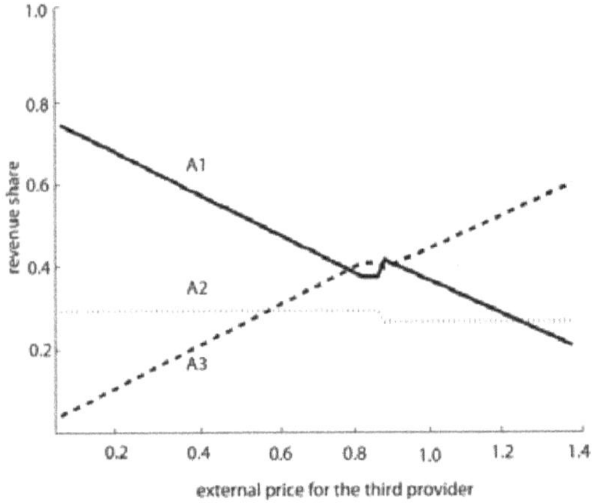

Figure 6. Dependence of revenue shares on the external opportunity price of the third provider.

how the investments shift over the development of different apps. In our model implementation, developer A3 was asked to make his/her highest on-platform contribution on app $P1$. Therefore, for a fixed revenue sharing scheme, $P1$ is the app yielding the least contribution to the Content Provider's profit. As a consequence, one can expect such a developer to reduce his/her investments on the considered app and move his/her investments off-platform.

8. Conclusions

In this chapter we have introduced and analyzed several novel models for the analysis of collaborative aggregator-led provisions of complex modular services under uncertainty. The approach has been twofold. Under discrete probability distribution settings, we have introduced a mixed integer linear model, which can readily account for different probability distributions over demand scenarios through sampling. Under the continuous distribution assumption, we have elaborated a non-linear modelling approach which has allowed us to account for risk aversion of each agent by explicitly characterizing the set of Nash equilibria. This has led to the extension of the portfolio theory of investment science to the single leader/multiple followers setting of the stochastic Stackelberg game of special structure. Numerical experiments and the case studies within the chapter confirm the validity of the theoretical results of this paper.

References

Abdel-Malek, L.L. and R. Montanari. 2005. An analysis of the multi-producer newsboy problem with a budget constraint. Int. J. Production Economics, 97: 296–307.

Avinadav, T.C. and Y. Perlman. 2015a. Consignment contract for mobile apps between a single retailer and competitive developers with different risk attitudes. European Journal of Operational Research, 246(33): 949–957.

Avinadav, T.C. and Y. Perlman. 2015b. The effect of risk sensitivity on a supply chain of mobile applications under a consignment contract with revenue sharing and quality investment. International Journal of Production Economics, 168: 31–40.

Ballon, P. and E.V. Heesvelde. 2011. ICT platforms and regulatory concerns in Europe. Telecommunications Policy, 35: 702–714.

Bard, J.F. 1998. Practical Bilevel Optimization. Kluwer Academic Publishers.

Bazaraa, M.S., H.D. Sherali and C.M. Shetty. 2006. Nonlinear Programming, Theory and Algorithms. 3rd Edition. Wiley-Interscience.

Birge, J. and F. Louveaux. 1997. Introduction to Stochastic Programming. Springer.

Cachon, G.P. and M.A. Lariviere. 2005. Supply chain coordination with revenue-sharing contracts: Strength and limitations. Management Science, Incentives and Coordination in Operations Management, 51(1): 30–44.

Chernonog, T., T. Avinadav, and T. Ben-Zvi. 2015. Pricing and sales-effort investment under bi-criteria in a supply chain of virtual products involving risk. European Journal of Operations Research, 246(2): 471–475.

Chung, C.S., J. Flynn and Ö. Kirca. 2008. A multi-item newsvendor problem with preseason production and capacitated reactive production. European Journal of Operations Research, 188: 775–792.

Crowther, J. 1964. Rationale for quantity discounts. Harvard Business Review, 42(2): 121–127.

Dance, C. and A.A. Gaivoronski. 2012. Stochastic optimization for real time service capacity allocation under random service demand. Annals of Operations Research, 193: 221–253.

Dempe, S. 2002. Foundations of Bilevel Programming. Kluwer Academic Publishers.

Dolan, R.J. 1987. Quantity discounts: Managerial issues and research opportunities. Marketing Science, 6(1): 1–22.

Erlebacher, S.J. 2000. Optimal and heuristic solutions for the multi-item newsvendor problem with a single capacity constraint. Production and Operations Management, 9: 303–318.

Fan, Y.F. and Y. Shou. 2020. A risk-averse and buyer-led supply chain under option contract: Cvar minimization and channel coordination. International Journal of Production Economics, 219: 66–81.

Gaivoronski, A.A. and J. Zoric. 2008. Business models for collaborative provision of advanced mobile data services: Portfolio theory approach. Operations Research/Computer Science Interfaces Series, 44: 356–383.

Gerchak, Y. and Y. Wang. 2004. Revenue-sharing vs. wholesale-price contracts in assembly systems with random demand. Production Oper. Management, 13(1): 23–33.

Govindan, K. and A. Malomfalean. 2019. A framework for evaluation of supply chain coordination by contracts under O2O environment. International Journal of Production Economics, 215: 11–23.

Goyal, S.K. 1976. An integrated inventory model for a single supplier - single customer problem. International Journal of Production Research, 15(1): 107–111.

Hadley, G. and T. Whitin. 1963. Analysis of Inventory Systems. Prentice Hall, Englewood Cliffs, NJ.

Hillegas, H. 2012. iOS app distribution options.

Joglekar, P. and S. Tharthare. 1990. The individually responsible and rational decision approach to economic lot sizes for one vendor and many purchasers. Decision Science, 21: 492–506.

Kall, P. and S.W. Wallace. 1994. Stochastic Programming. Wiley, Chichester.

Lal, R. and R. Staelin. 1984. An approach for developing an optimal discount pricing policy. Management Science, 30(12): 1524–1539.

Markowitz, H. 1991. Portfolio Selection. Blackwell, Second Edition.

Murati, E. 2021. What are digital platforms? An overview of definitions, typologies, economics, and legal challenges arising from the platform economy in EU. European Journal of Privacy Law & Technologies.

Parlar, M. and Q. Wang. 1994. Discounting decisions in a supplier/buyer relationship with a linear buyer's demand. IIE Transactions, 26(2): 34–41.

Pisciella, P. 2012. Methods for evaluation of business models for provision of advanced mobile services under uncertainty. Doctoral Thesis, Norwegian University of Science and Technology, Department of Industrial Economics and Technology Management.

Pisciella, P. and A.A. Gaivoronski. 2017. Stochastic programming bilevel models for service provision with a balancing coordinator. IMA Journal of Management Mathematics, 20(1): 131–152.

Pisciella, P., J. Zoric and A.A. Gaivoronski. 2009. Business model evaluation for an advanced multimedia service portfolio. Mobile Wireless Middleware, Operating Systems, and Applications - Workshops: Mobilware 2009 Workshops, Berlin, Germany, April 2009, Revised Selected Papers, pp. 23–32.

Reillier, L.C. and B. Reillier. 2017. Platform strategy: How to unlock the power of communities and networks to grow your business. Routledge.

Rockafellar, R.T. and S. Uryasev. 2000. Optimization of conditional value-at-risk. Journal of Risk, 2(99-4): 21–42.

Yang, H., J. Luo and H. Wang. 2017. The role of revenue sharing and first-mover advantage in emission abatement with carbon tax and consumer environmental awareness. International Journal of Production Economics, 193: 691–702.

Zenios, S.A. 2007. Practical Financial Optimization: Decision Making for Financial Engineers. Blackwell Publishing.

Appendix

Proof of Proposition 1.

Let us linearize complementarity constraint (39), introducing binary variables as in Bard (1998). Namely, constraint

$$f(y)g(y)=0$$

with bounded $f(y)$, $g(y)$ and $f(y) \geq 0$, $g(y) \geq 0$ is equivalent to pair of constraints

$$f(y) \leq Mz, g(y) \leq M - Mz, f(y) \geq 0, g(y) \geq 0. \tag{96}$$

where $z \in \{0, 1\}$ and $M > 0$ is a large number. This transforms (37)–(40) into

$$\max_{x,u} x^{\mathrm{T}} Rx + c^{\mathrm{T}} x$$

$$A_1 x \leq b_1 \tag{97}$$

$$Ex \leq M\overline{u} \tag{98}$$

$$b_1 - A_1 x \leq M\mathbf{1}_{m_1} - M\overline{u} \tag{99}$$

$$A_2 x = b_2 \tag{100}$$

$$x \geq 0, \overline{u} \in \{0,1\}^{m_1}$$

where is a column vector of all ones of length . This problem can be viewed as

$$\max_{u} \psi(\overline{u}) \tag{101}$$

$$\overline{u} \in \{0,1\}^{m_1}$$

where,

$$\psi(\overline{u}) = \max_{x \geq 0} x^{\mathrm{T}} Rx + c^{\mathrm{T}} x \tag{102}$$

subject to constraints (97)–(100).

Adding to (102) its Karush-Kuhn-Tucker conditions (Bazaraa and Shetty (2006)) will not change its solution, therefore

$$\psi(u) = \max_{x \geq 0} x^{\mathrm{T}} Rx + c^{\mathrm{T}} x \tag{103}$$

subject to constraints (97)–(100) and

$$\left(R + R^{\mathrm{T}}\right)x + c - A_1^{\mathrm{T}}\sigma_{11} - E^{\mathrm{T}}\sigma_3 + A_1^{\mathrm{T}}\sigma_{12} - A_2^{\mathrm{T}}\sigma_2 \leq 0$$

$$x^{\mathrm{T}}\left(-\left(R+R^{\mathrm{T}}\right)x-c+A_1^{\mathrm{T}}\sigma_{11}+E^{\mathrm{T}}\sigma_3-A_1^{\mathrm{T}}\sigma_{12}+A_2^{\mathrm{T}}\sigma_2\right)=0 \quad (104)$$

$$\sigma_{11}^{\mathrm{T}}\left(b_1-A_1x\right)=0 \quad (105)$$

$$\sigma_{12}^{\mathrm{T}}\left(M\mathbf{1}_{m_1}-M\bar{u}-b_1+A_1x\right)=0 \quad (106)$$

$$\sigma_3^{\mathrm{T}}\left(M\bar{u}-Ex\right)=0 \quad (107)$$

$$\sigma_2^{\mathrm{T}}\left(b_2-A_2x\right)=0 \quad (108)$$

$$x,\sigma_{11},\sigma_{12},\sigma_3\geq 0$$

Summing up complementarity conditions (104)–(108) yields:

$$2x^{\mathrm{T}}Rx=-c^{\mathrm{T}}x+b_1^{\mathrm{T}}\sigma_{11}-Mu^{\mathrm{T}}\sigma_{12}+M\mathbf{1}_{m_1}^{\mathrm{T}}\sigma_{12}-b_1^{\mathrm{T}}\sigma_{12}+b_2^{\mathrm{T}}\sigma_2+M\bar{u}^{\mathrm{T}}\sigma_3$$

Substituting this into (103) and the resulting problem into (101) we obtain the following equivalent to (37)–(40):

$$\max_{x,u,\sigma_{11},\sigma_{12},\sigma_2,\sigma_3}\frac{1}{2}\left(c^{\mathrm{T}}x+b_1^{\mathrm{T}}\sigma_{11}-M\bar{u}^{\mathrm{T}}\sigma_{12}+M\mathbf{1}_{m_1}^{\mathrm{T}}\sigma_{12}-b_1^{\mathrm{T}}\sigma_{12}+b_2^{\mathrm{T}}\sigma_2+M\bar{u}^{\mathrm{T}}\sigma_3\right) \quad (109)$$

$$A_1x\leq b_1 \quad (110)$$

$$Ex\leq M\bar{u} \quad (111)$$

$$b_1-A_1x\leq M\mathbf{1}_{m_1}-M\bar{u} \quad (112)$$

$$A_2x=b_2 \quad (113)$$

$$\left(R+R^{\mathrm{T}}\right)x+c-A_1^{\mathrm{T}}\sigma_{11}+A_1^{\mathrm{T}}\sigma_{12}-A_2^{\mathrm{T}}\sigma_2-E^{\mathrm{T}}\sigma_3\leq 0 \quad (114)$$

$$x^{\mathrm{T}}\left(-\left(R+R^{\mathrm{T}}\right)x-c+A_1^{\mathrm{T}}\sigma_{11}-A_1^{\mathrm{T}}\sigma_{12}+A_2^{\mathrm{T}}\sigma_2+E^{\mathrm{T}}\sigma_3\right)=0 \quad (115)$$

$$\sigma_{11}^{\mathrm{T}}\left(b_1-A_1x\right)=0 \quad (116)$$

$$\sigma_{12}^{\mathrm{T}}\left(M\mathbf{1}_{m_1}-M\bar{u}-b_1+A_1x\right)=0 \quad (117)$$

$$\sigma_3^{\mathrm{T}}\left(M\bar{u}-Ex\right)=0 \quad (118)$$

$$x,\sigma_{11},\sigma_{12},\sigma_3\geq 0,\bar{u}\in\{0,1\}^{m_1}$$

where we have dropped (108) due to its redundancy. Let us focus on the i-th row of constraints (110)–(112) and corresponding complementarity slackness conditions (115)–(117). Denoting a_i^1 and e_i the row vectors composed from the i-th row of matrices A_1 and E correspondingly, we obtain:

$$a_i^1 x \le b_{1i} \tag{119}$$

$$e_i x \le M\bar{u}_i \tag{120}$$

$$b_{1i} - a_i^1 x \le M - M\bar{u}_i \tag{121}$$

$$\sigma_{11i}\left(b_{1i} - a_i^1 x\right) = 0 \tag{122}$$

$$\sigma_{12i}\left(M - M\bar{u}_i - b_{1i} + a_i^1 x\right) = 0 \tag{123}$$

$$\sigma_{3i}\left(M\bar{u}_i - e_i x\right) = 0 \tag{124}$$

If $\bar{u}_i = 1$ then from (119) and (121) follows that $b_{1i} - a_i^1 x = 0$. Suppose now that $b_{1i} - a_i^1 x = 0$ and let us assume that in this case simultaneously $e_i x > 0$. This assumption restricts the feasible set of the problem (109)–(118), but it will not diminish its optimal value because $b_{1i} - a_i^1 x = 0$ and $e_i x = 0$ can not happen simultaneously at the optimal solution according to Condition 1 of Proposition. Due to (120) from $e_i x > 0$ follows $\bar{u}_i = 1$. Thus, $b_{1i} - a_i^1 x = 0$ if and only if $\bar{u}_i = 1$. Consequently, (122) is equivalent to

$$\sigma_{11i}\left(1 - \bar{u}_i\right) = 0 \tag{125}$$

We have seen already that if $\bar{u}_i = 1$ then $b_{1i} - a_i^1 x = 0$, from this follows $M - M\bar{u}_i - b_{1i} + a_i^1 x = 0$. Conversely, if $M - M\bar{u}_i - b_{1i} + a_i^1 x = 0$ then $\bar{u}_i$ can not be zero because M exceeds any feasible value of $b_{1i} - a_i^1 x$. Therefore $M - M\bar{u}_i - b_{1i} + a_i^1 x = 0$ if and only if $\bar{u}_i = 1$. Consequently, (123) is equivalent to

$$\sigma_{12i}\left(1 - \bar{u}_i\right) = 0 \tag{126}$$

If $\bar{u}_i = 0$ then from (120) $e_i x = 0$ and $M\bar{u}_i - e_i x = 0$. Conversely, if $M\bar{u}_i - e_i x = 0$ then inevitably $\bar{u}_i = 0$, $e_i x = 0$ because M exceeds any feasible value of $e_i x$ and therefore this equality can not be satisfied if $\bar{u}_i = 1$. Thus, $M\bar{u}_i - e_i x = 0$ if and only if $\bar{u}_i = 0$. Therefore (124) is equivalent to

$$\sigma_{3i}\bar{u}_i = 0 \tag{127}$$

From (127), (126) follows

$$M\bar{u}^{\mathrm{T}}\sigma_3 - M\bar{u}^{\mathrm{T}}\sigma_{12} + M\mathbf{1}_{m_1}^{\mathrm{T}}\sigma_{12} = 0. \tag{128}$$

Besides (125), (127), (126), are equivalent to

$$\sigma_{11i} \le M\bar{u}_i \tag{129}$$

$$\sigma_{12i} \le M\bar{u}_i \tag{130}$$

$$\sigma_{3i} \le M - M\bar{u}_i \tag{131}$$

correspondingly. Let us substitute now (128)–(130) into (109)–(118), utilizing equivalence between (131) and (124) and between (130) and (123).

$$\max_{x,u,v,\sigma_{11},\sigma_{12},\sigma_2,\sigma_3} \frac{1}{2}\left(c^{\mathrm{T}}x + b_1^{\mathrm{T}}(\sigma_{11}-\sigma_{12}) + b_2^{\mathrm{T}}\sigma_2\right) \tag{132}$$

subject to (110), (112), (113) and

$$\left(R+R^{\mathrm{T}}\right)x + c - A_1^{\mathrm{T}}(\sigma_{11}-\sigma_{12}) - A_2^{\mathrm{T}}\sigma_2 - E^{\mathrm{T}}\sigma_3 \le 0 \tag{133}$$

$$x^{\mathrm{T}}\left(-\left(R+R^{\mathrm{T}}\right)x - c + A_1^{\mathrm{T}}(\sigma_{11}-\sigma_{12}) + A_2^{\mathrm{T}}\sigma_2 + E^{\mathrm{T}}\sigma_3\right) = 0 \tag{134}$$

$$Ex \le M\bar{u} \tag{135}$$

$$\sigma_{11} \le M\bar{u} \tag{136}$$

$$\sigma_{12} \le M\bar{u} \tag{137}$$

$$\sigma_3 \le M\mathbf{1}_{m_1} - M\bar{u} \tag{138}$$

$$x,\sigma_{11},\sigma_{12},\sigma_3 \ge 0, \bar{u} \in \{0,1\}^{m_1}$$

Let us denote in (132)–(134)

$$\sigma_1 = \sigma_{11} - \sigma_{12} \tag{139}$$

and substitute constraints (136), (137) by

$$-M\bar{u} \le \sigma_1 \le M\bar{u}, \tag{140}$$

let us name the resulting problem Problem S1. Any feasible point of (132)–(138) will become a feasible point of S1 by applying (139), while the value of objective remains unchanged. Conversely, any feasible point of S1 can be transformed to feasible point of (132)–(138) by taking $\sigma_{11} = \max\{0, \sigma_1\}$, $\sigma_{12} = -\min\{0, \sigma_1\}$. Thus, substitution (139), (140) yields an equivalent problem to (132)–(138).

Let us make this substitution and express complementarity constraint (134) as a pair of "big M" constraints following (96). Then the problem becomes that of maximization of (132) subject to (113) and

$$-\left(R+R^{\mathrm{T}}\right)x + A_1^{\mathrm{T}}\sigma_1 + A_2^{\mathrm{T}}\sigma_2 + E^{\mathrm{T}}\sigma_3 \ge c \tag{141}$$

$$-\left(R+R^{\mathrm{T}}\right)x + A_1^{\mathrm{T}}\sigma_1 + A_2^{\mathrm{T}}\sigma_2 + E^{\mathrm{T}}\sigma_3 \le c + M\mathbf{1}_n - Mv \tag{142}$$

$$x \le Mv \tag{143}$$

$$b_1 - A_1 x \ge 0 \tag{144}$$

$$b_1 - A_1 x \le M\mathbf{1}_{m_1} - M\bar{u} \tag{145}$$

$$Ex \le M\bar{u} \tag{146}$$

$$-M\bar{u} \le \sigma_1 \le M\bar{u}, \tag{147}$$

$$\sigma_3 \le M\mathbf{1}_{m_1} - M\bar{u} \tag{148}$$

$$x, \sigma_3 \ge 0, \bar{u} \in \{0,1\}^{m_1}, v \in \{0,1\}^n$$

Due to Condition 1 for any element $(Ex)_i$ of vector Ex there exists k such that $(Ex)_i = x_k$ and corresponding constraint from (146) becomes

$$x_k \le M\bar{u}_i.$$

Let us consider together the k-th constraint from each of the constraint sets (141)–(143) and the i-th constraint from each of the constraint sets (144)–(148):

$$\left(-\left(R + R^{\mathrm{T}}\right)x + A_1^{\mathrm{T}}\sigma_1 + A_2^{\mathrm{T}}\sigma_2 + E^{\mathrm{T}}\sigma_3\right)_k \ge c_k \tag{149}$$

$$\left(-\left(R + R^{\mathrm{T}}\right)x + A_1^{\mathrm{T}}\sigma_1 + A_2^{\mathrm{T}}\sigma_2 + E^{\mathrm{T}}\sigma_3\right)_k \le c_k + M - Mv_k \tag{150}$$

$$x_k \le Mv_k \tag{151}$$

$$x_k \le M\bar{u}_i$$

$$b_{1i} - a_i^1 x \ge 0$$

$$b_{1i} - a_i^1 x \le M - M\bar{u}_i$$

$$-M\bar{u}_i \le \sigma_{1i} \le M\bar{u}_i,$$

$$\sigma_{3i} \le M - M\bar{u}_i \tag{152}$$

Let us take $\bar{u}_i = v_k = 0$. Then (149)–(152) reduces to

$$\left(-\left(R + R^{\mathrm{T}}\right)x + A_1^{\mathrm{T}}\sigma_1 + A_2^{\mathrm{T}}\sigma_2 + E^{\mathrm{T}}\sigma_3\right)_k \ge c_k \tag{153}$$

$$x_k = 0$$

$$b_{1i} - a_i^1 x \ge 0$$

$$\sigma_{1i} = 0,$$

$$\sigma_{3i} \geq 0 \tag{154}$$

If we take $\bar{u}_i = 0$, $v_k = 1$ then (149)–(152) yields:

$$\left(-\left(R + R^{\mathrm{T}}\right)x + A_1^{\mathrm{T}}\sigma_1 + A_2^{\mathrm{T}}\sigma_2 + E^{\mathrm{T}}\sigma_3\right)_k = c_k \tag{155}$$

$$x_k = 0$$

$$b_{1i} - a_i^1 x \geq 0$$

$$\sigma_{1i} = 0,$$

$$\sigma_{3i} \geq 0 \tag{156}$$

Thus, the set defined by constraints (155)–(156) is a subset of the set, defined by constraints (153)–(154). Similarly one can show that the set, defined by constraints (149)–(152) with $\bar{u}_i = 1$, $v_k = 0$ is a subset of the set, defined by constraints (149)–(152) with $\bar{u}_i = v_k = 1$. Therefore variable v_k is redundant and can be substituted by variable $\bar{u}_i$ in (149)–(151). Suppose now that k is such that $(Ex)_i \neq x_k$ for any i. Then v_k is not redundant and remains in constraints (149)–(151). Let us collect all such v_k in one vector and denote it by $\hat{u}$. Taking $u = (\hat{u}, \bar{u})$ and pruning redundant constraints from (141)–(148) we obtain (41)–(50). ■

Proof of Proposition 2. Let us select an arbitrary i. From (64) follows that $\bar{x}^i \leq \hat{x}^i$. Besides, $\hat{x}^i_j \in V^i$ as the solution of (65). Therefore $\bar{x}^i \in V^i$ due to Condition 2. Since $\hat{x}^i \in X^i$ as the solution of (65) then $F^i_k(\hat{x}^i) \leq 0$. Since $F^i_k(x^i)$ is nondecreasing on V^i and $\bar{x}^i \leq \hat{x}^i$ this yields $F^i_k(\bar{x}^i) \leq 0$ for all k. Thus, $\bar{x}^i$ is feasible for all i.

Suppose now that $\bar{x}$ is not Nash equilibrium. This means that for some i there exists $y^i \in X^i$ such that

$$F_0^i(y^i, z(\bar{x}_y)) > F_0^i(\bar{x}^i, z(\bar{x})), \tag{157}$$

where $\bar{x}_y = (\bar{x}^1, \ldots, \bar{x}^{i-1}, y^i, \bar{x}^{i+1}, \ldots, \bar{x}^N)$. Let us assume that there exist components $\bar{x}^i_j$ and y^i_j of vectors $\bar{x}^i$ and y^i such that $\bar{x}^i_j < y^i_j$ and take $\tilde{y}^i = (y^i_1, \ldots, y^i_{j-1}, \bar{x}^i_j, y^i_{j+1}, \ldots, y^i_n)$, $\tilde{x}_y = (\bar{x}^1, \ldots, \bar{x}^{i-1}, \tilde{y}^i, \bar{x}^{i+1}, \ldots, \bar{x}^N)$. Then $\tilde{y}^i \leq y^i$ componentwise and therefore $y^i \in X^i$ due to Conditions 1, 2. Besides, $z(\bar{x}_y) = z(\tilde{x}_y)$ because $\bar{x}_y$ and $\tilde{x}_y$ differ only in the j-th component of the strategy of agent i which in both cases is not smaller than this component of the strategies of other agents, and the strategies of other agents in both $\bar{x}_y$ and $\tilde{x}_y$ are equal. Thus, Condition 3 yields:

$$F_0^i(y^i, z(\bar{x}_y)) = F_0^i(y^i, z(\tilde{x}_y)) \leq F_0^i(\tilde{y}^i, z(\tilde{x}_y))$$

In other words, by substituting the point y^i by the point $\tilde{y}^i$ that coincides with y^i with the exception of component y^i_j that is changed to $\min\{y^i_j, \bar{x}^i_j\}$ the agent i will obtain the feasible strategy with at least as good value of his objective when the strategies of other agents do not change. Making a similar substitution for all other components $y^i_k : \bar{x}^i_k < y^i_k$ we shall obtain a feasible point $u^i = \min\{y^i, \bar{x}^i\}$ such that for $\bar{x}_u = (\bar{x}^1,..., \bar{x}^{i-1}, u^i, \bar{x}^{i+1},..., \bar{x}^N)$

$$F^i_0(y^i, z(\bar{x}_y)) \le F^i_0(u^i, z(\bar{x}_u))$$

and, consequently,

$$F^i_0(u^i, z(\bar{x}_u)) > F^i_0(\bar{x}^i, z(\bar{x})) \tag{158}$$

due to (157). Then there exist component u^i_j such that $u^i_j < \bar{x}^i_j$, otherwise we would have $u^i_j = \bar{x}^i_j$ and (158) will not be satisfied. Let us take $\tilde{u}^i = (u^i_1 ,..., u^i_{j-1}, \bar{x}^i_j, u^i_{j+1} ,..., u^i_n)$, $\tilde{x}_u = (\bar{x}^1,..., \bar{x}^{i-1}, \tilde{u}^i, \bar{x}^{i+1},..., \bar{x}^N)$. Again, $\tilde{u}^i \in X^i$ due to $\tilde{u}^i \le \bar{x}^i$ and Conditions 1, 2. Due to the definition of operator $z(x)$ from (63) and construction of points u^i, $\bar{x}_u$, $\tilde{x}_u$ we have $z(\bar{x}_u) = u^i$, $z(\tilde{x}_u) = \tilde{u}^i$ and, consequently, $F^i_0(u^i, z(\bar{x}_u)) = F^i_0(u^i, u^i)$, $F^i_0(\tilde{u}^i, z(\tilde{x}_u)) = F^i_0(\tilde{u}^i, \tilde{u}^i)$. In addition, $u^i_j \le \tilde{u}^i_j \le \hat{x}^i_j$. This, together with the unimodality from Condition 4 yields:

$$F^i_0(u^i, z(\bar{x}_u)) = F^i_0(u^i, u^i) \le F^i_0(\tilde{u}^i, \tilde{u}^i) = F^i_0(\tilde{u}^i, z(\tilde{x}_u))$$

In other words, by substituting the point u^i by the point $\tilde{u}^i$ that coincides with u^i with the exception of component $u^i_j < \bar{x}^i_j$ that is changed to $\bar{x}^i_j$ the agent i will obtain the feasible strategy with at least as good value of his objective. Performing this substitution for all components $u_k : \bar{x}^i_k > u_k$ we obtain

$$F^i_0(u^i, z(\bar{x}_u)) \le F^i_0(\bar{x}^i, z(\bar{x}))$$

which contradicts (158). The first assertion of this proposition is proved.

Let us take now the additional assumption of the Proposition and suppose that x' is some Nash equilibrium point that is not described by (64), (65). Let us assume that there exist i and k such that $x'^i_j > x'^k_j$ for some fixed j and define $\tilde{x}^i = (x'^i_1 ,..., x'^i_{j-1}, x'^k_j, x'^i_{j-1} ,..., x'^i_n)$, $\tilde{x} = (x'^1 ,..., x'^{i-1}, \tilde{x}^i, x'^{i+1} ,..., x'^N)$. Since $\tilde{x}^i \le x'^i$ we have $\tilde{x}^i \in X^i$ due to Conditions 1, 2. Due to (63) we have $z_j(\tilde{x}) = z_j(x')$, $z(\tilde{x}) = z(x')$ that together with Condition 3 where we have assumed the strict inequality yields:

$$F^i_0(x'^i, z(x')) < F^i_0(\tilde{x}^i, z(x')) = F^i_0(\tilde{x}^i, z(\tilde{x})).$$

This contradicts with the assumption that x' is the Nash equilibrium because x' and $\tilde{x}$ differ only in the strategies of the i-th actor. Thus, $x'^i_j = x'^k_j = x'_j$ for any Nash equilibrium point x', any i, k and fixed j. Therefore for operator $z(x)$ from (63) we have $z(x') = x'^i$ for any i.

Let us assume that there exists i such that $x'_j > \hat{x}^i_j$ for some j and any $\hat{x}^i$ that solves the problem (65). Selecting $\hat{x}^i$ as in additional assumption and taking $\tilde{x}^i = \left(x_1^{\prime i},...x_{j-1}^{\prime i},\hat{x}^i_j,x_{j+1}^{\prime i},...,x_n^{\prime i}\right)$, $\tilde{x} = \left(x^{\prime 1},...,x^{\prime i-1},\tilde{x}^i,x^{\prime i+1},...,x^{\prime N}\right)$ we obtain that $\tilde{x}^i \in X^i$ due to $\tilde{x}^i \le x^{\prime i}$ and Conditions 1, 2. In addition, $z(\tilde{x}) = \tilde{x}^i$ due to $x'_j > \bar{x}_j$ and definition of operator $z(x)$. Applying the additional assumption we obtain:

$$F_0^i(x',z(x')) = F_0^i(x',x^{\prime i}) < F_0^i(\tilde{x}^i,\tilde{x}^i) = F_0^i(\tilde{x}^i,z(\tilde{x}))$$

which shows again that x' can not be the Nash equilibrium. Thus, all Nash equilibria are described by (64), (65). ■

Proof of Proposition 4. The objective function in (71) is concave, therefore we can use necessary and sufficient conditions for global maximum from Bazaraa and Shetty (2006) to analyze its solution. Since d_j takes values from $[0, d_j^+]$ and has positive bounded density $h_j(d_j)$ on this interval then $H_j(d_j)$ is continuous strictly increasing function on $[0, d_j^+]$ with $H_j(0) = 0$, $H_j(d_j) = 1$. Therefore solution $H_j^{-1}(x)$ of equation $H_j(d_j) = x$ exists and is unique for any x : $0 < x < 1$. Two cases can occur.

1. Constraint (72) is not binding. Then the problem (71), (72) is decomposed into individual problems

$$\max_{x_j \ge 0}\left(a_j \mathbb{E}_{d_j} \min\{d_j,x_j\} - b_j x_j\right) \tag{159}$$

which objective function can be written as follows

$$a_j \int_0^\infty \min\{y,x_j\} h_j(y)dy - b_j x_j = a_j \int_0^{x_j} y h_j(y)dy + a_j x_j \int_{x_j}^\infty h_j(y)dy - b_j x_j.$$

Differentiating this with respect to x_j and equating the derivative with zero we obtain

$$a_j \int_{x_j}^\infty h_j(y)dy - b_j = 0 \tag{160}$$

Due to positivity of $h_j(y)$ for $y \in [0, d_j^+]$ this equation has unique solution $x_j(0)$ from (73). For this solution (74) is satisfied, because otherwise constraint (72) will be binding.

2. Constraint (72) is binding. Then the optimality conditions for the problem (71), (72) yield that in order to solve (71), (72) it is necessary to find solution $x_j(\rho)$ of the problem

$$\max_{x_j \ge 0} \sum_j \left(a_j \mathbb{E}_{d_j} \min\{d_j,x_j\} - (b_j + \rho c_j)x_j\right). \tag{161}$$

and find appropriate ρ by solving

$$\sum_j c_j x_j(\rho) = W. \tag{162}$$

In addition, solution of this equation will exist for $\rho \geq 0$. Similarly to the previous case, the problem (161) decomposes into problems

$$\max_{x_j \geq 0} \left(a_j \mathbb{E}_{d_j} \min\{d_j, x_j\} - (b_j + \rho c_j) x_j \right)$$

which similarly to (159) has solution $x(\rho)$ from (73). It will be unique due to positivity of $h_j(y)$ for $y \in [0, d_j^+]$ and for the same reason the solution of equation (162) will be unique too. ■

CHAPTER 9

Nash Equilibrium and its Modern Applications

*Vasyl Gorbachuk** and *Maxym Dunaievskyi*

1. Introduction

The strategic interaction is a key issue for such areas as international trade, comparative advantage, perfect and imperfect competition, market entry, innovation race, market power, organizational change, and others. It stems from the advanced equilibrium concepts and their applications. The concept of the Cournot–Stackelberg–Nash equilibrium is proposed in which the leader's payoff is no less than its payoff in the case of the Cournot–Nash equilibrium and the follower's payoff is no less than its payoff in the case of the Stackelberg–Nash equilibrium. The generalized equilibrium coincides with the Cournot–Nash and Stackelberg–Nash equilibria for extreme values of a parameter. The impact of a random production output on the expected market supply and price is investigated as well as the impact on the outputs and profits of companies. It is proved that the asymmetry of interactions between decision makers (DMs), in particular, the asymmetry of uncertainty can lead to the advantage of a leader and to generalized Cournot–Stackelberg–Nash equilibria. An DM with a better defined (determined) strategy always wins over an DM with a worse defined strategy. For example, in the interaction between the government and the central bank, it can be expected that the leader in the generalized Cournot–Stackelberg–Nash equilibrium will be the organization with a better defined strategy.

V.M. Glushkov Institute of Cybernetics, National Academy of Sciences of Ukraine.
* Corresponding author: GorbachukVasyl@netscape.net

2. Generalized Cournot–Stackelberg–Nash equilibrium

The concept of economic equilibrium introduced by A. Cournot (1801–1877) is a fundamental scientific concept. J. Nash (1928–2015) generalized it in such a manner (Nash 1950, Nash 1951) that it became one of the key concepts of applied mathematics and tools of modelling a trade cartel (J. Nash, R. Selten (1930–2016), and J. Harsanyi (1920–2000) shared the Nobel Prize in 1994 for their investigations of Cournot–Nash equilibria).

The concept of the Cournot equilibrium was simultaneously developed in the direction of decision theory, in particular, decision-making in the context of asymmetric information. H. von Stackelberg (1905–1946) applied this concept to the interaction between two companies, one of which (the leader) makes a decision earlier, knowing the reaction of the other company (the follower) (Von Stackelberg 1934). The obtained nontrivial result was highly commended by W. Leontief (1905–1999) (Leontief 1936), a Nobel Prize laureate in 1973, and experts in game theory. Economic applications are well known in which decision-makers are a government or a central bank (Ermoliev and Uryasev 1982, Gorbachuk 1991, Gorbachuk 2007a, Gorbachuk 2007b, Gorbachuk 2008a, Gorbachuk 2008b, Gorbachuk and Chumakov 2008, Gorbachuk 2010, Chikrii et al. 2012, Gorbachuk and Chumakov 2011, Gorbachuk 2014, Ermoliev et al. 2019, Gorbachuk et al. 2019, Borodina et al. 2020, Ermolieva et al. 2022, Gorbachuk et al. 2022). This part is an attempt to combine the concepts of the Cournot, Nash, and Stackelberg equilibria into one (Gorbachuk 2006). We assume that a homogeneous product is produced only by companies 1 and 2 and that the function of the inverse demand for this product is specified by the relationship

$$P = a - bQ, \qquad (1)$$

where P is the (positive) price of the product, Q is its quantity (volume) produced by the companies, and a, b are positive parameters (Nagurney 2009). The volume of output of the i-th company is $q_i \geq 0$ and its unit costs are c_i, $i = 1, 2$. Note that we have

$$Q = q_1 + q_2. \qquad (2)$$

The i-th company chooses a volume of output q_i, trying to maximize its (non-negative) payoff

$$\pi_i = Pq_i - c_i q_i \geq 0, \qquad (3)$$

whence, in view of relationships (1) and (2), we have

$$\pi_i = [a - b(q_1 + q_2)]q_i - c_i q_i = [a - c_i - b(q_1 + q_2)]q_i.$$

The Cournot–Nash equilibrium is understood to be a combination of outputs of companies q_1^C and q_2^C such that we have

$$\pi_1^C = \pi_1(q_1^C, q_2^C) \geq \pi_1(q_1, q_2^C) \;\; \forall q_1 \geq 0, \tag{4}$$

$$\pi_2^C = \pi_2(q_1^C, q_2^C) \geq \pi_2(q_1^C, q_2) \;\; \forall q_2 \geq 0. \tag{5}$$

Theorem 1. Under conditions (1)–(3) and also

$$a + c_2 > 2c_1, \tag{6}$$

$$a + c_1 > 2c_2, \tag{7}$$

the combination

$$q_1^* = (a - 2c_1 + c_2)/(3b), \tag{8}$$

$$q_2^* = (a - 2c_2 + c_1)/(3b) \tag{9}$$

is the Cournot–Nash equilibrium.

Proof. Conditions (2), (8), and (9) implies

$$Q(q_1, q_2^*) = q_1 + q_2^* = q_1 + (a - 2c_2 + c_1)/(3b),$$

$$Q(q_1^*, q_2) = q_1^* + q_2 = q_2 + (a - 2c_1 + c_2)/(3b),$$

whence

$$P(q_1, q_2^*) = a - bQ(q_1, q_2^*) = a - b[q_1 + (a - 2c_2 + c_1)/(3b)] = (2a + 2c_2 - c_1)/3 - bq_1,$$

$$P(q_1^*, q_2) = a - bQ(q_1^*, q_2) = a - b[q_1 + (a - 2c_1 + c_2)/(3b)] = (2a + 2c_1 - c_2)/3 - bq_2.$$

Thus, the payoff

$$\pi_1(q_1, q_2^*) = [P(q_1, q_2^*) - c_1]q_1 = [(2a + 2c_2 - c_1)/3 - bq_1 - c_1]q_1 = [2(a + c_2 - 2c_1)/3 - bq_1]q_1$$

is maximized when

$$2(a + c_2 - 2c_1)/3 - 2bq_1 = 0,$$

$$q_1 = (a - 2c_1 + c_2)/(3b) = q_1^*,$$

whence we obtain inequality (4) is true for $q_1^C = q_1^*$ and $q_2^C = q_2^*$. Similarly, the payoff

$$\pi_2(q_1^*, q_2) = [P(q_1^*, q_2) - c_2]q_2 = [(2a + 2c_1 - c_2)/3 - bq_2 - c_2]q_2 = [2(a + c_1 - 2c_2)/3 - bq_2]q_2$$

is maximized when

$$2(a + c_1 - 2c_2)/3 - 2bq_2 = 0,$$

$$q_2 = (a - 2c_2 + c_1)/(3b) = q_2^*,$$

whence it follows that inequality (5) holds for $q_1^C = q_1^*$ and $q_2^C = q_2^*$.

By the Stackelberg–Nash equilibrium we understand a combination of the outputs of the company's q_1^S and q_2^S that is such that we have

$$\pi_1^S = \pi_1(q_1^S, q_2^S) \geq \pi_1(q_1, q_2^S) \quad \forall q_1 \geq 0, \tag{10}$$

where $q_1^S = q_2^S(q_1)$ is determined from

$$\pi_2 = \pi_2(q_1, q_2^S) \geq \pi_1(q_1, q_2) \quad \forall q_1, q_2 \geq 0. \tag{11}$$

In this case, we call company 1 the leader with the strategy q_1 and company 2 the follower with the strategy $q_1(q_1)$.

Theorem 2. Under conditions (1)–(3), (6), and (7), the combination

$$q_1^l = (a - 2c_1 + c_2)/(2b),$$
$$q_1^f = (a + 2c_1 - 3c_2)/(4b), \tag{12}$$

is the Stackelberg–Nash equilibrium.

Proof. Taking into account conditions (1)–(3), the payoff

$$\pi_2(q_1, q_2) = (P - c_2)q_2 = (a - bQ - c_2)q_2 = [a - b(q_1 + q_2) - c_2]q_2$$

is maximized with respect to q_2 when the best response of the follower is as follows:

$$0 = a - b(q_1 + q_2 - c_2 - bq_2^{**}) = a - c_2 - bq_1 - 2bq_2^{**},$$
$$q_2^{**} = (a - c_2 - bq_1)/(2b) = q_2^S(q_1).$$

Note that if $q_1 = q_1^l$, then we have

$$q_2^{**} = [a - c_1 - b(a - 2c_1 + c_2)/(2b)]/(2b) =$$
$$= (2a - 2c_2 - a + 2c_1 - c_2)/(4b) = (a + 2c_1 - 3c_2)/(4b) = q_2^f,$$

i.e., the equalities in conditions of Theorem 2 satisfy the equation above.

It follows from relationships (2), (10)–(12) that we have

$$Q(q_1, q_2^{**}) = q_1 + q_2^{**} = q_1 + (a - c_1 - bq_1)/(2b) = (bq_1 + a - c_1)/(2b),$$

$$P(q_1, q_2^{**}) = a - bQ(q_1, q_2^{**}) = a - (bq_1 + a - c_2)/2 = (a + c_2 - bq_1)/2.$$

Thus, the payoff

$$\pi_1(q_1, q_2^{**}) = [P(q_1, q_2^{**}) - c_1]q_1 = (a + c_2 - bq_1 - 2c_1)q_1/2$$

is maximized when

$$0 = a + c_2 - 2c_1 - 2bq_1^{**},$$
$$q_1^{**} = (a - 2c_1 + c_2)/(2b) = q_1^l.$$

Thus, inequalities (10) and (11) hold when we have $q_1^S = q_1^l$, $q_2^S = q_2^f$.

By virtue of inequality (3), for the parameters $a = 2$, $b = 4$, and $c_1 = c_2 = 1$, we have

$$q_1^l = (2 - 2\times 1 + 1)/(2\times 4) = 1/8,$$

$$q_2^f = (2 + 2\times 1 - 3\times 1)/(4\times 4) = 1/16 = q_2^S(q_1^l).$$

First of all,

$$\pi_2(q_1, q_2^S) = \{1 - 4[q_1 + (1 - 4q_1)/8]\}(1 - 4q_1)/8 = (1 - 4q_1)^2/16,$$

$$\pi_2(q_1, q_2) = [1 - 4(q_1 + q_2)]q_2,$$

$$(1 - 4q_1)^2/16 - (1 - 4q_1)q_2 + 4(q_2)^2 = [(1 - 4q_1)/4 - 2q_2]^2 \geq 0 \ \ \forall q_1, q_2 \geq 0,$$

whence it follows that relationship (11) holds.

Next, we have

$$\pi_1(q_1^l, q_2^f) = [1 - 4(1/8 + 1/16)]/8 = 1/32,$$

$$\pi_1(q_1, q_2^S(q_1)) = \{1 - 4[q_1 + (2 - 1 - 4q_1)/(2\times 4)]q_1 = (1 - 4q_1)q_1/2 \leq 32 \ \ \forall q_1 \geq 0,$$

i.e., the combination of q_1^l and q_2^f satisfies the relationship (10).

Theorem 3. Under conditions (1)–(3), (6), and (7) and also the condition

$$c_1 \geq c_2, \tag{13}$$

we have

$$\pi_1^S > \pi_1^C, \tag{14}$$

$$\pi_2^S < \pi_2^C. \tag{15}$$

Proof. From relationships (2), (8), and (9) we have

$$Q^C = Q(q_1^C, q_2^C) = q_1^C + q_2^C = (2a - c_1 - c_2)/(3b),$$

whence, by virtue of equation (1), we obtain

$$P^C = P(Q^C) = a - bQ^c = (a + c_1 + c_2)/3.$$

Then, according to equality (3), we have

$$\pi_1^C = (a - 2c_1 + c_2)^2/(9b), \tag{16}$$

$$\pi_2^C = (a - 2c_2 + c_1)^2/(9b). \tag{17}$$

From relationships (2), (12), and (13) we have

$$Q^S = Q(q_1^S, q_2^S) = q_1^S + q_2^S = (3a - 2c_1 - c_2)/(9b),$$

whence, by virtue of equation (1), we obtain

$$P^S = P(Q^S) = a - bQ^S = (a + 2c_1 + c_2)/4.$$

Then, according to equality (3), we have

$$\pi_1^S = (a - 2c_1 + c_2)^2 /(8b), \tag{18}$$

$$\pi_2^S = (a + 2c_1 - 3c_2)^2 /(16b). \tag{19}$$

Equations (16) and (18) and condition (6) imply inequality (14).

In order to derive inequality (15), we use equations (17) and (19):

$$\pi_2^S - \pi_2^C = (7a + 10c_1 - 17c_2)(2c_1 - a - c_1)/(144b).$$

Note that, by virtue of condition (6), we have

$$2c_1 - a - c_2 < 0,$$

and, according to condition (7), we have

$$7a + 7c_1 - 14c_2 > 0,$$

whence, in view of condition (13), we obtain

$$7a + 10c_1 - 17c_2 > 0,$$

and this inequality together with the latter equation implies inequality (15).

Lemma 1. If conditions (1)–(3) and the inequality

$$q_1 > 0, \tag{20}$$

are true, then the volume of output q_1 of company 1 is of the form

$$q_1^{Gl} = \gamma(a - c_1)/b, \;\; \gamma \in (0,\, 1). \tag{21}$$

Proof. Based on relationships (1) and (2), we have

$$q_1 = Q - q_2 = (a - P)/b - q_2 \leq (a - P)/b,$$

and, by virtue of inequalities (3) and (20), we obtain

$$P > c_1,$$

$$q_1^{Gl} < (a - c_1)/b.$$

Theorem 4. If conditions (1)–(3) and (21) are true and inequality (13) is satisfied as the equality

$$c = c_1 = c_2, \tag{22}$$

$$q_2^{Gf} = (a - c)(1 - \gamma)/(2b), \tag{23}$$

then the Cournot–Nash equilibrium is obtained when $\gamma = 1/3$ and the Stackelberg–Nash equilibrium is obtained when $\gamma = 1/2$.

Proof. Note that conditions (1)–(3) and (21) imply the condition

$$c_1 < P < a,$$

that, together with equality (22), gives inequalities (6) and (7).
When $\gamma = 1/3$ and equality (22) is true, condition (21) implies the quantity

$$q_1^{Gl} = (a - c)/(3b),$$

that coincides with the quantity q_1^* calculated by the formula (8), and condition (23) implies

$$q_2^{Gf} = (a - c)(1 - 1/3)/(2b),$$

that coincides with the quantity q_2^* calculated by the formula (9).
When $\gamma = 1/2$ and equality (22) holds, condition (21) implies the quantity

$$q_1^{Gl} = (a - c)/(2b),$$

that coincides with the quantity q_1^l calculated by the formula (12) and condition (23) implies the quantity

$$q_2^{Gf} = (a - c)(1 - 1/2)/(2b),$$

that coincides with the quantity q_2^f calculated by the formula (13).

By the generalized Cournot–Stackelberg–Nash equilibrium we understand a combination of outputs of companies q_1^G and q_2^G in which the payoff of the leader is no less than its payoff in the Cournot–Nash equilibrium and the payoff of the follower is no less than its payoff in the Stackelberg–Nash equilibrium,

$$\pi_1^G = \pi_1(q_1^G, q_2^G) \geq \pi_1^C, \tag{24}$$

$$\pi_2^G = \pi_2(q_1^G, q_2^G) \geq \pi_2^S, \tag{25}$$

where $q_2^G = q_2^G(q_1)$ is determined from the inequality

$$\pi_2(q_1, q_2^G) \geq \pi_2(q_1, q_2) \;\; \forall q_1, q_2 \geq 0. \tag{26}$$

Theorem 5. If conditions (1)–(3) and (22) are true, then the following combination of conditions

$$q_1^{Gl} = \gamma(a - c)/b, \;\; \gamma \in (0, 1), \tag{21}$$

$$q_2^{Gf} = (a - c)(1 - \gamma)/(2b) \tag{23}$$

is the generalized Cournot–Stackelberg–Nash equilibrium under the condition

$$\gamma \in [1/3,\ 1/2]. \tag{27}$$

Proof. By conditions (2) and (21)–(23), we have

$$Q^G = q_1^{Gl} + q_2^{Gf} = (1+\gamma)\,(a-c)/(2b),$$

whence, because of relationship (1), we obtain

$$P^G = a - bQ^G = [a + c - \gamma(a-c)]/2. \tag{28}$$

According to conditions (16), (21), (22), and (28), to prove inequality (24), it suffices to prove the inequality

$$(a-c)^2/(9b) \le [a + c - \gamma(a-c)]/2 - c\}\gamma(a-c)/b = \gamma(1-\gamma)(a-c)^2/(2b),$$

$$9\gamma\,(1-\gamma) \ge 2,$$

$$9\gamma^2 - 9\gamma + 2 \le 0,$$

$$(\gamma - 1/3)\,(\gamma - 2/3) \le 0,$$

$$1/3 \le \gamma \le 2/3. \tag{29}$$

By conditions (19), (22), (23), and (28), to prove inequality (25), it suffices to prove the inequality

$$(a-c)^2/(16b) \le \{[a = c - \gamma(a-c)]/2 - c\}(1-\gamma)(a-c)/(2b) = (1-\gamma)^2(a-c)^2/(4b),$$

$$1 - \gamma \ge 1/2,$$

$$\gamma \le 1/2. \tag{30}$$

Inequalities (29) and (30) follow from condition (27). To prove inequality (26), it suffices to show that, when equality (22) holds, the quantity q_2^{**} determined by condition (12) coincides with that determined by condition (23) for the value of q_1 determined by condition (21):

$$q_2^{**} = [a - c - b\gamma(a-c)/b]/(2b) = (1-\gamma)(a-c)/(2b).$$

Thus, for the considered case, it is possible to parametrically relate the Cournot–Nash and Stackelberg–Nash equilibria by a parameter. In a more general case, the problem of parametrization of these equilibria remains open, as well as the parametrization of other well-known equilibria. Such a parameter is not simply a control variable but also the selection of the objective function itself and a way of behavior since a greater value of $\gamma \in [1/3, 1/2]$ corresponds to a greater degree of leadership and a smaller one corresponds to a greater degree of competition (Gorbachuk 2006).

3. An asymmetric Cournot–Nash equilibrium under uncertainty as a generalized Cournot–Stackelberg–Nash equilibrium

In solving economic problems, game theory models and methods, including game theory models under uncertainty, are often used. The importance of uncertainty is obvious after the fall of stocks of Internet companies in 2000, the terrorist attack on September 11, 2001, oil price increase up to a threshold above 70 dollars per barrel, and drastic climatic change. Therefore, in the New Millennium, the need to further develop stochastic optimization methods arises. Here the influence of randomness in a production output on the expected commercial manufacture and price and also on the outputs and profits of companies is investigated (Gorbachuk 2007a).

We assume that a homogeneous product is manufactured by two decision makers (DMs) only, namely, by the 1-st company and the 2-nd company. The situation, in which the values of q_1 and q_2 are deterministic (are not random), was investigated in detail in many works. The randomness of one of these values can influence the strategies of both companies. Let the value of q_2 be deterministic, and let the value of q_1 be uniformly distributed over a (nondegenerate) segment $[d_1, D_1]$, i.e., let that density function value be

$$f(q_1) = \{1/(D_1 - d_1), q_1 \in [d_1, D_1]; 0, q_1 \notin [d_1, D_1]\}.$$

In particular, it is assumed in that the oil price is uniformly distributed between 17 and 51 dollars per barrel (De Wolf and Smeers 1997). The 2-nd company chooses the volume q_2 of its production to maximize the expected value of its profit

$$E(\pi_2) = E[(P - c_2)q_2],$$

whence, because of formulas (1) and (2), we have

$$\begin{aligned} E(\pi_2) &= E\{[a - b(q_1 + q_2)]q_2 - c_2 q_2\} = E\{[a - c_2 - b(q_1 + q_2)]q_2\} \\ &= (a - c_2)q_2 - b(q_2)^2 - bq_2\bar{q}_1, \end{aligned} \tag{31}$$

where the expected value of the quantity q_1 equals

$$\bar{q}_1 = E(q_1) = (d_1 + D_1)/2. \tag{32}$$

If the largest possible expected profit of a company is negative, this company can choose zero output, i.e., it will not bring its product to market. Since the strategy of one DM is deterministic and the strategy of the other DM is random, we operate under asymmetric uncertainty.

The 1-st company chooses the expected volume $\bar{q}_1$ of its output to maximize the expected value of its profit

$$E(\pi_1) = E(Pq_1 - c_1 q_1),$$

whence, because of formulas (1) and (2), we obtain

$$E(\pi_1) = E\{[a - b(q_1 + q_2)]q_1 - c_1 q_1\} = E\{[a - c_1 - b(q_1 + q_2)]\, q_1\}$$
$$= (a - c_1)\bar{q}_1 - bE[(q_1)^2] - bq_2\bar{q}_1, \tag{33}$$

where, by relationship (32), we have

$$E[(q_1)^2] = \int_{-\infty}^{\infty} (q_1)^2 f(q_1)dq_1 = \int_{d_1}^{D_1} (q_1)^2 dq_1 /(D_1 - d_1) = (q_1)^3 \big|_{d_1}^{D_1} /[3(D_1 - d_1)] =$$

$$= [(D_1)^3 - (d_1)^3]/[3(D_1 - d_1)] = [(D_1)^2 + D_1 d_1 + (d_1)^2]/3$$

$$= [(D_1 + d_1)^2 - D_1 d_1]/3 = [4(\bar{q}_1)^2 - D_1 d_1]/3.$$

Thus, we have

$$E(\pi_1) = (a - c_1)\bar{q}_1 - bq_2\bar{q}_1 - b[4(\bar{q}_1)^2 - D_1 d_1]/3.$$

Then, for a random output of the 1-st company (under uncertainty), we call a Cournot–Nash equilibrium a combination of an output q_2^{C1} of the 2-nd company and expected output of the 1-st company

$$\bar{q}_1^{C1} = (d^{C1} + D^{C1})/2$$

such that we have

$$E(\pi_1^{C1}) = E[\pi_1(\bar{q}_1^{C1}, d_1^{C1}, D_1^{C1}, q_2^{C1})] \geq E[\pi_1(\bar{q}_1, d_1, D_1, q_2^{C1})] \quad \forall \bar{q}_1 \geq 0; \tag{34}$$

$$E(\pi_2^{C1}) = E[\pi_2(\bar{q}_1^{C1}, q_2^{C1})] \geq E[\pi_1(\bar{q}_1^{C1}, q_2)] \quad \forall q_2 \geq 0. \tag{35}$$

Next, we assume that the lower bound d_1 for the values of the quantity q_1 is a constant (such a constant can always be, for example, zero). It is natural to assume that the volume of a deterministic Cournot–Nash equilibrium output q_2^{C1} of the 1-st company belongs to the segment $[d_1, D_1]$.

Theorem 6. Given formulas (1), (2), (31)–(33) and also the relationships

$$d_1 \leq q_1^C = (a - 2c_1 + c_2)/(3b), \tag{36}$$

$$a + c_1 \geq 2c_2, \tag{37}$$

$$a \geq c_2 + bd_1, \tag{38}$$

the combination

$$\overline{q}_1^* = [3(a - 2c_1 + c_2) + 4bd]/(13b),$$

$$q_2^* = (5a - 8c_2 + 3c_1 - 2bd)/(13b)$$

is a Cournot–Nash equilibrium for a random output of the 1-st company.

The unstrict inequality (37) corresponds to a strict inequality of Theorem 1 and, when $d_1 > 0$, another strict inequality of Theorem 1 follows from inequality (36).

Let us take advantage of the concavity of the expected profit concerning q_2 to maximize $E(\pi_2)$ with respect to q_2:

$$0 = \partial E(\pi_2)/\partial q_2 = a - c_2 - 2bq_2 - b\overline{q}_1. \tag{39}$$

Let us take advantage of the concavity of the expected profit $E(\pi_1)$ with respect to $\overline{q}_1$ to maximize $E(\pi_1)$ with respect to q_1:

$$0 = \partial E(\pi_1)/\partial\overline{q}_1 = a - c_1 - bq_2 - b[8\overline{q}_1 - \partial(D_1 d_1)/\partial\overline{q}_1]/3. \tag{40}$$

In a view of equality (32) and the deterministic character of d_1, we obtain the equality

$$\partial(D_1 d_1)/\partial\overline{q}_1 = D_1\partial d_1/\partial\overline{q}_1 + d_1\partial D_1/\partial\overline{q}_1 = 2d_1$$

and substitute it into equation (40):

$$0 = a - c_1 - bq_2 - b(8\overline{q}_1 - 2d_1)/3. \tag{41}$$

Equation (39) implies the expression

$$q_2 = (a - c_2 - b\overline{q}_1)/(2b), \tag{42}$$

that is substituted into equality (41):

$$0 = a - c_1 - (a - c_2 - b\overline{q}_1)/2 - b(8\overline{q}_1 - 2d_1)/3,$$

$$0 = 6(a - c_1) - 3(a - c_2 - b\overline{q}_1) - 4b(4\overline{q}_1 - d_1) = 3(a - 2c_1 + c_2) + 4bd_1 - 13b\overline{q}_1$$

$$\overline{q}_1 = [3(a - 2c_1 + c_2) + 4bd_1]/(13b) = \overline{q}_1^{C1}. \tag{43}$$

In a view of equality (32) and the inequality $D_1 \geq d_1$, we have

$$d_1 \leq (d_1 + D_1^C)/2 = [3(a - 2c_1 + c_2) + 4bd_1]/(13b),$$

$$13bd_1 \leq 3(a - 2c_1 + c_2) + 4bd_1,$$

which is guaranteed by condition (36).

From relationships (42) and (43), we finally obtain

$$q_2 = \{a - c_2 - [3(a - 2c_1 + c_2) + 4bd_1]/13\}/(2b) =$$
$$= (13a - 13c_2 - 3a + 6c_1 - 3c_2 - 4bd_1)/(26b) =$$
$$= (10a + 6c_1 - 16c_2 - 4bd_1)/(26b) = (5a + 3c_1 - 8c_2 - 2bd_1)/(13b) = q_2^{C1}.$$

Note that, by inequalities (37) and (38), the inequality $q_2^{C1} \geq 0$ is fulfilled.

Corollary 1. By Theorem 6, we have

$$D_1^C = [6(a - 2c_1 + c_2) - 5bd_1]/(13b).$$

The statement follows from the equalities

$$(d_1 + D_1^C)/2 = \overline{q}_1^{C1} = [3(a - 2c_1 + c_2) + 4bd_1]/(13b).$$

Corollary 2. If condition (36) is satisfied as the equality, we have $D_1 = d_1$, i.e., the output of the 1-st company is deterministic.

Thus, the randomness of the output of the 1st company leads to the change in the Cournot–Nash equilibrium from q_2^C to q_2^{C1} and also from q_1^C to $\overline{q}_1^{C1}$. The answer concerning the direction of these changes is given by the theorem formulated below.

Theorem 7. Under the conditions of Theorem 6, we have

$$\overline{q}_1^{C1} \leq q_1^C, \tag{44}$$

$$q_2^{C1} \geq q_2^C. \tag{45}$$

We have

$$\overline{q}_1^{C1} - q_1^C = [3(a - 2c_1 + c_2 + 4bd_1]/(13b) - (a - 2c_1 + c_2)/(3b) =$$
$$= [9(a - 2c_1 + c_2) + 12bd_1 - 13(a - 2c_1 + c_2)/(39b) =$$
$$= (9a - 18c_1 + 9c_2 + 12bd_1 - 13a + 26c_1 - 13c_2)/(39b) =$$
$$= (-4a + 8c_1 - 4c_2 + 12bd_1)/(39b) = 4(2c_1 + 3bd_1 - a - c_2)/(39b) \leq 0;$$

$$q_2^{C1} - q_2^C = (5a + 3c_1 - 8c_2 - 2bd_1)/(13b) - (a - 2c_2 + c_1)/(3b) =$$
$$= [3(5a + 3c_1 - 8c_2 - 2bd_1)/(13b) - 13(a - 2c_2 + c_1)/(39b) =$$
$$= (15a + 9c_1 - 24c_2 - 6bd_1 - 13a + 26c_2 - 13c_1)/(39b) =$$
$$= (2a - 4c_1 + 2c_2 - 6bd_1)/(39b) = 2(a + c_2 - 2c_1 - 3bd_1)/(39b) \geq 0.$$

Inequalities (36)–(38) mean that the maximally possible price a must be sufficiently high:

$$a \geq \max\{2c_2 - c_1; c_2 + bd_1; 2c_1 - c_2 + 3bd_1\}.$$

If the cost per unit for the 1-st company with a random output is no less than the cost per unit for the 2-nd company with a deterministic output,

$$c_1 \geq c_2, \tag{46}$$

then inequality (36) implies condition (37) of Theorem 6.

Corollary 3. Under conditions (1)–(33), (38), and (46), inequality (36) is the necessary and sufficient condition of inequalities (44) and (45).

Theorem 7 testifies that, under the conditions of Theorem 6, the randomness of q_1 leads to a decrease in the expected equilibrium value of $\bar{q}_1^{C1}$ in comparison with that of q_1^C and also to an increase in q_2^{C1} in comparison with q_2. Let us compare the total equilibrium commercial manufacture Q^C with the total expected equilibrium commercial manufacture,

$$\bar{Q}^{C1} = \bar{q}_1^{C1} + q_2^{C1} = (3a - 6c_1 + 3c_2 + 4bd_1 + 5a + 3c_1 - 8c_2 - 2bd_1)/(13b) =$$
$$= (8a - 3c_1 - 5c_2 + 2bd_1)/(13b).$$

Theorem 8. Under the conditions of Theorem 6, we have $\bar{Q}^{C1} \leq Q^C$.

We have

$$\bar{Q}^{C1} - Q^C = (24a - 9c_1 - 15c_2 + 6bd_1 - 26a + 13c_1 + 13c_2)/(13b) = (-2a + 4c_1 - 2c_2 + 6bd_1)/(13b) \leq 0.$$

Corollary 4. According to the conditions of Theorem 6, we have

$$P^C \leq \bar{P}^{C1} = a - b\bar{Q}^{C1}.$$

By Theorem 8, under the conditions of Theorem 6, the randomness of q_1 leads to a decrease in the expected equilibrium commercial manufacture $\bar{Q}^{C1}$ in comparison with Q^C.

Theorem 9. By Theorem 6, the 2-nd company whose output is deterministic and whose DM knows the expected output of the 1-st company obtains an additional payoff

$$E(\pi_2^{C1}) - \pi_2^C = (G^2 - F^2)/b \geq 0,$$

where

$$F = (a - 2c_2 + c_1)/3,$$

$$G = (5a + 3c_1 - 8c_2 - 2bd_1)/13.$$

The additional payoff of the 2-nd company is specified by the expression

$$(\bar{P}^{C1} - c_2)q_2^{C1} - F^2/b = (5a + 3c_1 + 5c_2 - 2bd_1 - 13c_2)(5a + 3c_1 - 8c_2 - 2bd_1)/(13^2 b) - F^2/b =$$
$$= G^2/b - F^2/b = (G + F)(G - F)/b.$$

Since we have $F \geq 0$, $G \geq 0$, the multiplier $(F + G)$ is nonnegative. The other multiplier is nonnegative by inequality (36):

$$G - F = 15a + 9c_1 - 24c_2 - 6bd_1 - 13a + 26c_2 - 13c_1 =$$
$$= 2a - 4c_1 + 2c_2 - 6bd_1 = 2(a + c_2 - 2c_1 - 3bd_1) \geq 0.$$

As a random output of the 1-st company increases the profit of its competitor (the 2-nd company with a deterministic output), the question about motivations of the 1-st company concerning its random output (the question about changes in the profit of the 1-st company) arises.

Theorem 10. The expected equilibrium profit of the 1-st company under the conditions of Theorem 6 is no more than its profit under the conditions of a Cournot–Nash equilibrium:

$$E(\pi_1^{C1}) \leq \pi_1^C.$$

We denote $H = a - 2c_1 + c_2$. Then, by Corollary 1, we have

$$E(\pi_1^{C1}) = (a - c_1 - bq_2^{C1})\overline{q}_1^{C1} - b[4(\overline{q}_1^{C1})^2 - d_1 D_1^C]/3 =$$
$$(13a - 13c_1 - 5a - 3c_1 + 8c_2 + 2bd_1)(3H + 4bd_1)/(13^2 b) -$$
$$-b[4(3H + 4bd_1)^2/(13b)^2 - (6H - 5bd_1)d_1/(13b)]/3 = 2(4H + bd_1)(3H + 4bd_1)/(13^2 b) -$$
$$-[4(3H + 4bd_1)^2 - 13b(6H - 5bd_1)d_1]/(13^2 b) =$$
$$= [6(4H + bd_1)(3H + 4bd_1) - 4(3H + 4bd_1)^2 + 13bd_1(6H - 5bd_1)]/(13^2 3b) =$$
$$= \{(3H + 4bd_1)[6(4H + bd_1) - 4(3H + 4bd_1)] + 13bd_1(6H - 5bd_1)\}/(13^2 3b) =$$
$$= [2(3H + 4bd_1)(12H + 3bd_1 - 6H + 8bd_1) + 13bd_1(6H - 5bd_1)]/(13^2 3b) =$$
$$= (6H - 5bd_1)(6H + 8bd_1 + 13bd_1)/(13^2 3b) = (6H - 5bd_1)(2H + 7bd_1)/(13^2 b) =$$
$$= [12H^2 + 42bd_1 H - 10bd_1 H - 35(bd_1)^2]/(13^2 b) = [12H^2 + 32bd_1 H - 35(bd_1)^2]/(13^2 b).$$

Thus, we obtain

$$\pi_1^C - E(\pi_1^{C1}) = H^2/(9b) - [12H^2 + 32bd_1 H - 35(bd_1)^2]/(13^2 b) =$$
$$= [169H^2 - 108H^2 - 288bd_1 H + 315(bd_1)^2]/(13^2 9b) = [61H^2 - 288bd_1 H + 315(bd_1)^2]/(13^2 9b).$$

To estimate the sign of the numerator, we find its roots,

$$bd_1 = [288H \pm H(288^2 - 4 \times 61 \times 315)^{0.5}/(2 \times 315) = (288 \pm 78)H/630.$$

Thus, the sign of the numerator coincides with the sign of the expression

$$(bd_1 - 366\ H/630)(bd_1 - 210H/630) = (bd_1 - 61H/105)(bd_1 - H/3).$$

Since both multipliers of the latter expression are nonpositive by the Theorem 10 conditions, this expression is nonnegative. Theorem 10 is proved.

Theorems 9 and 10 assert that the randomness of the output of the 1-st company provides an additional payoff for its competitor (for the 2-nd company with a deterministic output) but not for the 1-st company itself since the determinacy of its output is preferable for its DM than the randomness. This conclusion explains enhanced attention to prediction and planning.

Thus, the asymmetry of uncertainty leads to the leadership of the DM with a deterministic output and the DM with a random output becomes its follower (Gorbachuk 2007b). The DM with a deterministic output does not know the random volume of the output of the other DM, but he knows the expected volume of the random output of the competitor.

Theorem 11. Under the conditions of Theorem 6, a Cournot–Nash equilibrium for a random q_1 is a generalized C0ournot–Stackelberg–Nash equilibrium.

By Theorem 9, a Cournot–Nash equilibrium under uncertainty is a generalized Cournot–Stackelberg–Nash equilibrium if, when $c = c_1 = c_2$, we have

$$E(\pi_1^{C1}) \geq (a-c)^2/(16\text{b}).$$

Let us take into account the equality above,

$$E(\pi_1^{C1}) - H^2/(16b) = [12H^2 + 32bd_1H - 35(bd_1)^2]/(13^2 b) - H^2/(16b) =$$
$$= -[560(bd_1)^2 - 512Hbd_1 - 13H^2]/(13^2 16b).$$

To prove the positivity of the latter expression, we find the roots of the square-law function in the numerator:

$$bd_1 = [512H \pm (512^2H^2 + 4\times 560\times 23H^2)^{0.5}]/(2\times 560) = (512\pm 8\sqrt{4901})H/1120.$$

From this, we obtain that the sign of the numerator coincides with the sign of the product

$$= -[bd_1 - (512+8\sqrt{4901})H/1120][bd_1 - (512-8\sqrt{4901})H/1120],$$

or, with sufficiently high accuracy, with the sign of the following product:

$$= -[bd_1 - (512+8\times 70)\ H/1120][bd_1 - (512-8\times 70)H/1120] =$$
$$= (1078H/1120 - bd_1))(bd_1 + 48H/1120).$$

Given condition (36), this sign is positive, which is what had to be proved.

The results obtained show that the asymmetry of the interaction of DMs, in particular, the asymmetry of uncertainty can provide an advantage for a leader and can lead to generalized Cournot–Stackelberg–Nash equilibria.

4. The Cournot–Nash equilibria under mutual uncertainty

In high-technology industries, producing new products or offering new services, a small number of companies, capable of producing such products or services, are engaged into a market. In addition, markets of new products or services are characterized by uncertainty in demand or output (Gorbachuk 2008b). The importance of uncertainty after the apparent collapse of shares of Internet companies during 2000, terrorist attacks in the United States on September 11, 2001, exceeding a threshold of 130 dollars per barrel by oil prices, and abrupt climate change is obvious. The uncertainty often constrains the business activity of people accustomed to the so-called centralized planning. In this work, the influence of random outputs on the expected market output and price, as well as on the company's outputs and profits, is investigated.

Let us suppose that a homogeneous product is produced only by two DMs – companies 1 and 2. Let the function of inverse demand for this product be given by the equation (1).

The case, when the values q_1 and q_2 are deterministic (i.e., are not random), has been studied in detail. The randomness of one of these values can affect the company's strategy. Let us investigate the situation where both of these values are independent random values. Let us assume q_1 and q_2 that have a uniform distribution on the (nondegenerate) segments $[d_1, D_1]$ and $[d_2, D_2]$ respectively, whereas the density functions are determined by such equations:

$$f(q_1) = \{1/(D_1 - d_1), q_1 \in [d_1, D_1]; 0, q_1 \notin [d_1, D_1]\},$$

$$f(q_2) = \{1/(D_2 - d_2), q_2 \in [d_2, D_2]; 0, q_2 \notin [d_2, D_2]\}.$$

For example, it was assumed that the price of oil is uniformly distributed in the range from 17 to 51 dollars per barrel (by the way, this scenario is of interest today, 35 years later the year of publication by De Wolf and Smeers 1997), but analytical estimations are not obtained.

The expected value of q_1 is

$$\overline{q}_1 = E(q_1) = (d_1 + D_1)/2, \tag{47}$$

and that of q_2 is

$$\overline{q}_2 = E(q_2) = (d_2 + D_2)/2. \tag{48}$$

Company 1 chooses the expected value $\overline{q}_1$ of output, seeking to maximize the expected value of its profit:

$$E(\pi_1) = E(Pq_1 - c_1 q_1) \geq 0.$$

From this expression, taking into account (1) and (2), one gets

$$E(\pi_1) = E\{[a - b(q_1 + q_2)]q_1 - c_1 q_1\} = E\{[a - c_1 - b(q_1 + q_2)]q_1\} =$$
$$= (a - c_1)\overline{q}_1 - bE[(q_1)^2] - bq_2\overline{q}_1,$$

where due to relationship (47)

$$E[(q_1)^2] = \int_{-\infty}^{\infty} (q_1)^2 f(q_1)dq_1 = \int_{d_1}^{D_1} (q_1)^2 dq_1 /(D_1 - d_1) = (q_1)^3 |_{d_1}^{D_1} /[3(D_1 - d_1)] =$$
$$= [(D_1)^3 - (d_1)^3]/[3(D_1 - d_1)] = [(D_1)^2 + D_1 d_1 + (d_1)^2]/3 =$$
$$= [(D_1)^2 + 2D_1 d_1 + (d_1)^2 - D_1 d_1]/3 = [(D_1 + d_1)^2 - D_1 d_1]/3 = [4(\overline{q}_1)^2 - D_1 d_1]/3.$$

Due to the independence of random variables q_1 and q_2

$$E(q_1 q_2) = E(q_1)E(q_2) = \overline{q}_1\overline{q}_2 .$$

Thus,

$$E(\pi_1) = (a - c_1)\overline{q}_1 - b\overline{q}_1\overline{q}_2 - b[4(\overline{q}_1)^2 - D_1 d_1]/3. \tag{49}$$

Similarly, taking into account the symmetry,

$$E(\pi_2) = (a - c_2)\overline{q}_2 - b\overline{q}_2\overline{q}_1 - b[4(\overline{q}_2)^2 - D_2 d_2]/3. \tag{50}$$

If the highest possible expected profit of the company is negative, then the company may choose zero output, that is not to enter the market. Since the strategies of both DMs are random, then there is a mutual (symmetric) uncertainty.

In this case, while random outputs of both companies (in conditions of mutual uncertainty), the combination of the expected outputs

$$\overline{q}_1^C = (d_1^C + D_1^C)/2,$$

$$\overline{q}_2^C = (d_2^C + D_2^C)/2,$$

that

$$E(\pi_1)^C \equiv E[\pi_1(\overline{q}_1^C, d_1^C, D_1^C, d_2^C, D_2^C, \overline{q}_2^C)] \geq E[\pi_1(\overline{q}_1, d_1, D_1, d_2^C, D_2^C, \overline{q}_2^C)]$$

for any nonnegative $\overline{q}_1$, as well as

$$E(\pi_2)^C \equiv E[\pi_1(\overline{q}_1^C, d_1^C, D_1^C, d_2^C, D_2^C, \overline{q}_2^C)] \geq E[\pi_2(\overline{q}_1^C, d_1^C, D_1^C, d_2, D_2, \overline{q}_2)]$$

for any nonnegative $\overline{q}_2$, is called the Cournot–Nash equilibrium.

Next, assume the lower limit d_i for the values of magnitude q_i is some constant (for example, zero can always be such constant), $i = 1, 2$. It is natural to assume that the value of the determined equilibrium by Cournot–Nash output q_i^C of the company i belongs to the segment $[d_i, D_i]$, $i = 1, 2$. Then the market output of the companies under uncertainty is not less than the sum of the lower boundaries of random outputs of these companies;

$$Q^C \equiv q_1^C + q_2^C \geq d_1 + d_2.$$

Theorem 12. Under conditions (1), (2), (47)–(50), as well as while

$$d_1 \leq q_1^C = (a - 2c_1 + c_2)/(3b), \tag{51}$$

$$d_2 \leq q_2^C = (a - 2c_2 + c_1)/(3b), \tag{52}$$

the combination

$$\overline{q}_1^* = [3(5a + 3c_2 - 8c_1) + 2b(8d_1 - 3d_2)]/(55b), \tag{53}$$

$$\overline{q}_2^* = [3(5a + 3c_1 - 8c_2) + 2b(8d_2 - 3d_1)]/(55b), \tag{54}$$

is a Cournot–Nash equilibrium under mutual uncertainty.

Proof. For $d_i > 0$, $i = 1, 2$, the inequalities (51) and (52) yield strict inequalities (6) and (7) from the work (Gorbachuk 2006) respectively.

Since the function of expected profit $E(\pi_1)$ is concave by $\overline{q}_1$, maximization of $E(\pi_1)$ by $\overline{q}_1$ gives

$$0 = \partial E(\pi_1)/\partial \overline{q}_1 = a - c_1 - b\overline{q}_2 - 2b(4\overline{q}_1 - d_1)/3, \tag{55}$$

where the equality (47) and fixity of d_1 are taken into account:

$$\partial(D_1 d_1)/\partial \overline{q}_1 = D_1 \partial d_1/\partial \overline{q}_1 + d_1 \partial D_1/\partial \overline{q}_1 = 2d_1.$$

Similarly, because of the symmetry

$$0 = \partial E(\pi_2)/\partial \overline{q}_2 = a - c_2 - b\overline{q}_1 - 2b(4\overline{q}_2 - d_2)/3,$$

where from we get

$$\overline{q}_1 = [3(a - c_2) - 2b(4\overline{q}_2 - d_2)]/(3b).$$

Substitute the last expression into equality (55):

$$0 = 3(a - c_1 - b\overline{q}_2) - 2b(4\overline{q}_1 - d_1) = 3(a - c_1 - b\overline{q}_2) -$$

$$-2b\{4[3(a - c_2) - 2b(4\overline{q}_2 - d_2)]/(3b) - d_1\},$$

$$0 = 9(a - c_1 - b\overline{q}_2) - 2[12(a - c_2) - 8b(4\overline{q}_2 - d_2) - 3bd_1] =$$

$$= 9(a - c_1) - 9b\overline{q}_2 - 24(a - c_2) + 64b\overline{q}_2 - 16bd_2 + 6bd_1,$$

$$55b\overline{q}_2 = 15a + 9c_1 - 24c_2 + 2b(8d_2 - 3d_1). \tag{56}$$

The latter equation yields the relationship (54).

Note that in the light of conditions (51), (52) we have

$$2a - 4c_1 + 2c_2 - 6bd_1 \geq 0,$$

$$13a + 13c_1 - 26c_2 + 16bd_2 \geq 0$$

respectively. The sum of these two inequalities shows that the right side of equation (56) is nonnegative, i.e., $\overline{q}_2^*$ is nonnegative.

Similarly, we get the relationship (53). Theorem 12 is proved.

Corollary 5. Under the conditions of Theorem 12

$$D_1^C = [(5a + 3c_2 - 8c_1) - b(23d_1 + 12d_2)]/(55b),$$

$$D_2^C = [(5a + 3c_1 - 8c_2) - b(23d_2 + 12d_1)]/(55b).$$

that follows from the equality

$$(d_1 + D_1^C)/2 = \overline{q}_1^C = [3(5a + 3c_2 - 8c_1) + 2b(8d_1 - 3d_2)]/(55b),$$

$$(d_2 + D_2^C)/2 = \overline{q}_2^C = [3(5a + 3c_1 - 8c_2) + 2b(8d_2 - 3d1)]/(55b).$$

Corollary 6. If conditions (51) and (52) become equalities, then $D_1^C = d_1$ and $D_2^C = d_2$, that is, the outputs of both companies are determinate.

So, the randomness of the outputs of companies leads to changes in the Cournot–Nash equilibrium from q_1^C (Gorbachuk 2007a) to $\overline{q}_1^C$ and from q_2^C to $\overline{q}_2^C$. The direction of these changes is determined by the following theorem.

Theorem 13. Let the conditions of Theorem 12 be met. If the inequality

$$c_2 - c_1 + b(d_2 - d_1) \geq 0, \tag{57}$$

holds, then

$$\overline{q}_2^C \leq q_1^C; \tag{58}$$

if the inequality (57) does not hold, then

$$\overline{q}_2^C < q_2^C. \tag{59}$$

Proof. Really,

$$q_1^C - \overline{q}_1^C = (a - 2c_1 + c_2)/(3b) - [3(5a + 3c_2 - 8c_1) + 2b(8d_1 - 3d_2)]/(55b) =$$
$$= [55(a - 2c_1 + c_2) - 9(5a + 3c_2 - 8c_1) - 6b(8d_1 - 3d_2)]/(165b) =$$
$$= (55a - 110c_1 + 55c_2 - 45a - 27c_2 + 72c_1 - 48bd_1 + 18bd_2)/(165b) =$$
$$= (10a - 38c_1 + 28c_2 - 48bd_1 + 18bd_2)/(165b). \tag{60}$$

From the inequality (51) one gets

$$10a - 20c_1 + 10c_2 - 30bd_1 \geq 0,$$

and from the inequality (57) –

$$18c_2 - 18c_1 + 18bd_2 - 18bd_1 \geq 0.$$

The value of the sum of the latter two inequalities proves the nonnegativeness of the numerator in expression (60), which yields the inequality (58).

Similarly, we find

$$q_2^C - \overline{q}_2^C = (a - 2c_2 + c_1)/(3b) - [3(5a + 3c_1 - 8c_2) + 2b(8d_2 - 3d_1)]/(55b) =$$
$$= (10a - 38c_2 + 28c_1 - 48bd_2 + 18bd_1)/(165b). \tag{61}$$

From the inequality (52) one gets

$$10a - 20c_2 + 10c_1 - 30bd_2 \geq 0,$$

and from the negation of inequality (57) –

$$-18c_2 + 18c_1 - 18bd_2 + 18bd_1 > 0.$$

The sum of the latter two inequalities shows the nonnegativeness of the numerator in expression (61); therefore, inequality (59) holds. Theorem 13 is proved.

Note that condition (57) is fulfilled for

$$c = c_1 = c_2, \tag{62}$$

$$d_1 = d_2. \tag{63}$$

The condition (57) means that if the product cost of company 1 is higher, then the lower limit of a random output of this company is to be lower:

$$d_2 - d_1 \geq (c_1 - c_2)/b.$$

Lemma 2. Let the conditions of Theorem 12 and inequality (10) of Theorem 1 from the work (Gorbachuk 2007a)

$$a \geq c_2 + bd_1. \tag{64}$$

is met. Then, if the inequality

$$a \geq c_1 + b(2d_1 + d_2), \tag{65}$$

is fulfilled, then the upper limit D_1^C of the random output of company 1 under mutual uncertainty is not less than the upper limit D_1 of the output of this company under asymmetric uncertainty (Gorbachuk 2007a):

$$D_1^C = [6(5a + 3c_2 - 8c_1) - b(23d_1 + 12d_2)]/(55b) \geq$$
$$\geq [6(a - 2c_1 + c_2) - 5bd_1]/(13b) = D_1 .$$

Proof. Really,

$$715b(D_1^C - D_1) = 13[6(5a + 3c_2 - 8c_1) - b(23d_1 + 12d_2)] -$$
$$-55[6(a - 2c_1 + c_2) - 5bd_1] =$$
$$= 390a + 234c_2 - 624c_1 - 299bd_1 - 156bd_2 - 330a + 660c_1 - 330c_2 + 275bd_1 =$$
$$= 60a - 96c_2 + 36c_1 - 24bd_1 - 156bd_2 =$$
$$= 12(5a - 8c_2 + 3c_1 - 2bd_1 - 13bd_2) . \tag{66}$$

Condition (52) yields the inequality

$$4a - 8c_2 + 4c_1 - 12bd_2 \geq 0,$$

adding it up with inequality (65), we get the nonnegativeness of the numerator in expression (66). Lemma 2 is proven.

Note that for relations (62), (63) condition (65) follows from inequality (51) or (52).

Corollary 7. Under the conditions of Theorem 3 and relations (64), (65) the expected output $\overline{q}_1^C$ of company 1 under mutual uncertainty is not less than the expected output $\overline{q}_1^{C1}$ (Gorbachuk 2007a) under asymmetric uncertainty.

Corollary 7 yields that the degree of uncertainty of output of corporation 1 orders equilibrium values of its outputs:

$$q_1^C \geq \overline{q}_1^C \geq \overline{q}_1^{C1}.$$

Conditions (51), (52), (64), (65) indicate that the maximum possible price a should be large enough:

$$a \geq \max\{2c_1 - c_2 + 3bd_1; 2c_2 - c_1 + 3bd_2; c_2 + bd_1; c_1 + b(2d_1 + d_2)\}.$$

Theorem 14. Under the conditions of Theorem 12 the expected market output

$$\overline{Q}^C \equiv \overline{q}_1^C + \overline{q}_2^C$$

in the case of mutual uncertainty is not greater than the market output Q^C (Gorbachuk 2006) while the Cournot–Nash equilibrium.

Proof. Indeed, because of equations (53) and (54)

$$\bar{Q}^C = [3(5a + 3c_2 - 8c_1) + 2b(8d_1 - 3d_2) + 3(5a + 3c_1 - 8c_2) +$$
$$+2b(8d_2 - 3d_1)]/(55b) =$$
$$= [(30a - 15c_1 - 15c_2 + 10b(d_1 + d_2)]/(55b) = [3(2a - c_1 - c_2) + 2b(d_1 + d_2)]/(11b)\cdot$$

Furthermore,

$$Q^C - \bar{Q}^C = (2a - c_1 - c_2)/(3b) - [2(2a - c_1 - c_2) + 2b(d_1 + d_2)]/(11b) =$$
$$= [11(2a - c_1 - c_2) - 9(2a - c_1 - c_2) - 6b(d_1 + d_2)]/(33b) =$$
$$= 2(2a - c_1 - c_2 - 3bd_1 - 3bd_2)/(33b).$$

Adding up inequalities (51) and (52), we conclude that the numerator in the latter expression is nonnegative. Theorem 14 is proved.

Theorem 14 implies that mutual uncertainty results in reducing the expected market output.

Theorem 15. Under conditions of Theorem 12 and fulfilment of inequalities (64), (65) the expected market output $\bar{Q}^C$ under mutual uncertainty is not greater than the expected market output

$$\bar{Q}^{C1} = (8a - 3c_1 - 5c_2 + 2bd_1)/(13b)$$

under asymmetric uncertainty (Gorbachuk 2007a).

Proof. Really.

$$\bar{Q}^C - \bar{Q}^{C1} = [3(2a - c_1 - c_2) + 2b(d_1 + d_2)]/(11b) - (8a - 3c_1 - 5c_2 + 2bd_1)/(13b) =$$
$$= [39(2a - c_1 - c_2) + 26b(d_1 + d_2) - 11(8a - 3c_1 - 5c_2 + 2bd_1)]/(143b) =$$
$$= (78a - 39c_1 - 39c_2 + 26bd_1 + 26bd_2 - 88a + 33c_1 + 55c_2 - 22bd_1)/(143b) =$$
$$= -2(5a + 3c_1 - 8c_2 - 2bd_1 - 13bd_2)/(143b). \quad (67)$$

By the condition (52) let us write the inequality

$$4a - 8c_2 + 4c_1 - 12bd_2 \geq 0.$$

Adding it with the inequality (65), we get the negativeness of the numerator in expression (67). Theorem 15 is proved.

Theorem 15 implies that the equilibrium values of market outputs depend upon the degree of determinacy of the outputs of companies:

$$Q^C \geq \bar{Q}^{C1} \geq \bar{Q}^C.$$

Corollary 8. Under the conditions of Theorem 15 we have

$$P^C \le \overline{P}^{C1} \le \overline{P}^C \equiv a - b\overline{Q}^C = a - b[3(2a - c_1 - c_2) + 2b(d_1 + d_2)]/(11b) =$$
$$= [5a + 3c_1 + 3c_2 - 2b(d_1 + d_2)]/11.$$

Theorem 16. If the conditions of Theorem 12, conditions (62) are fulfilled, and

$$\Delta \equiv d_1 - d_2 \ge 5(a-c)(\sqrt{183} - 12)/(39b) > 0.195(a-c)/b, \tag{68}$$

then

$$E(\pi_1)^C \ge \pi_1^C.$$

Proof. Note that inequality (68) implies the negation of inequality (57). From equation (49) we find

$$E(\pi_1)^C = (a-c)\overline{q}_1^C - b\overline{q}_2^C\overline{q}_1^C - b(\overline{q}_1^C)^2 + b[D_1^C d_1 - (\overline{q}_1^C)^2]/3 =$$
$$= (\overline{P}^C - c)\overline{q}_1^C + b[D_1^C d_1 - (\overline{q}_1^C)^2]/3. \tag{69}$$

On account of corollary 8

$$\overline{P}^C - c = [5a + 6c - 2b(d_1 + d_2) - 11c]/11 = [5(a-c) - 2b(d_1 + d_2)]/11,$$

where from, taking into account equation (53), we get

$$(\overline{P}^C - c)\overline{q}_1^C = [5(a-c) - 2b(d_1 + d_2)][15(a-c) + 2b(8d_1 - 3d_2)]/(11^2 5b). \tag{70}$$

According to equation (53) and corollary 5,

$$(\overline{q}_1^C)^2 - D_1^C d_1 = [15(a-c) + 2b(8d_1 - 3d_2)]^2/(55^2 b^2) - d_1[30(a-c) - b(23d_1 + 12d_2)]/(55b) =$$
$$= \{[15(a-c) + 2b(8d_1 - 3d_2)]^2 - 55bd_1[30(a-c) - b(23d_1 + 12d_2)]\}/(11^2 5^2 b^2). \tag{71}$$

Substitute relationships (70) and (71) into equality (69):

$$E(\pi_1)^C = \{15[5(a-c) - 2b(d_1 + d_2)][15(a-c) + 2b(8d_1 - 3d_2)] -$$
$$-[15(a-c) + 2b(8d_1 - 3d_2)]^2 + 55bd_1[30(a-c) - b(23d_1 + 12d_2)]\}/(11^2 5^2 3b) =$$
$$= \{[15(a-c) + 2b(8d_1 - 3d_2)]\{15[5(a-c) - 2b(d_1 + d_2)] -$$
$$-[15(a-c) + 2b(8d_1 - 3d_2)]\} + 55bd_1[30(a-c) - b(23d_1 + 12d_2)]\}/(11^2 5^2 3b) =$$
$$= \{[15(a-c) + 2b(8d_1 - 3d_2)][75(a-c) - 30b(d_1 + d_2)] -$$
$$-15(a-c) - 2b(8d_1 - 3d_2)] + 55bd_1[30(a-c) - b(23d_1 + 12d_2)]\}/(11^2 5^2 3b) =$$
$$= [30(a-c) - b(23d_1 + 12d_2)]\{2[15(a-c) + 2b(8d_1 - 3d_2)] + 55bd_1\}/(11^2 5^2 3b) =$$
$$= [30(a-c) - 12bd_2 - 23bd_1][30(a-c) - 12bd_2 + 87bd_1]/(11^2 5^2 3b). \tag{72}$$

The nonnegativeness of $E(\pi_1^C)$ is guaranteed by conditions (51) and (52). Comparing the latter expression with the equality

$$\pi_1^C = (a-c)^2/(3^2 b),$$

we get

$$11^2 5^2 3^2 b[E(\pi_1)^C - \pi_1^C] =$$

$$= 3[30(a-c) - 12bd_2 - 23bd_1][30(a-c) - 12bd_2 + 87bd_1] - 3025(a-c)^2 =$$

$$= 3[30(a-c) + 12b\Delta - 35bd_1][30(a-c) + 12b\Delta + 75bd_1] - 3025(a-c)^2 =$$

$$= -325(a-c)^2 + 1080(a-c)b\Delta + 6750(a-c)bd_1 + 1080(a-c)b\Delta + 432(b\Delta)^2 +$$

$$+2700(bd_1)(b\Delta) - 3150(a-c)bd_1 - 1260(bd_1)(b\Delta) - 7875(bd_1)^2 = -325(a-c)^2 +$$

$$+2160(a-c)b\Delta + 3600(a-c)bd_1 + 1440bd_1b\Delta + 432(b\Delta)^2 - 7875(bd_1)^2,$$

where from, taking into account the inequality

$$b\Delta \le bd_1 \le (a-c)/3, \tag{73}$$

it follows

$$11^2 5^2 3^2 b[E(\pi_1)^C - \pi_1^C] \ge$$

$$\ge -325(a-c)^2 + 2160(a-c)b\Delta + 3600(a-c)b\Delta + 1440(b\Delta)^2 +$$

$$+432(b\Delta)^2 - 7875(a-c)^2/9 = -1200(a-c)^2 + 5760(a-c)b\Delta + 1872(b\Delta)^2 =$$

$$= 48[-25(a-c)^2 + 120(a-c)b\Delta + 39(b\Delta)^2]. \tag{74}$$

To estimate the sign of the latter expression, we will find its roots:

$$b\Delta = \{-120(a-c) \pm [120^2(a-c)^2 + 3900(a-c)^2]^{0.5}\}/78 =$$

$$= (a-c)(-120 \pm 10\sqrt{183})/78 = 5(a-c)(\pm\sqrt{183} - 12)/39.$$

The sign of expression (74) coincides with the sign of product

$$[b\Delta - 5(a-c)(\sqrt{183} - 12)/39][b\Delta + 5(a-c)(\sqrt{183} + 12)/39],$$

and for nonnegativity of expression (68), it is sufficient the fulfilment of inequality (74). Theorem 16 is proved.

Theorem 17. Under the conditions of Theorem 16 the Cournot–Nash equilibrium under mutual uncertainty is a generalized Cournot–Stackelberg–Nash equilibrium.

Proof. Theorem 16 implies that the Cournot–Nash equilibrium under mutual uncertainty is a generalized Cournot–Stackelberg–Nash equilibrium (Gorbachuk 2006) if

$$E(\pi_2)^C \geq (a-c)^2/(16b).$$

Because of symmetry

$$E(\pi_2)^C = [30(a-c) - 12bd_1 - 23bd_2][30(a-c) - 12bd_1 + 87bd_2]/(11^2 5^2 3b) =$$

$$= [30(a-c) + 23b\Delta - 35bd_1][30(a-c) - 87b\Delta + 75bd_1]/(11^2 5^2 3b).$$

From this expression, taking into account inequalities (73), we get:

$$16bE(\pi_2)^C - 11^2 5^2 3b(a-c)^2 =$$

$$= 16b[900(a-c)^2 - 2610(a-c)b\Delta + 2250(a-c)bd_1 + 690(a-c)b\Delta - 2001(b\Delta)^2 +$$

$$+1725bd_1b\Delta - 1050(a-c)bd_1 + 3045bd_1b\Delta - 2625(bd_1)^2] - 9075b(a-c)^2 =$$

$$= 16b[900(a-c)^2 - 1920(a-c)b\Delta + 1200(a-c)bd_1 + 4770bd_1b\Delta -$$

$$-2001(b\Delta)^2 - 2625(bd_1)^2] - 9075b(a-c)^2 \geq$$

$$\geq 16b[900(a-c)^2 - 1920(a-c)b\Delta + 1200(a-c)b\Delta + 4770(b\Delta)^2 -$$

$$-2001(b\Delta)^2 - 2625(a-c)^2/9] - 9075b(a-c)^2 =$$

$$= 16b[(900 - 875/3)(a-c)^2 - 720(a-c)b\Delta + 2769(b\Delta)^2] - 9075b(a-c)^2 =$$

$$= b[(a-c)^2 1975/3 - 4^2 720(a-c)b\Delta + 4^2 2769(b\Delta)^2]. \qquad (75)$$

To estimate the sign of the latter expression, let us find the roots of its quadratic function:

$$b\Delta = \{4^2 720(a-c) \pm [4^4 720^2 (a-c)^2 -$$

$$-4^3 1975 \times 2759(a-c)^2/3]^{0.5}\}/(2 \times 4^2 2769) =$$

$$= (a-c)^2[4^2 720 \pm 8(144^2 5^2 4 - 5^2 79 \times 923)^{0.5}]/(2^5 2769) =$$

$$= (a-c)(4^2 720 \pm 40\sqrt{10027})(2^5 2769) = (a-c)(1440 \pm 5\sqrt{10027})/(2^2 2769).$$

It implies that the sign of expression (75) coincides with the sign of product

$$[b\Delta - (a-c)(1440 + 5\sqrt{10027})(2^2 2769)][b\Delta - (a-c)(1440 - 5\sqrt{10027})/(2^2 2769)],$$

or, with a very high probability, with the sign of a product

$$[b\Delta-(a-c)(1440+500)(2^2 2769)][b\Delta-(a-c)(1440-500)/(2^2 2769)]=$$

$$=[b\Delta-(a-c)1940/(2^2 2769)][b\Delta-(a-c)940/(2^2 2769)]=$$

$$=[b\Delta-(a-c)485/2769][b\Delta-(a-c)235/2769].$$

Given condition (68) this sign is positive. Theorem 17 is proved.

Example 1. The values

$$d_1 = 0.2(a-c)/b,\ d_2 = 0, \tag{76}$$

satisfy the conditions (51), (52), (68). Based on Theorem 12

$$\overline{q}_1^C = 18.2(a-c)/(55b), \tag{77}$$

$$\overline{q}_2^C = 13.8(a-c)/(55b) < (a-c)/(3b) = q_2^C, \tag{78}$$

While equalities (62) and (76) hold true, inequalities (57) and (59) do hold true by theorem 13.

Taking into account corollary 5, we get

$$D_1^C = 25.4(a-c)/(55b), \tag{79}$$

$$D_2^C = 27.6(a-c)/(55b) > D_1^C,$$

and corollary 8 stipulates that the equality

$$\overline{P}^C = (4.6a+6.4c)/11. \tag{80}$$

While equalities (69), (76), (77), (79), (80) are fulfilled,

$$E(\pi_1)^C = (\overline{P}^C - c)\overline{q}_1^C + b[D_1^C d_1 - (\overline{q}_1^C)^2]/3 =$$

$$=(4.6a+6.4c-11c)25.4(a-c)/(55b11)+$$

$$+b[25.6(a-c)0.2(a-c)/(55b^2)-18.2^2(a-c)^2/(55^2b^2]/3=$$

$$=4.6(a-c)^2 25.4/(55b11)+[55(a-c)25.6(a-c)0.2-18.2^2(a-c)^2]/(55^2 3b)=$$

$$=[15(a-c)^2 116.84+11(a-c)^2 25.6-18.2^2(a-c)^2]/(55^2 3b)=$$

$$=[1752.6(a-c)^2+281.6(a-c)^2-331.24(a-c)^2]/(9075b)=$$

$$=1702.96(a-c)^2/(9075b) > 0.187(a-c)^2/b > (a-c)^2/(9b) = \pi_1^C,$$

which confirms the result of Theorem 16.

Similarly, because of symmetry

$$E(\pi_2)^C = (\overline{P}^C - c)\overline{q}_2^C + b[D_2^C d_2 - (\overline{q}_2^C)^2]/3,$$

where from, taking into account equalities (76), (78), (80), we get

$$E(\pi_2)^C = (4.6a + 6.4c - 11c)13.8(a-c)/(55b11) - 13.8^2(a-c)^2/(55^2 3b) =$$

$$= [15(a-c)4.6(a-c)13.8 - 13.8^2(a-c)^2]/(55^2 3b) =$$

$$= 13.8(a-c)^2(69 - 13.8)/(55^2 3b) =$$

$$= 761.76(a-c)^2/(9075b) > 0.083(a-c)^2/b > (a-c)^2/(16b) = \pi_2^S.$$

Thus, the result of Theorem 17 is also confirmed.

Lemma 3. Under the conditions of Theorem 12 and fulfilment of equality (62) negation of inequality (57) results in

$$D_2^C - D_1^C = (d_1 - d_2)/5 > 0.$$

Indeed, because of corollary 5 and equality (62)

$$D_1^C = [30(a-c) - b(23d_1 + 12d_2)]/(55b),$$

$$D_2^C = [30(a-c) - b(23d_2 + 12d_1)]/(55b),$$

$$D_2^C - D_1^C = (23d_1 + 12d_2 - 23d_2 - 12d_1)/55 = (d_1 - d_2)/5 > 0.$$

Theorem 18. Suppose the value q_i has a discrete distribution with the average

$$\overline{q}_i = p_i d_i + (1 - p_i)D_i, \ i = 1, 2,$$

where p_i is a given probability on the interval [0, 1]. Let equality (62) and the inequalities

$$2bp_i d_i < (a-c)(1+p_i) + bp_j d_j(1-p_i), \ i, j = 1, 2, \ i \neq j,$$

be fulfilled. Then for $i, j = 1, 2, i \neq j$,

$$D^C = [(1-p_j)(bp_i d_i - a + c) + 2(a - c - bp_j d_j)]/\{[4 - (1-p_i)(1-p_j)]b\}.$$

Proof of Theorem 18 is similar to the proof of Theorem 12. It is not difficult to see that under the conditions of Theorem 18 the assertion of Lemma 3 that $D_2^C \geq D_1^C$ remains valid.

Lemma 4. While the fulfilment of the conditions of Theorem 18 and inequalities

$$d_2 \leq d_1 \leq (a-c)/(2b)$$

the inequality $D_2^C \geq D_1^C$ holds.

In general, the case, to obtain estimates computer modeling is used (Tramontana et al. 2011). Analytical estimates are usually obtained for certain types of distributions (Prokopovych and Yannelis 2014).

Thus, when the unit costs of companies for output are the same, then, while the Cournot–Nash equilibrium, they have the same outputs and profits. If the output of one of the companies is random in the class of uniform distributions, then, while the generalized Cournot–Stackelberg–Nash equilibrium, this company, for rather general conditions, becomes a follower, a company with determinate output – a leader. If the output of each company is random, then, while the generalized Cournot–Stackelberg–Nash equilibrium, a company with a wider range of random values of outputs (under condition (68)) is a follower, a company with a narrower range of random values of outputs – a leader. So, a company with a more definite output in all the cases receives priority over the company with less certain outputs (Gorbachuk 2008b).

Similarly, while the interaction of government and the central bank it can be expected that a leader while the generalized Cournot–Stackelberg–Nash equilibrium will be an organization with a more definite strategy.

References

Borodina, O.M., S.V. Kyryziuk, O.V. Fraier, Y.M. Ermoliev, T.Y. Ermolieva, P.S. Knopov and V.M. Horbachuk. 2020. Mathematical modeling of agricultural crop diversification in Ukraine: Scientific approaches and empirical results. Cybernetics and Systems Analysis, 56(2): 213–222.

Chikrii, A., N. Denisova, V. Gorbachuk, K. Gromaszek, Y. Krivonos, V. Lytvynenko, I. Matychyn, V. Osypenko, S. Smailova and W. Wojcik. 2012. Current problems in information and computational technologies. Wojcik, W. and J. Sikora (eds.). V. 2, Lublin: Politechnika Lubelska.

De Wolf, D. and Y. Smeers. 1997. A stochastic version of a Stackelberg–Nash–Cournot model. Management Science, 43(2): 190–197.

Ermoliev, Yu.M. and S.P. Uryasev. 1982. Nash equilibrium in n-person games. Cybernetics and Systems Analysis, 18(3): 367–372.

Ermoliev, Yu., T. Ermolieva, T. Kahil, M. Obersteiner, V. Gorbachuk and P. Knopov. 2019. Stochastic optimization models for risk-based reservoir management. Cybernetics and Systems Analysis, 55(1): 55–64.

Ermolieva, T., P. Havlik, S. Frank, T. Kahil, J. Balkovic, R. Skalsky, Y. Ermoliev, P.S. Knopov, O.M. Borodina, and V.M. Gorbachuk. 2022. A risk-informed decision-making framework for climate change adaptation through robust land use and irrigation planning. Sustainability, 14(3). https://doi.org/10.3390/su14031430.

Gorbachuk, V. 1991. Methods for Nash Equilibria Search. Nonsmooth analysis and its applications to mathematical economics. Baku, Azerbaijan: Ac. Sci. USSR, p. 65.

Gorbachuk, V. 2010. Cournot–Nash equilibria and Bertrand–Nash equilibria for a heterogenous duopoly of differentiated products. Cybernetics and Systems Analysis, 46(1): 25–33.

Gorbachuk, V. and B. Chumakov. 2008. Who should lead: A producer or a supplier? Modelare Matematica, Optimizare si Tehnologii Informationale. Chisinau, Moldova: Academia de transporturi, informatica si comunicatii, pp. 122–130.

Gorbachuk, V. and B. Chumakov. 2011. Local Cournot–Nash equilibria for duopoly and triopoly with linear and quadratic costs. Transport systems and logistics. Chisinau, Moldova: Academia de Transporturi, Informatică şi Comunicaţii, pp. 233–242.

Gorbachuk, V., M. Dunaievskyi and S.-B. Suleimanov. 2019. Modeling of agency problems in complex decentralized systems under information asymmetry. IEEE Conference on Advanced Trends in Information Theory (December 18–20, Kyiv, Ukraine). Kyiv: Taras Shevchenko National University of Kyiv, pp. 449–454.

Gorbachuk, V.M. 2006. Generalized Cournot–Stackelebrg–Nash equilibrium. Cybernetics and Systems Analysis, 42(1): 25–33.

Gorbachuk, V.M. 2007a. An asymmetric Cournot–Nash equilibrium under uncertainty as a generalized Cournot–Stackelberg–Nash equilibrium. Cybernetics and Systems Analysis, 43(4): 471–477.

Gorbachuk, V.M. 2007b. The mixed strategy of cooperation and generalized leadership for outputs of symmetric duopoly. Journal of Automation and Information Sciences, 39(7): 68–74.

Gorbachuk, V.M. 2008a. The cartel optimum and the reasonable Cournot–Nash equilibrium for fractional objective functions. Journal of Automation and Information Sciences, 40(12): 61–69.

Gorbachuk, V.M. 2008b. The Cournot–Nash equilibrium under mutual uncertainty. Journal of Automation and Information Sciences, 40(7): 59–72.

Gorbachuk, V.M. 2014. Equilibria of international public goods. Education and science and their role in social and industrial progress of society. Kyiv: Alexander von Humboldt Foundation, pp. 13–14.

Gorbachuk, V.M., M.S. Dunaievskyi, A.A. Syrku and S.-B. Suleimanov. 2022. Substantiating the diffusion model of innovation implementation and its application to vaccine propagation. Cybernetics and Systems Analysis, 58(1): 84–94.

Leontief, W. 1936. Stackelberg on monpolistic competiton. Journal of Political Economy, 44(4): 554–559.

Nagurney, A. 2009. Oligopolistic market equilibrium. pp. 2691–2694. *In*: Floudas, C.A., P.M. Pardalos (eds.). Encyclopedia of Optimization. 2-nd ed. Springer.

Nash, J.F. 1950. Equilibrium points in n-person games. Proceedings of the National Academy of Sciences of the United States of America, 36(1): 48–49.

Nash, J.F. 1951. Noncooperative games. Annals of Mathematics, 54(2): 286–295.

Prokopovych, P. and N.C. Yannelis. 2014. On the existence of mixed strategy Nash equilibria. Journal of Mathematical Economics, 52(C): 87–97.

Tramontana, F., L. Gardini and T. Puu. 2011. Mathematical properties of a discontinuous Cournot–Stackelberg model. Chaos, Solitons & Fractals, 44(1): 58–70.

Von Stackelberg, H. 1934. Marktform und Gleichgewicht. Vienna: Julius Springer.

CHAPTER 10

On the Vector Optimal Control of Risk Processes

Bogdan V Norkin,[1,*] *Vladimir I Norkin*[1] and *Roman V Sukach*[2]

1. Introduction

In this section, optimization, and vector optimal control problems for stochastic risk processes are considered. Mathematical models in the form of risk processes describe stochastic systems with a wholly ordered state space, i.e., for each pair of different states, one is better or worse than the other. Typical examples of such systems are banking, financial and insurance systems, stochastic business processes, inventory management systems, and energy storage. The goal of control is to transfer the system from the initial state to better states while avoiding bad states. This goal is formalized as a vector optimality criterion, in which some indicators characterize the effectiveness of management while others characterize the risk of getting into unfavorable states. Formally, the problem of vector optimal control is constructing an effective frontier in the space "efficiency - risk" and finding the corresponding Pareto-optimal controls.

In the paper, several approaches are proposed for the numerical implementation of this scheme.

[1] V.M. Glushkov Institute of Cybernetics of the National Academy of Sciences & National Technical University of Ukraine "Igor Sikorsky Kyiv Polytechnic Institute".
[2] National Technical University of Ukraine "Igor Sikorsky Kyiv Polytechnic Institute".
Emails: vladimir.norkin@gmail.com; majorissue@outlook.com
* Corresponding author: bogdan.norkin@gmail.com

The basic approach consists of finding the optimal control among some parametric controls. This approach is supported by the fact that many stochastic optimal control problems have analytical solutions up to unknown numerical parameters. Thus, the original problem is reduced to the finite-dimensional vector stochastic programming problem concerning these parameters. The values of efficiency and risk indicators under fixed control can be estimated by multiple process simulations. In addition, it is shown that the discounted indicators of efficiency and risk under a fixed control satisfy some integral equations, which can be effectively solved by a successive approximation method. Thus, the problem of vector stochastic optimal control is reduced to the problem of stochastic vector optimization on a finite-dimensional set of parameters.

In the case of a small number of parameters, this problem is solved by finite deterministic or random approximation of a continuous set of parameters with a discrete one. For a finite number of vectors in the criteria space, finding the Pareto-optimal points is carried out by directed enumeration and is not a significant problem. In parallel, the corresponding Pareto-optimal parametric controls are found, which give an approximate solution to the original problem.

In the case of a large number of parameters, to find Pareto optimal solutions, it is possible to optimize the convolutions of the criteria. Since optimal strategies are non-convex, non-smooth, or discontinuous functions of state variables and parameters, the corresponding stochastic programming problems turn out to be non-convex, non-smooth, or even discontinuous. For its approximate solution, stochastic generalized gradient directed random search, and successive smoothing methods are used.

One more approach formalizes the control problem as a finite-dimensional vector stochastic programming problem with fixed unknown control. The efficiency indicator is the gain from the decisions made, and the risk indicator is the probability of falling into unfavorable states. The obtained optimal solution as a function of the initial state of the process can be used in a rolling horizon manner, i.e., at each new current state, the decision is recalculated. In the case of a finite number of process evolution scenarios, the problem is reduced to high-dimensional mixed-integer programming problems.

Thus, the following results are displayed in the work.

The methodology of two-stage and vector (multi-criteria) optimal control of risk processes is developed.

For controlled risk processes, integral efficiency and risk functionals are formulated.

Linear integral equations are derived to calculate the values of the efficiency and risk functionals for a particular control strategy.

Successive approximation methods for solving these integral equations are developed and substantiated.

A scheme is developed for reducing a vector optimal control problem with a parametrically specified family of admissible controls to vector stochastic programming problems.

For the vector stochastic optimization problems, an approximation method for constructing the Pareto-optimal frontier in the space "efficiency - risk" and for finding the corresponding Pareto-optimal set of controls is developed and substantiated.

The developed approaches to solving vector optimal control problems are applied to the problems of managing an insurance company's reserves and inventory control problems.

2. Stochastic risk processes in insurance and inventory control theory

The general theory of single-criteria optimal control of Markov decision-making processes is described, for example, in Dynkin and Yushkevich 1979, Bertsekas and Shreve 1996, Feinberg and Shwartz 2002, Powell 2011. In this paper, we consider a particular class of processes, the so-called controlled risk processes describing the stochastic evolution of some assets. The management of such processes is considered in terms of efficiency and risk. Risk is understood as the possibility of getting into some undesirable region of the state space. Thus, the management of these processes is two-criteria or multicriteria. In this case, usually, one criterion is chosen as the objective function, and the rest are used in the constraints. The Lagrange multiplier method, the dynamic programming method, and the Pontryagin maximum principle are usually used to solve the problem. The difficulties of these approaches are related to the nonconvexity of optimal control problems, the nonsmoothness of the Bellman function, the necessity to find viscosity solutions for Bellman equations, and the curse of dimensionality. In this paper, we propose a different approach to the multicriteria optimal control of (discrete-time) risk processes, namely, its reduction to a multicriteria stochastic programming problem by using parametric control strategies.

2.1 Risk processes in insurance mathematics

This section considers stochastic optimization models of insurance mathematics and methods for their solution from the point of view of the methodology of stochastic programming and stochastic optimal control with vector optimality criteria. The evolution of the capital of an insurance company is considered in discrete time. The main random variables that influence this evolution are payment levels, which are ratios of paid claims to the corresponding premiums per unit of time. The main optimization variables are the structure of the insurance portfolio (the structure of the total premium) and the number of dividends. The performance criteria are profitability indicators of the insurance business, and the risk criteria are the probability of ruin and the capital necessary to prevent the ruin. Optimization aims to build efficient frontiers and efficient (Pareto) feasible solutions. Methods for solving these problems are proposed.

The traditional insurance theory is based on the expected utility theory (see, e.g., Bowers et al. 1997). An alternative approach is based on the optimization of return and/or risk (see de Finetti 1957, Borch 1974, Gerber 1979, Beard et al. 1894, Nakonechnyi 1996, Lyubchenko and Nakonechnyi 1998, Bayraktar and Young 2008, Schmidli 2008, Norkin 2011, Ascue and Muler 2014). The probability of ruin (probability of insolvency) is most often used as a risk indicator (see Asmussen and Albrecher 2010, Korolev et al. 2011, Ascue and Muler 2014). The difficulty of the arising optimization problems is explained by the fact that they belong to the class of nonconvex problems of stochastic programming or stochastic optimal control under probabilistic constraints. This paper shows how insurance optimization tasks can be posed and numerically solved according to several criteria. In contrast to the classical Cramer-Lundberg collective risk model (see, e.g., Bowers et al. 1997), in this paper, the stochastic evolution of the capital of an insurance company is considered in discrete time, which is justified by discrete-time (quarterly, annual) reporting on a company activity results (see, e.g., Forinshurer 2022). The primary source of randomness in insurance is insurance claims that appear at random times and have random sizes. In the discrete-time models considered below, the primary random variable is the level of insurance payments, which is the ratio of claim payments to premiums per unit of time (year, quarter) (see Norkin 2014, Ermoliev et al. 2020).

Discrete-time dynamic actuarial risk models. Let the reserves of an insurance company evolve over (discrete) time according to the following law:

$$X^{t+1} = X^t + c(y) - u^t - \xi^t(y), \quad X^0 = x, \quad t = 0,1,\dots,T \leq \infty, \tag{1}$$

where $y = (y_1,\dots,y_n) \in Y$ is a fixed deterministic vector parameter, which determines insurance premiums $c(y)$ and influence on the insurance claims $\xi^t(y)$; $u^t \in [0, X^t]$; is the collected dividends in the time interval $[t, t+1)$.

For example, vector y can describe the insurance portfolio with components $y_i \geq 0$, $\sum_{i=1}^{n} y_i \leq 1$; $c(y)$ is the net deterministic premiums per unit of time from the insurance portfolio y, e.g., $c(y) = (1 - v)\sum_{i=1}^{n} y_i$, where $v \in (0, 1)$ is the level of operational expenses; $\xi^t(y)$ is realizations of the level of insurance claims under the portfolio y, for example, $\xi^t(y) = \sum_{i=1}^{n} y_i \xi_i^t$, where ξ_i^t is random payment level for insurance type i in the time interval $[t, t+1)$, y_i is the premium from the i-th type of insurance in this time interval.

In another example, y may describe a stop-loss reinsurance contract with net premiums $c(y) = c_0 - c_1 y$ and bounded claims $\xi^t(y) = \min\{y, \xi^t\}$ in the time interval $[t, t+1)$, where c_0 and ξ^t are the premiums and claims without reinsurance, c_1 is the cost of reinsurance of the stop-loss level y. Process (1) is called a discrete-time risk process. Under the event of ruin (or insolvency) of the risk process, one means such realization of the process (1), when $X^t < 0$ for some $t > 0$.

A similar approach to modelling investment risk insurance was used in Norkin VI 2007, and Ermol'yev et al. 2001.

The expected accumulated dividends as performance indicators. Let us denote the first random moment τ when the process (1) stops, i.e., when the company's reserves become negative:

$$\tau = \sup\left\{t \in [0, T] : \min_{0 \leq k < t} X^k \geq 0\right\}.$$

The dividend optimal control problem (under the fixed portfolio y) has the form:

$$F(u) = \mathbf{E}\sum_{t=0}^{\tau-1} u^t \rightarrow \max_{u:u^t \in [0, X^t]}, \tag{2}$$

where $\{X^t\}$ satisfy equation (1), E denotes mathematical expectation.

The paradox (see de Finetti 1957) for this problem is that under the optimal dividend strategy the probability of ruining the risk process can be equal to one. Therefore, it is necessary to limit the set of admissible controls and/or take into account additional control criteria. This problem has attracted considerable attention of researchers (see Piunovsky 1997, Sethi 1997, Hipp 2003, Paulsen 2003, Dickson and Waters 2004, Gerber et al. 2006, Thonhauser and Albrecher 2007, Bayraktar and Young 2008, Avanzi 2009, Ascue and Muler 2014).

Parametric control strategies. In this paper, we consider this problem as a multiobjective stochastic optimal control one and solve it approximately by using parametric control strategies. For example, one can use positional parametric control strategies of the form $u^t = u(X^t, z)$, where $u(\cdot,\cdot)$ is a given function, and $z \in Z$ is a vector of parameters from the set Z. In particular, $\{u^t\}$ can be a deterministic strategy $\{u^t = z^t,\ t = 0, 1, ..., T < \infty\}$. An indicator of the effectiveness of the management strategy is the expected accumulated dividends for the time τ before the ruin:

$$D(x, y, z) = \mathbf{E}\sum_{t=0}^{\tau-1} u(X^t, z).$$

It appears that in optimal control actuarial problems, the optimal controls often have the so-called band structure (see Chapter 5 in Ascue and Muler 2014 and references therein). This means that there is a band partition $\{B_k,\ k = 1,2,...\}$ of the state space $R_+ = \cup_k B_k$, where $\{B_k\}$ are disjoint subsets with boundary points $b = \{0 = b_0 < b_1 < ... < b_k < ...\}$. For each partition set B_k there is a corresponding analytically known control rule $B_k(x, b,...)$ depending on the state variable x, partition points b, and maybe some other parameters. The whole control strategy has the form

$$u(x,b) = \{u_k(x,b,...),\quad x \in B_k,\ k = 1,2,...\}. \tag{3}$$

The class of band controls includes some other particular controls like barrier control, (B, b)-control, proportional control and other known rules. Substituting the control $u^t = u(X^t, b,...)$ into the original control problem (2) transforms it into a finite-dimensional stochastic programming problem concerning the unknown parameters $(b,...)$. For example, in the case of barrier control, $(b > 0)$,

$$u(x,b) = \begin{cases} x - b, & x > b, \\ 0, & 0 \le x \le b; \end{cases}$$

in the case of (B, b)-control, $(B > b)$,

$$u(x,B,b) = \begin{cases} x - b, & x > B, \\ 0, & 0 \le x \le B; \end{cases}$$

In the case of proportional control

$$u(x,b) = \{\alpha_k x,\ x \in B_k,\ k = 1,2,...\},\quad \alpha_k \in [0,1],\ \cup_k B_k = R_+.$$

Remark that the control function (3) can be nonsmooth and even discontinuous on its variables, so the output stochastic programming problem may occur nonconvex, nonsmooth, and multiexremal.

The probability of ruin is the risk indicator. Let $\psi^t(x)$ denote the probability of ruin, and $\varphi^t(x)$ be the probability of non-ruin of process (1) in the time interval $[0, t]$ and with the initial capital $X^0 = x$:

$$\psi^t(x) = \Pr\left\{\exists k \in \{0,1,\dots,t\} : X^k < 0;\ X^0 = x\right\},$$

$$\varphi^t(x) = 1 - \psi^t(x) = \Pr\left\{X^k \geq 0,\ 0 \leq k \leq t,\ X^0 = x\right\},$$

where $\Pr\{\cdot\}$ denotes the probability of the event in the curly braces. Functions ψ^t, φ^t depend not only on x, but also on other parameters (y, z) of the process (1), but in this context, this dependence is not explicitly indicated.

Assumption 1. All $\{\xi^t \in R^1\}$ are independent identically distributed random variables, distributed as a random variable $\xi \in R^1$ with the cumulative distribution function $\Phi(w) = \Pr\{\xi \leq w\}$; $\xi^t(y) = y\xi^t$ with $y \in R^1$.

Under Assumption 1, the sequence of functions $\{\varphi^t(\cdot), t = 0, 1, \dots\}$ satisfies the following integral relations:

$$\begin{aligned}\varphi^1(x) &= \Pr\{x + c(y) - u(x,z) - y\xi \geq 0\} = \\ &= \Pr\{\xi \leq (x + c(y) - u(x,z))/y\} = \\ &= \Phi\big((x + c(y) - u(x,z))/y\big),\end{aligned}$$

$$\varphi^{t+1}(x) = \mathbf{E}_\xi \varphi^t(x + c(y) - u(x,z) - y\xi), \quad t = 1, 2, \dots,$$

where $\mathbf{E}_\xi$ denotes the expectation operator (Lebesgue integral over the measure induced by the random variable ξ); by definition, $\varphi^t(x) = 0$, $x < 0$, for all t.

The non-ruin probability of the process (1) on the infinite time interval

$$\varphi(x) = \Pr\{X^t \geq 0\ \ \forall t \geq 0, X^0 = x\},$$

as a function of the initial state x, satisfies under Assumption 1 for fixed y, z the (integral) equation

$$\begin{aligned}\varphi(x) &= \mathbf{E}_\xi \varphi\big(x + c(y) - u(x,z) - y\xi\big) = \\ &= \int_{\{w: x + c(y) - u(x,z) - yw \geq 0\}} \varphi\big(x + c(y) - u(x,z) - yw\big) d\Phi(w),\end{aligned} \tag{4}$$

where, by definition, $\varphi(x) = 0$ for $x < 0$. This is a linear integral equation that always has a zero solution. However, we are interested in conditions under which there exists a nondecreasing solution $\varphi(\cdot)$ such that

$$0 \leq \varphi(x) \leq 1, \qquad \lim_{\{x \to +\infty\}} \varphi(x) = 1. \tag{5}$$

Such solutions can be found by the successive approximations method:

$$\varphi^{k+1}(x) = \mathrm{E}_{\xi}\varphi^{k}\left(x + c(y) - u(x,z) - \xi y\right), \quad \varphi^{0} \equiv 1, \qquad k = 0,1,\dots .$$

The existence of solutions to the problem (4), (5) and the convergence conditions for the successive approximations method were studied by Norkin 2007 and Norkin 2008.

Discounted risk functionals in insurance mathematics. For the process (1), we consider several additive utility and risk functionals. Let, at each step, the process (1) be characterized by an indicator $r(\cdot, \xi(y))$, then define the additive functional

$$V(x;y,z) = \mathbf{E}\sum_{t=0}^{\tau-1}\gamma^{t} r\left(X^{t}, \xi^{t}(y)\right),$$

where $\gamma \in (0,1)$ is the discount factor, the mathematical expectation $\mathbf{E}$ is taken over all possible trajectories of the process, and τ is a random moment of the ruin of the process, i.e., $\tau = \sup\{t : \min_{0\le k<t} X^{k} \ge 0\}$.

If function $r(\cdot)$ grows no faster than a linear one, then the function $V(\cdot; y, z)$ satisfies the equation (see Norkin 2014b):

$$V(x;y,z) = \mathrm{E}_{\xi} r\left(x, \xi(y)\right) + \gamma \mathrm{E}_{\xi} V\left(x + c(y) - u(x,z) - \xi(y); y, z\right), \qquad (6)$$

where $V(x; y, z)$ for $x < 0$. Note that, in contrast to the standard Bellman equation, the right-hand side of (6) does not contain any supremum or infimum operations.

If $r(x, \xi) = r(x)$ is upper semicontinuous and $c(y)$ and $u(x, z)$ are continuous in their arguments, then $V(x; y, z)$ is upper semicontinuous and, for each fixed pair (y, z), its values can be found by the successive approximations method (see Norkin 2014b and Norkin 2014c):

$$V^{k+1}(x;y,z) = r(x) + \gamma \mathrm{E}_{\xi} V^{k}\left(x + c(y) - u(x,z) - \xi(y); y, z\right), \;\; V^{0}(u;x,y) = 0, \;\; k = 0,1,\dots .$$

For example,

$$r_0(X^t) = \begin{cases} u(X^t, z), & X^t \ge 0, \\ 0, & X^t < 0, \end{cases}$$

then the mean discounted dividends until the ruin are expressed as follows

$$V_0(x;y,z) = \mathrm{E}\sum_{t=0}^{\tau-1}\gamma^{t} u(X^{t}, z),$$

and function $V_0(\cdot; y, z)$ satisfies the equation

$$V_0(x;y,z) = u(x,z) + \gamma \mathrm{E}_{\xi} V_0\left(x + c(y) - u(x,z) - \xi(y); y, z\right).$$

If

$$r_1(X^t) = \begin{cases} 1, & X^t \geq 0, \\ 0, & X^t < 0, \end{cases}$$

then the discounted ruin probability

$$V_2(x, y, z) = \mathbf{E}\sum_{t=0}^{\tau-1} \gamma^t \mathbf{1}_{\{X^t + c(y) - u(X^t, z) - \xi^t(y) < 0\}}$$

under Assumption 1 satisfies the equation

$$\begin{aligned} V_2(x; y, z) &= \mathrm{E}_\xi \mathbf{1}_{\{x + c(y) - u(x,z) - \xi(y) < 0\}} + \gamma \mathrm{E}_\xi V_2\left(x + c(y) - u(x, z) - \xi(y); y, z\right) = \\ &= (1 - \Phi((x + c(y) - u(x, z))/y) + \gamma \mathrm{E}_\xi V_2(x + c(y) - u(x, z) - y\xi; y, z). \end{aligned}$$

Consider another risk functional for a non-stop process (1). Let us introduce a reward function for the one-time step of the form (see Norkin 2014b and Norkin 2014c):

$$r_3(x, y, z, \xi(y)) = u(x, z) + \lambda \min\{0, x + c(y) - u(x, z) - y\xi(y)\},$$

which combines the gain (dividends) $u(x, z)$ and the risk factor (capital deficit) with the penalty factor $\lambda \geq 0$. Let us denote the discounted function of net income under the control strategy $u(\cdot, z)$:

$$V_3(x, y, z) = \mathbf{E}\left[\sum\nolimits_{t=0}^{\infty} \gamma^k r_3\left(X^t, y, z, \xi^t(y)\right)\right],$$

where γ is the discounting factor, $0 < \gamma \leq 1$.

All listed indicators $V_i(x;\, y,\, z)$, $i = 0, 1, 2, 3$, as functions of the initial state x at a fixed value of parameters (y, z), satisfy integral equations of the form $V_i(x;\, y,\, z) = f(x;\, y,\, z) + A_i(x;\, y,\, z) \circ V_i(\cdot)$, where $A_i(x;\, y,\, z)$ is some integral operator, $f(x;\, y,\, z)$ is a given function. Similar integral equations were considered in Borch 1974, and Gerber 1979. Therefore, the indicator functions $V_i(x;\, y,\, z)$ can be found by the successive approximation's method:

$$V_i^{k+1}(x; y, z) = f(x; y, z) + A_i(x; y, z) \circ V_i^k(\cdot), \quad V_i^0(x; y, z) = 0, \quad k = 0, 1, \ldots. \tag{7}$$

Details of the implementation of the successive approximation method for these equations are described in Norkin 2007, Norkin 2008, Norkin 2014b, and Norkin 2014c. Thus all indicators $V_i(x;\, y,\, z)$ can be efficiently calculated for any fixed control $u(\cdot,\, z)$. An alternative method for calculating the values of the indicators $V_i(x;\, y,\, z)$ is the method of statistical simulations (the Monte Carlo method) (see Norkin 2014a).

2.2 Stochastic risk processes in inventory theory

Stochastic inventory theory provides another example of risk processes (see, e.g., Prabhu 1998). The evolution of the stock X^t in a discrete-time follows the equation

$$X^{t+1} = X^t + u^t - g(X^t + u^t, \xi^t), \quad X^0 = x, \quad t = 0,1,\ldots,T \le \infty, \tag{8}$$

where $t = 0,1,\ldots$ is the discreet time, X^t is the value of the stock at the beginning of the time interval $[t, t + 1)$; u^t is the replenishment of the stock at the beginning of the time interval $[t, t + 1)$; ξ^t is the demand of the good in the t-th time interval; $g(X^t + u^t, \xi^t)$ is the actual satisfying demand in the time interval t. Obviously, $g(X^t + u^t, \xi^t) \le \xi^t$. The storage capacity is bounded, $X^t \in [0, Q]$, $t = 0,1,\ldots$.

There may be two models, (a) with possibly negative and (b) with nonnegative stock.

(a) The stock can be negative, $X^{t+1} = X^t + u^t - \xi^t$, $t = 0,1,\ldots$, i.e., $g(X^t + u^t, \xi^t) = \xi^t$.

The deficit at the time $t + 1$ is equal to $B_{t+1} = \max\{0, -X^{t+1}\} = -\min\{0, X^t + u^t - \xi^t\}$.

(b) The physical stock in the warehouse can only be non-negative, $X^{t+1} = \max\{0, X^t + u^t - \xi^t\} = \{X^t + u^t - \xi^t\}_+$, $t = 0,1,\ldots$,

i.e., $g(X^t + u^t, \xi^t) = \min\{X^t + u^t, \xi^t\}$. The deficit at the time $t + 1$ is equal to

$$D_{t+1} = \max\{0, -Z_{t+1}\} = -\min\{0, Z_t + u_{t+1} - \xi_{t+1}\}.$$

An alternative variant of this equation is the following:

$$X^{t+1} = \left(X^t - \xi^t\right)_+ + u^t = \max\{0, X^t - \xi^t\} + u^t, \quad t = 0,1,\ldots,T,$$

The stock replenishment at the beginning of the time interval $[t, t + 1)$ can be selected as a function of the amount of stock X^t. For example, the so-called (s, S)-replenishment strategy is determined by two parameters, $(s, S) = z$, namely,

$$u(x,z) = \begin{cases} 0, & x \ge s, \\ S - x, & x < s. \end{cases}$$

Probability of deficit. Let the random variables (demand) $\{\xi^t\}$ be independent and identically distributed with a common distribution function Φ. Let the stock replenishment strategy depends only on the stock level $u^t = u(X^t)$, i.e., the equation of stock dynamics has the form:

$$X^{t+1} = X^t + u(X^t) - \xi^t, \quad t = 0,1,\ldots.$$

Denote the probability of debt (ruin)

$$\psi(x) = P\left\{ \exists t \geq 0 : X^t < 0 \mid X^0 = x \right\}$$

and the non-ruin (survival) probability

$$\varphi(x) = 1 - \psi(x) = P\left\{ X^t \geq 0, \forall t \geq 0 \mid X^0 = x \right\}.$$

The latter probability satisfies the (integral) equation

$$\varphi(x) = \int_{\{\omega \in \Omega : \omega \leq x + u(x)\}} \varphi\left(x + u(x) - \omega\right) dF(\omega) = \mathbf{E}\mathbf{1}_{\{\omega \leq x + u(x)\}} \varphi\left(x + u(x) - \omega\right),$$

where $\mathbf{1}_{\{\omega \leq x + u(x)\}}$ is the indicator function of the event in parentheses, which is equal to one if the event occurred, and zero otherwise.

Efficiency indicator. The expected total losses for shortages and surpluses of goods in the stock under a parametric control strategy $u^t = u(X^t, z)$ are as follows,

$$V(x, z) = \mathbf{E}\sum_{t=0}^{\infty} (\lambda)^t \max\left\{ d^+(X^t + u^t - \xi^t), d^-(\xi^t - X^t - u^t) \right\},$$

where **E** is the mathematical expectation sign; λ is a discount factor, $\lambda \in (0, 1]$; d^+ denotes the cost of storing a unit of stock; d^- is the costs due to stock shortage. It is natural to assume that the warehouse has a limited volume Q, i.e., $u^t \in [0, Q]$. If all random variables ξ^t are independent and equally distributed (like some random variable ξ, for example, discretely distributed), then function $V(x, z)$ under fixed parameter z satisfies the following equation:

$$V(x, z) = \mathbf{E}_\xi \max\{ d^+(x + u(x, z) - \xi), d^-(\xi - x - u(x, z)) \} + \\ + \lambda \mathbf{E}_\xi V\left(x + u(x, z) - g(x + u(x, z) - \xi), z\right),$$

which can be solved by the numerical successive approximation method.

Estimation of performance indicators by simulation and Monte Carlo method. Consider an approach to the inventory management problem based on the search for parametric control strategies. Often the solution to the inventory optimal control problem takes the form of the so-called (s, S)-strategy (see Prabhu 1998, Daduna et al. 1999):

$$u^t = u_{X^t}(s, S), \qquad t = 0, 1, \ldots, T, \qquad u_x(s, S) = \begin{cases} \varnothing, & s > S, \\ 0, & s \leq S, s < x, \\ S - x, & s \leq S, s \geq x. \end{cases} \quad (9)$$

where pare (s, S) are parameters of the strategy, $0 \leq s < S \leq Q$. Remark that function $u_x(s, S)$ is discontinuous in both variables x and (s, S).

A cost function associated with each time interval $[t, t+1)$: $r^t = r(x^t, u^t, \xi^t)$, consists of three terms: $c(X^t, u^t, \xi^t)$ is storage cost of goods, $d(u^t)$ is the cost of buying goods, and $e(X^t, u^t, \xi^t)$ is the revenue from selling goods, where

$$r(x,u,\xi) = c(x,u,\xi) + d(u) - e(x,u,\xi),$$

$$c(x,u,\xi) = \max\{\alpha(x+u-\xi), 0\} = \begin{cases} \alpha(x+u-\xi), & x+u > \xi, \\ 0, & x+u \le \xi; \end{cases}$$

$$d(u) = \begin{cases} 0, & u = 0, \\ a+bu, & u > 0; \end{cases} \quad e(x,u,\xi) = \begin{cases} \beta\xi, & x+u > \xi, \\ \beta(x+u), & x+u \le \xi; \end{cases} \quad \alpha, \beta, a, b \ge 0.$$

Then the cost function $f_{x,\xi}(s, S) = r(x, u_x(s, S), \xi)$, for the simplest (s, S) restocking strategy takes on the form:

$$f_{x,\xi}(s,S) = r(x, u_x(s,S), \xi) = \begin{cases} \varnothing, & s > S; \\ \alpha(x-\xi) - \beta\xi, & s \le S,\ s \le x,\ x > \xi; \\ -\beta x, & s \le S,\ s \le x,\ x \le \xi; \\ a + b(S-x) + \alpha(S-\xi) - \beta\xi, & s \le S,\ s > x,\ S > \xi; \\ a + b(S-x) - \beta S, & s \le S,\ s > x,\ S \le \xi. \end{cases}$$

The cost function $f_{x,\xi}(s, S)$ is defined on the set $\{(s, S) \in \mathrm{R}^2 : 0 \le s \le S \le Q\}$. Remark that functions $d(u)$, $r(x, u, \xi)$ and $f_{x,\xi}(s, S)$ are discontinuous. Other parametric restocking strategies are considered in Knopov and Norkin 2022.

The evolution of process (8) on the time interval $[0, T]$ under the (s, S) restocking strategy (9) can be characterized by the following cost and ruin probability criteria:

$$V_0(s,S) = \mathbf{E}_\xi \frac{1}{T} \sum_{t=0}^{T-1} (\gamma)^t f_{X^t,\xi^t}(s,S), \tag{10}$$

and

$$V_1(s,S) = \Pr\{\exists t \in \{0,T\} : X^t + u_{X^t}(s,S) < \xi^t\} = \mathbf{E}_\xi \sum_{t=0}^{T-1} \mathbf{1}_{\{X^t + u_{X^t}(s,S) < \xi^t\}}, \tag{11}$$

subject to dynamic constraints (8), (9). Here $\mathbf{E}_\xi$ denotes mathematical expectation with respect to random variable $\xi = (\xi^0, \xi^1, \ldots, \xi^{T-1})$; γ is a discounting factor, $0 < \gamma \le 1$; $\mathbf{1}_{\{\cdot\}}$ is the indicator function of the event in the brackets.

The vector stochastic optimization problem for selecting the optimal restocking (s, S)-strategy is as follows:

$$\vec{V}(s,S)=\{V_0(s,S),V_1(s,S)\}\rightarrow \min\nolimits_{0\le s\le S\le Q}. \tag{12}$$

In general, the function $f_{x,\xi}$ (s, S) is discontinuous on the set $\{(s, S) : 0 \le s \le S \le Q\}$ both due to the discontinuity of the control u and cost d. The indicator function $I_{X^t,\xi^t}(s,S)=\mathbf{1}_{\left\{X^t+u_{X^t}(s,S)<\xi^t\right\}}$ is also discontinuous, and these circumstances make the problem particularly difficult.

Mathematical expectations in (10), (11) can be estimated by sample average values (see Shapiro 2003, Pagnoncelli et al. 2009, Knopov and Norkin 2020):

$$F_N(s,S)=\frac{1}{N}\sum_{n=1}^{N}\frac{1}{T}\sum\nolimits_{t=0}^{T-1}(\gamma)^t f_{X_n^t,\xi_n^t}(s,S)\rightarrow \min\nolimits_{0\le s\le S\le Q},$$

$$F_N(s,S)=\frac{1}{N}\sum_{n=1}^{N}\sum\nolimits_{t=0}^{T-1} I_{X_n^t,\xi_n^t}(s,S)\rightarrow \min\nolimits_{0\le s\le S\le Q},$$

where $\xi_n = (\xi_n^1,\dots,\xi_n^{T-1})$ are independent samples ξ_n^t of the random demand in independent series $n = 1,\dots,N$; $\{X_n^t,\ t = 0,\dots,T-1\}$ is an evolution of the store under sample ξ_n according to dynamic equations (8); $u_n=\left\{u_n^0=u_{X_n^0}(s,S),\dots,u_n^{T-1}=u_{X_n^{T-1}}(s,S)\right\}$ is a sequence of controls, which corresponds to samples of the demand ξ_n. Functions $F_N(s, S)$ can be discontinuous and multi-extremal. In the case of barrier strategies, they may be continuous but still nonsmooth nonconvex and multiextremal. For their optimization, one can apply the so-called stochastic smoothing method from Norkin 2020, Knopov and Norkin 2022.

3. Multicriteria risk process control and optimization

3.1 Reduction to vector optimization

The problem of vector optimal control of the risk process (1) is to find non-dominant values of the vector indicator $\vec{V}(x; y, z) = \{V_i(x; y, z), i = 0, 1,\dots\}$:

$$\vec{V}(x;y,z)\rightarrow \mathrm{extr}_{x\in X,\ y\in Y,\ z\in Z}, \tag{13}$$

and the corresponding Pareto-optimal values of the parameters (x, y, z), where $x \in X$ indicates an initial state of the risk process (1), $y \in Y$ describes a structure and the volume of the insurance portfolio, and the parameter

$z \in Z$ is responsible for choosing a control strategy. For example, from heuristic reasons for the numerical approximation of the Pareto-optimal frontier, the so-called dividend barrier-proportional control strategies are considered, i.e., $u(x, z) = z_1 \max\{0, x - z_2\}$, $z = (z_1, z_2)$, $z_1 \in [0, 1]$, $z_2 \geq 0$, .

The complexity of this problem lies in the fact that, firstly, the indicators $V_i(x; y, z)$ themselves are not known explicitly but are solutions of corresponding integral equations; secondly, these indicators can be non-convex functions; and thirdly, the corresponding Pareto-optimal set can have a rather complex structure. Function values $V_i(x; y, z)$ can be found by successive approximations (7) or by statistical simulations, particularly by their parallel versions (see Norkin 2008, Norkin 2014a, Norkin 2014b).

For small dimensions of the parameter vector (x, y, z), problem (13) can be approximately solved by discrete approximation of sets X, Y, Z by finite N-point sets X^N, Y^N, Z^N and solving a discrete vector optimization problem:

$$\vec{V}(x; y, z) \to \text{extr}_{x \in X^N,\ y \in Y^N,\ z \in Z^N}.$$

A similar method for solving vector optimization problems was considered by Statnikov and Matusov 2002, and its asymptotic convergence under $N \to \infty$ was studied in Norkin 2015a and Norkin 2015b.

A similar approach can be applied to the two-criteria inventory control problem (12) for searching the (s, S)-strategy and other parametric control strategies.

Remark. As a very particular case, one can consider in (1) just a deterministic and fixed control strategy $u(\cdot, z) = z \in R^1$. Then the stochastic optimal control problem turns out into a vector optimization problem with a complex performance criteria $\{V_i(\cdot)\}$, which still can be calculated by the successive approximation method or by the Monte Carlo method. The obtained optimal controls $z^*(x)$, $x \in X$, as functions of the initial state x can then be applied for the risk process control dynamically by using controls $u^t = z^*(X^t)$.

Results of numerical experiments on the application of the proposed methodology to solving insurance business optimization are presented in Norkin 2008, Norkin 2014a, Norkin, B.V. 2014b.

3.2 Solution of the problem in case of a finite number of scenarios

Consider the following stochastic programming model for finding a deterministic optimal control strategy $\{u^t\}$ (see Norkin 2014a):

$$\sum_{t=0}^{T-1} u^t \to \max_{y,u}, \tag{14}$$

$$X^{t+1} = X^t + c(y) - u^t - \sum_{i=1}^{n} \xi_i^t y_i, \quad X^0 = x, \quad t = 0,1,...,T-1, \quad j = 1,...,m; \quad (15)$$

$$\sum_{i=1}^{n} y_i \le c, \quad y_i \ge 0, \quad i = 1,...,n; \ u^t \in \left[0, X^t\right], \ t = 0,1,...,T-1; \quad (16)$$

$$\mathrm{P}\left\{X^t \ge a^t, i = 1,...,T\right\} \ge 1 - \alpha. \quad (17)$$

Here, the variable $y = (y_1,...,y_n)$ specifies the (unchanged) structure of the company's portfolio of insurance contracts; the variable $u = (u^0, u^1,...,u^{T-1})$ specifies planned deductions from premiums unrelated to claim payments (for example, investments, dividends, etc.); the random variable $X = (X^0, X^1,...,X^T)$ describes the dynamics of the company's capital. The parameter c specifies the total premium under insurance contracts; x is the initial capital; $c(y) = (1 - \nu)\sum_{i=1}^{n} y_i$; $\nu \in [0,1)$, $a^t \ge 0$, $\alpha \in (0,1)$ are the company's reliability parameters ($\alpha \in [0, 1]$); T is the planned time horizon (the number of quarters or years taken into account to assess the probability of ruin). Thus, the values $\xi^t = (\xi_1^t,..., \xi_n^t)$ specify (random) loss per unit of premium in various types $i = 1,...,n$ of insurance in the time interval $[t, t + 1)$ (a quarter, a year), i.e., $\xi_i^t y_i$ is the total insurance claims from the i-th type of insurance in the period $[t, t + 1)$. The function

$$f(y,u) = \sum_{t=0}^{T-1} u^t$$

is the total reward,

$$g(x,y,u) = \Pr\left\{X^t \ge a^t, i = 1,...,T\right\}$$

is the probability that the reserves fall below the prescribed levels a^t, $t = 0,1,...,T$. If $a^t \equiv 0$, then $g(x, y, u)$ describes the probability of the company not going bankrupt over periods T under a constant total premium c, portfolio structure y and deductions $\{u^t\}$.

Problem (14)–(17) is a complex stochastic programming model with a probabilistic constraint (17). In the case of a finite set of independent scenarios $(\xi^1, \xi^2,...,\xi^T)_j$, $j = 1,...,m$ for trajectories $\{\xi^1, \xi^2,...,\xi^T\}$ of a vector random variable ξ occurring with probabilities p_j, problem (14)–(17) can be reduced to a partially integer linear programming problem.

Let us denote ξ_{ij}^t the value of the quantity ξ_i^t in the j-th scenario. Assume that all ξ_{ij}^t do not exceed some maximum value M. Let for each scenario

j, there corresponds a trajectory of the reserves $\{x_j^1, x_j^2, \ldots, x_j^T\}$. Note that $-cMt \le x_j^t \le x + ct$ for all t. Let us introduce Boolean variables μ_j that take on the value 1 if at least one value $x_j^t - a^t$, $t = 1, 2, \ldots, T$, in scenario j is negative, i.e., a ruin has occurred in this scenario. Consider the restrictions:

$$x_j^t - a^t \ge \left(-cMt - a^t\right)\mu_j, \quad t = 1, \ldots, T. \tag{18}$$

If at least one $x_j^{t'} - a^{t'}$, $t' \in \{1, 2, \ldots, T\}$, in (18) is negative, then the corresponding constraint $x_j^{t'} - a^{t'} \ge (-cMt' - a^{t'})\mu_j$ implies that μ_j is equal to one. Relations (18), in which $x_j^t - a^t \ge 0$, do not impose restrictions on the values of $\mu_j \in \{0, 1\}$. When restrictions (18) are satisfied for all scenarios j, i.e.,

$$x_j^t - a^t \ge (-cMt - a^t)\mu_j, \; t = 1, \ldots, T, \; j = 1, \ldots, m,$$

the value $\sum_{j=1}^{m} p_j \mu_j$ gives an upper estimate for the ruin probability over all scenarios, so constraint (17) can be written as:

$$\sum_{J=1}^{m} p_j \mu_j \le \alpha, \qquad \mu_j \in \{0,1\}, \qquad j = 1, 2, \ldots, m.$$

Thus, problem (14)–(17) in the case of a discrete set of scenarios can be equivalently represented in the following form:

$$\sum_{t=0}^{T-1} u^t \to \max_{y,x,u,\mu},$$

$$x_j^{t+1} = x_j^t + c(y) - u^t - \sum_{i=1}^{n} \xi_{ij}^t y_i, \quad x_j^0 = x, \qquad t = 0, 1, \ldots, T, \qquad j = 1, \ldots, m;$$

$$\sum_{i=1}^{n} y_i \le c, \qquad y_i \ge 0, \qquad i = 1, \ldots, n;$$

$$x_j^t - a^t \ge (-cMt - a^t)\mu_j, \; t = 1, \ldots, T, \; j = 1, \ldots, m,$$

$$\sum_{J=1}^{m} p_j \mu_j \le \alpha, \qquad\qquad \mu_j \in \{0,1\}, \qquad j = 1, 2, \ldots, m.$$

This model is a large-scale mixed integer linear programming problem and can be solved numerically by standard optimization methods. Obviously, it has nontrivial solutions for $\alpha \ge \min_j p_j$.

4. Conclusions

The article proposes a new approach to practical multicriteria (vector) optimization and control of risk processes in discrete time. These processes describe a stochastic evolution of some asset, for example, the evolution of reserves of an insurance company or the state of stock in a warehouse under conditions of random demand. In the case of the insurance model, the so-called level of payments acts as the main random factor, i.e., the ratio of insurance claim payments to premiums under relevant contracts. In inventory theory, the primary random variable is the demand for a product. Control affects the parameters of the risk process, such as the rate of replenishment or consumption of an asset. The quality of management is evaluated according to utility and risk criteria. The expected asset accumulation level, expected cumulative dividends, expected satisfied demand, and other additive indicators can be used as utility criteria. The risk criteria are, for example, the probability of ruin, i.e., the probability that the process falls into some risk area, as well as the expected cost of ruin prevention. The optimal control is sought in the class of parametric strategies, i.e., in the class of functions depending on the state of the process and a finite number of unknown deterministic parameters. The heuristic basis for choosing the class of parametric control strategies are theoretical results on analytical forms of optimal controls, for example, results on the barrier or band structure of optimal controls. When such a parametric control is substituted into the control criteria, the problem of stochastic optimal control is transformed into a finite-dimensional stochastic optimization problem with complex probabilistic criteria in the form of mathematical expectations and ruin event probabilities. This paper shows that there are two options for calculating optimization criteria, these are the solution of some integral equations and the Monte Carlo method. To find the Pareto-optimal frontier, it is proposed to use a deterministic or random discrete approximation of a finite-dimensional region of optimization parameters. As a result, the original vector optimal control problem is reduced to a discrete finite-dimensional vector optimization problem, which is solved by enumeration algorithms. Numerical experiments show the efficiency of this approach.

Acknowledgement

The contribution by Norkin V.I. was supported by Ukrainian-Norwegian project CPEA-LT-2016/10003 "Advanced Collaborative Program for Research-based Education on Risk Management in Industry and Services under Global Economic, Technological and Environmental Changes: Enhanced Edition" funded by the Norwegian Agency for International Cooperation and Quality Enhancement in Higher Education (Diku) and by a grant of Volkswagen Foundation (2022, #9C090).

References

Ascue, P. and N. Muler. 2014. Stochastic Optimization in Insurance. A Dynamic Programming Approach. Springer, New York.

Asmussen, S. and H. Albrecher. 2010. Ruin Probabilities. Second Edition, World Scientific, London.

Avanzi, B. 2009. Strategies for dividend distribution: A review. North American Actuarial Journal, 13(2): 217–251. DOI: 10.1080/10920277.2009.10597549.

Bayraktar, E. and V. Young. 2008. Maximizing utility of consumption subject to a constraint on the probability of lifetime ruin. Finance and Research Letters, 5(4): 204–212. DOI: 10.1016/j.frl.2008.08.002.

Beard, R.E., T. Pentikainen and E. Pesonen. 1984. Risk Theory. The Stochastic Basis of Insurance. 3rd Ed., Chapman and Hall, London, New York.

Bertsekas, D.P. and S.E. Shreve. 1996. Stochastic Optimal Control: The Discrete-time Case, Athena Scientific, Belmont, Massachusett.

Borch, K. 1974. The Mathematical Theory of Insurance, Lexington Books, Lexington, Massachusetts.

Bowers, N.L., H.U. Gerber, J.C. Hickman, D.A. Jones and C.J. Nesbitt. 1997. Actuarial Mathematics, Society of Aactuaries, Chicago.

Daduna, H., P.S. Knopov and L.P. Tur. 1999. Optimal strategies for an inventory system with cost function of general form. Cybernetics and Systems Analysis, 35(4): 601–618. DOI: https://doi.org/10.1007/BF02835856.

de Finetti, B. 1957. Su un'impostazione alternativa della teoria collettiva del rischio. Transactions of the XV-th International Congress of Actuaries, 2: 433–443, Actuarial Society of America, New York.

Dickson, D.C.M. and H.R. Waters. 2004. Some optimal dividend problems. ASTIN Bulletin, 34(1): 49–74. DOI: https://doi.org/10.1017/S0515036100013878.

Dynkin, E.B. and A.A. Yushkevich. 1979. Controlled Markov Processes, Springer, New York.

Ermoliev, Y.M., V.I. Norkin and B.V. Norkin. 2020. Stochastic optimization models of actuarial mathematics. Cybernetics and Systems Analysis, 56(1): 58–67. DOI: https://doi.org/10.1007/s10559-020-00221-0.

Ermol'yev, Y.M., T.Y. Ermol'yeva, G. McDonald and V.I. Norkin. 2001. Problems on insurance of catastrophic risks. Cybernetics and Systems Analysis, 37(2): 220–234. DOI: https://doi.org/10.1023/A:1016798903215.

Feinberg, E.A. and A. Shwartz. (eds.). 2002. Handbook of Markov Decision Processes. Methods and Applications, Kluwer Academic Publishing, Dordrecht.

Forinshurer. 2022. Online magazine about insurance in Ukraine. Statistics of the insurance market of Ukraine. Access mode (01.01.2022): http://forinsurer.com/stat.

Gerber, H.U. 1979. An Introduction to Mathematical Risk Theory, Huebner Foundation Monographs, Philadelphia.

Gerber, H.U., E.S.W. Shiu and N. Smith. 2006. Maximizing dividends without bankruptcy. ASTIN Bulletin, 36(1): 5–23. DOI: https://doi.org/10.1017/S0515036100014392.

Hipp, C. 2003. Optimal dividend payment under a ruin constraint: Discrete time and state space. Blätter der DGVFM (Blätter der Deutschen Gesellschaft für Versicherungs-und Finanzmathematik e.V.), 26(2): 255–264. DOI:10.1007/bf02808376.

Knopov, P.S. and V.I. Norkin. 2020. Convergence conditions for the observed mean method in stochastic programming. Cybernetics and Systems Analysis, 54(1): 45–59. DOI: https://doi.org/10.1007/s10559-018-0006-3.

Knopov, P.S. and V.I. Norkin. 2022. Stochastic optimization methods for the stochastic storage process control. pp. 79–111. *In*: Blondin, M.J. et al. (eds.). Intelligent Control and Smart Energy Management, Springer Optimization and its Applications 181. Springer Nature Switzerland AG. https://doi.org/10.1007/978-3-030-84474-5_3.

Korolev, V.Y., V.E. Bening and S.Y. Shorgin. 2011. Mathematical Basis of Risk Theory. 2nd ed., Fizmatlit, Moscow. (In Russian).

Lyubchenko, G.I. and A.N. Nakonechnyi. 1998. Optimization methods for compound poisson risk processes. Cybernetics and Systems Analysis, 34(2): 230–237. DOI: https://doi.org/10.1007/BF02742072.

Nakonechnyi, A.N. 1996. Optimization of risk processes. Cybernetics and Systems Analysis, 32(5): 641–646. DOI: https://doi.org/10.1007/BF02367767.

Norkin, B.V. 2007. On solution of the basic actuarial integral equation by a successive approximation method. Ukrainian Mathematical Journal, 59: 1689–1698. DOI: https://doi.org/10.1007/s11253-008-0033-8.

Norkin, B.V. 2008. Stochastic successive approximation method for the assessment of insolvency risk of an insurance company. Cybernetics and Systems Analysis, 44(6): 892–905. DOI: https://doi.org/10.1007/s10559-008-9065-1.

Norkin, B.V. 2011. Mathematical models for insurance business optimization. Cybernetics and Systems Analysis, 47(1): 117–133. DOI: https://doi.org/10.1007/s10559-011-9295-5.

Norkin, B.V. 2014a. Systems simulation analysis and optimization of insurance business. Cybernetics and Systems Analysis, 50(2): 260–270. DOI: https://doi.org/10.1007/s10559-014-9613-9.

Norkin, B.V. 2014b. Stochastic optimal control of risk processes with lipschitz payoff functions. Cybernetics and Systems Analysis, 50(5): 774–787. DOI: https://doi.org/10.1007/s10559-014-9668-7.

Norkin, B.V. 2014c. On stochastic optimal control of discrete-time risk processes. Journal of Automation and Information Sciences, 46(10): 30–44. DOI: 10.1615/JAutomatInfScien.v46.i10.40.

Norkin, B.V. 2015a. Statistical approximation of multicriteria stochastic optimization problems. Reports of the National Academy of Sciences of Ukraine, 2015(4): 35–41. DOI: https://doi.org/10.15407/dopovidi2015.04.035.

Norkin, B.V. 2015b. On the approximation of vector optimization problems. Cybernetics and Computer Engineering, 179: 35–42. Access (31.05.2019): DOI: https://doi.org/10.15407/kvt179.01.035.

Norkin, V.I. 2007. Self-insurance of investor under repeating catastrophic risks. Cybernetics and Systems Analysis, 43(3): 377–383. DOI: https://doi.org/10.1007/s10559-007-0059-1.

Norkin, V.I. 2020. A stochastic smoothing method for nonsmooth global optimization. Cybernetics and Computer Technologies, 2020(1): 5–14. DOI: https://doi.org/10.34229/2707-451X.20.1.1.

Pagnoncelli, B.K., S. Ahmed and A. Shapiro. 2009. Sample average approximation method for chance constrained programming: Theory and applications. Journal of Optimization Theory and Applications, 142: 399–416. DOI: 10.1007/s10957-009-9523-6.

Paulsen, J. 2003. Optimal dividend payouts for diffusions with solvency constraints. Finance Stochastics, 7(4): 457–473. DOI:10.1007/s007800200098.

Piunovsky, A.B. 1997. Optimal Control of Random Sequences in Problems with Constraints, Springer, Dordrecht.

Powell, W.B. 2011. Approximate Dynamic Programming. Solving the Curses of Dimensionality, 2nd ed., Wiley, Hoboken, NJ.

Prabhu, N.U. 1998. Stochastic Storage Processes. 2nd ed., Springer, New York. DOI: https://doi.org/10.34229/2707-451X.20.1.1.
Schmidli, H. 2008. Stochastic Control in Insurance. Springer-Verlag, London.
Sethi, S.P. 1997. Optimal Consumption and Investment with Bankruptcy, Kluwer Academic Publishers, Boston, Dordrecht, London.
Shapiro, A. 2003. Monte Carlo sampling methods. pp. 353–425. *In*: Ruszczyński, A. and A. Shapiro (eds.). Handbooks in OR&MS, 10. Elsevier, Amsterdam.
Statnikov, R.B. and J.B. Matusov. 2002. Multicriteria Analysis in Engineering. Kluwer Academic Publishers, Dordrecht.
Thonhauser, S. and H. Albrecher. 2007. Dividend maximization under consideration of the time value of ruin. Insurance: Mathematics and Economics, 41: 163–184. DOI:10.1016/j.insmatheco.2006.10.013.

CHAPTER 11

The Type-Variety Principle in Ensuring the Reliability, Safety and Resilience of Critical Infrastructures

*Volodymyr A Zaslavskyi** and *Oleh A Horbunov*

1. Introduction

Ensuring the security of the country, regions, critical objects, and infrastructures is an important scientific and technical problem. The solution to it defines directions, requirements, and resources to ensure system security, formation and implementation of sustainable development of the country (Lewis 2015, Zaslavskii and Greindl 1998, IIASA 2016, Ushakov 2006, Zaslavsky 2011).

Effective support for the functioning of critical infrastructure (CI) facilities (government agencies, military facilities, nuclear power plants, oil, and gas production platforms, etc.), protecting them from external influences and counteracting cyber-attacks, efficient work of personnel ensuring the reliable functioning of systems requires constant attention to the problem of security to identify and eliminate sources of threats and develop adequate ways to overcome them (Ushakov 2006, Zaslavsky 2011, Zaslavsky and Pasichna 2019a, Norkin et al. 2018).

The reliability and safety of CI object functioning depend on the reliability of the technical and information as well communication components, including security systems and the human factor - decision-making (DM) personnel (human operator, platform worker, pilot, etc.), as well as various external and environmental influences (Taylor 2014).

Taras Shevchenko National University of Kyiv, Faculty of Computer Sciences and Cybernetics, Kyiv, Ukraine.

* Corresponding author: zas@unicyb.kiev.ua

CI and its security systems consist of many various highly reliable functional components and monitoring, telemetry, and control systems for detecting, collecting, processing and tracking various information. CI security support is based on monitoring data for the purpose of continuous information support of the processes of generation, coordination, and decision-making in a real situation, considering the activation of various threats, such as terrorist actions, as well as ensuring the safety of personnel and their normal physical and psychological state (Ushakov 2006, Taylor 2014). Violation of the functioning of the CI can lead to severe negative social and economic consequences for the state, national security and defence (Lewis 2015, Tagarev and Pavlov 2007, Biryukov and Kondratov 2015). The close relationship of CI with government, organizational and informational interaction, increased social and economic dependence on the services of CI set new requirements in extreme situations for their safety and reliability of operation using innovative developments. In the process of creating critical systems, topical problems of reliability optimization (optimal redundancy) are considered, which are associated with the choice of the composition of the components (various types and implementations of systems, subsystems, nodes, blocks, and elements) and their integration, taking into account various efficiency criteria and resource constraints (Volkovich et al. 1992, Afanasiev et al. 1993, Zaslavsky and Ievgiienko 2010). The human factor, i.e., professional qualifications of personnel, their reliability in performing a list of highly important operations, normal physical condition and working capacity are also very important components of the safety and reliability of the CI functioning with human-machine and cybernetic systems (Taylor 2014, Rasmussen and Taylor 1976).

The global pandemic of the coronavirus, COVID-19 has led to a restriction of interaction between countries, business companies, and education systems, mass illness of staff, high psychological stress on people and teams in organizations and the transition to online mode (Tsai et al. 2020, Galea et al. 2020). Significant health problems and stress for the population were brought about by the unleashed large-scale war of Russia against Ukraine. The presence of military personnel and citizens in the zones of violent hostilities, genocide against the civilian population in the occupied territories, rocket attacks and often announced alarms throughout Ukraine have caused a significant long-term deterioration in the health of people, various physical disabilities, and psychological disorders (Podkovka et al. 2021, Panchenko et al. 2015).

Identification with a certain probability of human functional states that are dangerous in terms of maintaining the level of professional performance and reliability is an important task. The solution to it makes it possible to prevent the occurrence of errors and accidents in critical systems.

In Sections 1 and 2, mathematical models and an algorithm of the type-variety principle are considered in ensuring the reliability of critical systems during their design and the detection of defects during operation. Based on mathematical models and the algorithm of the type-variety principle, Section 3 proposes an approach to solving the problem of minimizing human errors through the purposeful organization of a variety of physical exercises that help eliminate physical deficiencies that, for example, can be associated with human pain. Their elimination increases the reliability of the human factor while ensuring the safety of critical systems.

2. Application of the type-variety principle in ensuring the reliability and security of critical infrastructures

2.1 Type-variety principle and its applications

One of the fundamental concepts in nature, natural science, mathematics, cybernetics and systems analysis is the concept of "diversity", meaning that either two materials, things of a process, element, model, algorithm or service are significantly different, or they have changed and continue to change over time. The concept of diversity proposed by Ashby in biology, command and control systems are currently widely used by researchers in various fields of knowledge and applied fields (Ashby 1956). This concept initiated the emergence of problem statements, mathematical models, and optimization methods in the economy, which significantly affect changes in organizations, business processes, education, and the formation of innovation processes in various applied fields (Ottosson 2013, Zaslavsky 2014).

The practical application by the authors of the idea of diversity concepts for solving various applied problems (Volkovich et al. 1992, Afanasiev et al. 1993, Zaslavsky2014, Volkovich and Zaslavsky 1986, Zaslavsky and Strizak 2006a), made it possible to formulate the type-variety principle in the study of complex systems as the principle of system analysis in the following formulation (Zaslavsky 2006b).

Type-variety principle - is focused on the usage of different nature (the principle of operation, construction) components (e.g., systems, subsystems, elements, technologies, raw materials of different origins, various types of models, algorithms, software components, experts, tools, etc.) that perform the same functions, each of them may be used separately (independently), but their simultaneous integration (consolidation, combination, utilization, application) and interaction provide better solutions for emerging issues (e.g., providing a highly reliable system operation, exception of a possible repetition of the failure in general reason, defect detection, analysis of projects, developing of innovative products, business processes, etc.).

The ideas of the type-variety principle were used in solving a number of applied problems: designing critical facilities and CI, forming a water purification system in a river basin, forming a complex of various non-destructive testing devices to detect defects in complex systems, building mathematical models and algorithms to detect fraudulent transactions in the payment system, formation of a diverse energy portfolio of generating companies, the security of information systems (Zaslavsky 2011, Zaslavskyi and Pasichna 2019a, Zaslavsky and Strizak 2006a, Zaslavskyi and Pasichna 2019b, Zaslavsky 2016, Kadenko et al. 1999. Zaslavsky 2001).

In connection with the intensification of terrorist activity, it is necessary to review measures and develop new approaches to ensure the systemic security of critical infrastructures, provision the physical protection of critical facilities, and introduce new methods of countering terrorism (Ushakov 2006, Norkin et al. 2018, Harris 2004). Section 3 describes the mathematical model and algorithm for optimizing the structures of protection systems in conditions of limited resources using typical redundancy, and Chapter 4 describes the application of this idea to detect defects in security systems.

2.2 *Type-variety principle and reliability optimization*

Models and methods for reliability optimization and optimal redundancy are effective methods for increasing the reliability of critical infrastructures, protection systems, and complex technical and information systems, by selecting and including additional (redundant) elements and links compared to the minimum required to perform the specified functions under certain conditions trouble-free operation (Ushakov 2006). At the same time, the main devices (element, block, subsystem) necessary to ensure the functioning of the system are distinguished in the system design, and the elements intended to ensure the device's operability in the case of failure of the main elements are called backup (Volkovich et al. 1992, Afanasiev et al. 1993, Ushakov 1969, Barlow and Proschan 1975). The problems of optimal redundancy of complex systems were considered in numerous works by various authors (Ushakov 1969, Barlow and Proschan 1975). However, when implementing redundancy, as a rule, the same type (identical to the main) reserve elements were considered. The study of problems of multi-type redundancy, analysis, formalization, and algorithm for solving the problem of optimal redundancy with type-variety elements was developed and presented in works (Volkovich et al. 1992, Afanasiev et al. 1993, Volkovich and Zaslavsky 1986). An important factor in the analysis of this applied optimization problem is the following: different types of elements that are used in the construction of modules for subsystems of a complex system have different technical and economic characteristics but are identical in their functionality. With this method of redundancy (diverse redundancy), a common cause failure is eliminated, which increases the time of active existence of complex systems, CI, and the stage of their life cycle - operation, and this approach also allows finding new optimal design solutions that have never been considered before.

The problem of CI safety is associated with the high cost of their failure, which requires the presence of redundant elements in the system structure that ensure risk diversification and failure prevention due to the common cause of failure based on various types of redundancy. An important component in ensuring the safety and fault tolerance of CI is the human factor. In the following sections, the formulations, and the algorithm for solving problems of ensuring reliability for CI, considering into account the human factor, are considered, in which the mechanism for generating and analysing options is used as an implementation of the type-variety principle.

2.3 *Mathematical models and methods for ensuring high reliability and safety of critical systems*

In this chapter, we consider the formulation of the reliability optimization (redundancy) problem to optimization structure of critical infrastructure protection systems under resource constraints. The critical infrastructure protection system is considered structurally a monotonic (coherent) system (Barlow and Proschan 1975). Redundancy in such a system, as distinct from traditional redundancy, is achieved by using elements of different types (the type-variety redundancy) (Volkovich et al. 1992, Afanasiev et al. 1993, Volkovich and Zaslavsky 1986). Let the CI protection system consists of n subsystems $u_j, j \in J = \{1,2,...,n\}$, that implements different types of protection methods. A subsystem j may be implemented in alternative versions v_j with the usage of a set $U_j = \{u_{jk}, k \in K_j\}$ of elements u_{jk} with different types, where $K_j = \{1,2,...,k_j^*\}$ is the index set of element types. Elements of different types have different rates of techno-economic characteristics $p_j(u_{jk})$, $g_{ij}(u_{jk})$, $j \in J$, $i \in I = \{1,2,...,m\}$ but are identical in their functional purpose. We define a method for the formation of the subsystem structure from elements of different types. Let $R_j = \{1,2,...,r_j,..., k_j^*\}$ be the set of lengths of the alternative version of the subsystem j and set of combinations of r_j indexes from the set K_j

$$M_{K_j}^{r_j} = \left\{k_{l_j}(r_j) = \left\{k_1^{l_j},...,k_p^{l_j},...,k_{r_j}^{l_j}\right\}\right\}, k_p^{l_j} \in K_j, \tag{1}$$

where $l_j = 1,2,...,C_{|Kj|}^{rj}$, $C_{|Kj|}^{rj} = \dfrac{k_j^*!}{rj!(k_i^* - rj)!}$, l_j – index of combination in set M_{Kj}^{rj}.

For each combination $k_{l_j}(r_j)$ of length r_j the set $V_{jk(rj)}$ of possible combinations is defined:

$$v_{jk(rj)} = \left(u_{jk_1},...,u_{jk_p},...,u_{jk_{r_j}}\right). \tag{2}$$

Let $V_j^{r_j} = \bigcup_{k(r_j)\in M_{K_j}^{r_j}} V_{jk(r_j)}$ be the set of alternatives of subsystem j of length r_j.

Then $V_j = \bigcup_{r_j \in R_j} V_j^{r_j}$ is the set of possible variants and $v = \left(v_{jk(r_1)},...,v_{jk(r_j)},...,v_{jk(r_n)}\right)$ is the implementation of subsystem j, and $V = \prod_{j\in J} V_j$ is the set of possible implementations of the complex system.

The number of possible implementations of subsystem j is given by:

$$\left|V_j\right| = \sum_{r_j\in R_j} \sum_{k(r_j)\in M_{K_j}^{r_j}} \left|V_{jk(r_j)}\right|, \left|V_{jk(r_j)}\right| = C_{|K_j|}^{r_j}.$$

The reliability of the CI protection subsystem and the resource indexes on the alternative $v_{jk(r_j)}$ is given by:

$$p_j(v_{jk(r_j)}) = 1 - \prod_{k_p \in k(r_j)} (1 - p_j(u_{jk})),\ g_{ij}(v_{jk(r_j)}) = \sum_{k_p \in k(r_j)} g_{jk}(u_{jk_p}), i \in I.$$

We have to choose the implementation of the complex CI protection system v, which has the maximum reliability score for the given constraints b_i, $i \in I$ on resources. The formulated problem of protection system structure optimization is a two-level problem of discrete monotonic programming, which general notation is the following (Problem A):

Maximize:

$$P(v) = \sum_{x\in X} Y(x) \prod_{j\in J} p_j(v_{jk(r)})^{x_j} (1 - p_j(v_{jk(r)}))^{1-x_j}, \tag{3}$$

under constraints:

$$g_i(v) = \sum_{j\in J} g_{ij}(v_{jk(r_j)}) \le b_i, i \in I, \tag{4}$$

$$v = (v_{jk(r_1)},...,v_{jk(r_j)},...,v_{jk(r_n)}) \in V = \prod_{j\in J} V_j, \tag{5}$$

where $P(v)$, $g_i(v)$, $i \in I$ are, respectively, the reliability and technical-economic indexes of the alternative v. Note that each j-th subsystem may be in one of two states:

$$x_j = \begin{cases} 1, \text{ operational,} \\ 0, \text{ failed.} \end{cases}$$

and if boolean structure function $Y(x) = 1$ in state $x = (x_1,...,x_j,...,x_n)$ then CI protection system is operational, otherwise if $Y(x) = 0$ protection system is failed. The algorithm for solving the problem is based on the Sequential Analysis and Sifting of Variants technique (Volkovich et al. 1992, Volkovich and Zaslavsky 1986).

Problem (3)–(5) can also be considered to study systems with serial connection of elements (serial structure), then the system reliability indicator is calculated by the formula $P(v) = \prod_{j \in J} p_j(v_{jk(r)})$, and the scheme for solving the problem is considered as a special case of criterion (3) and remains unchanged.

The approach to solve problems (3)–(5) includes the calculation of tolerances on two levels (Volkovich et al. 1992, Volkovich and Zaslavsky 1986). The first level tolerances on resources should be defined for a set of variants $v_{jk(r_j)}$, and the second level tolerances should be defined for sets of elements U_j that are used to create variants of the subsystem.

The procedures Ψ_1 and Ψ_2 are the basis of the algorithm, we compute the tolerances of the first and second levels, improve them, and reduce the length of subsystems variants, i.e., reduce the problem dimension. Rules for calculation tolerances of the first level are established by the following theorem.

Theorem 1. Value

$$d_{ij} = b_i - \sum_{\ell \in J, \ell \neq j} \min_{v_{jk(r_\ell)} \in V_\ell} \{g_{i\ell}(v_{jk(r_\ell)})\}, \; i \in I, j \in J, \tag{6}$$

is a first-level tolerance for the set V_j, by the i-th constraint.

Analysis of the sets V_j by the tolerances d_{ij} is performed using the following constraint subsystems:

$$g_{ij}(v_{jk(r_j)}) \leq d_{ij}, i \in I, \tag{7}$$

$$v_{jk(r_j)} \in V_j. \tag{8}$$

Let $Q_j = \{d_{ij}, i \in I\}$ be the set of tolerances for the constraint subsystems (7)–(8), $Q = \bigcup_{j \in J} Q_j$ is a "tolerance system". The set of tolerances $Q_j = \{d_{ij}, i \in I\}$ is preferred to set of tolerances $Q'_j = \{d'_{ij}, i \in I\}$ (notation $Q \geq Q'$) if $d_{ij} \leq d'_{ij}$ and at least for one $i \in I$ inequality is strict. The tolerance system Q is preferred to the tolerance system $Q' = \bigcup_{j \in J} Q'_j$ if at least for one $j \in J$ the preference $Q_j \geq Q'_j$ is strict. The tolerance system Q is computed and improved by the procedure Ψ_1 both by improving the values and by restricting the set V_j. The constraint subsystems (10)–(11) with the tolerances d_{ij} are applied for the following purposes in the analysis of the sets of elements by second-level tolerances performed by the procedure Ψ_2:

(1) to eliminate the types of elements $k \in K_j$, thus reducing the number of combinations $k(r_j) \in M_{Kj}^{rj}$ and eliminating the sets $V_{jk(r_j)}$;

(2) to reduce $r_j \in R_j$, i.e., the lengths of the subsystem alternatives.

Theorem 2. Let $v_{jk(rj-1)}^{(i)} = \arg\min\limits_{k(r_j-1)\in M_{K_j}^{r_j-1}} g_{ij}(v_{jk(r_j-1)}), i \in I$. Then

$$d_{ij}^{r_j} = d_{ij} - g_{ij}(v_{jk(r_j-1)}^{(i)}), r_j \in R_j, \tag{9}$$

is the second-level tolerance for the set of elements $u_{jk} \in U_j$ by the i-th constraint (7) in the analysis of the alternatives $v_{jk(r_j)}$ of length rj of the j-th subsystem. Note that for $r_j = 1$ we analyse alternatives consisting of elements of one type and then $g_{ij}(v_{jk(rj-1)}) = 0$.

Proposition 1. Let d_{ij}^{rj} be the tolerance (9). If for at least one $i \in I$

$$g_{ij}(u_{jk}) > d_{ij}^{rj}, \tag{10}$$

then the k-th type element is not used in the construction of the alternative $v_{jk(rj)}$ of length r_j.

From this Proposition 1 follows that if

$$|K'_j| < r_j, \tag{11}$$

where K'_j is the set of element types remaining in K_j after a check of the condition (13), then there exist no alternatives of length r_j of the subsystem j which satisfy (3)–(4). Thus, analysis of the conditions (7)–(8) produces a new set of alternatives for subsystem j:

$$V'_j = \bigcup_{r_j \in R_j} V_j'^{r_j} \subseteq V_j, j \in J,$$

where

$$V_j'^{r_j} = \bigcup_{k(r_j) \in M_{K_j}'^{r_j}} V_{jk(r_j)}, M_{K_j}'^{r_j} \subseteq M_{K_j}^{r_j}, R_j \subseteq R_j.$$

Note that for $R'_j = \emptyset$ exists no alternative realizations of subsystem j satisfying the tolerance d_{ij}. Clearly, if $V'_j = \emptyset$, i.e., the application of the procedure Ψ_2 has eliminated the entire set of alternatives V_j, the problem is infeasible. If the number $|V'_j|, j \in J$ of elements in the sets $V'_j, j \in J$ is small (not exceeding some number N determined by the computer resources), then an aggregated problem $A_{(0)}$ is generated and solved. The optimal solution to this aggregated problem, to be considered below, is also the optimal solution to the original problem. If the set $V'_j, j \in J$ is large, the search for the optimal solution is continued by introducing a supplementary constraint on the reliability of the protection system:

$$P(v) \geq P^*, \tag{12}$$

where P^* is defined on $[P_{\min}, P_{\max}]$, $P_{\min} = \min_{v \in V} P(v)$, $P_{\max} = \max_{v \in V} P(v)$.

We represent $P(v)$ in the form $P(v) = \varphi(p_1,...,p_j,...,p_n)$, where the reliability function $\varphi(p_1,...,p_n)$ is defined on $\prod_{j \in J} Hp_j(V_j)$. Here $Hp_j(V_j)$ is the ordered value set of the function p_j.

The first-level reliability tolerances (constraint 12) are determined by the following theorem.

Theorem 3.

$$d_j = \min_{\prod_{j \in J} Hp_j(V_j)} \varphi(\tilde{p}_1,..., \tilde{p}_{j-1}, p_j, \tilde{p}_{j+1},..., \tilde{p}_n) \geq P^*, \tag{13}$$

where $\tilde{p}_j = \max_{v_j \in V_j} p_j(v_j), j \in J$ is a first-level reliability tolerance for the set V_j.

We supplement the constraint subsystem (7)–(8) with the set of tolerances Q_j with the constraint

$$p_j(v_{jk(r_j)}) \geq d_j. \tag{14}$$

Then $Q = \cup_{j \in J} Q_j$ is a complete tolerance system for the constraint subsystem (7)–(8), (14) where $\bar{Q}_j = Q_j \cup \{d_j\}$. Using these constraints subsystems, we apply the procedure Ψ_2 to continue our analysis of the set of elements.

Theorem 4. Let d_j be the tolerance (11),

$v_{jk(r_j-1)} = \left(u_{jk_1},..., u_{jk_p},..., u_{jk_{r_j-1}}\right) = \arg\max_{k(r_j-1) \in M_{K_j}^{r_j-1}} p_j(v_{jk(r_j-1)})$. Then

$$d_j^{r_j} = \lg(1-d_j) - \sum_{k_p \in k(r_j-1)} \lg\left(1 - p_j\left(u_{jk_p}\right)\right), r_j \in R_j, \tag{15}$$

is a second-level reliability tolerance for the set of elements in the analysis of the alternatives $v_{jk(r_j)}$ of the length r_j of subsystem j.

Proposition 2. Let $d_j^{r_j}$ be the tolerance (14). If

$$\lg(1 - p_j(u_{jk})) > d_j^{r_j},$$

then the element u_{jk} of type k is not used in the alternatives $v_{jk(r_j)}$ of length r_j. If

$$0 <] d_j^{r_j} * \lg(1 - p_j(u_{jk}))^{-1}[, \tag{16}$$

then the element u_{jk} of type k may not be used in alternatives $v_{jk(r_j)}$ of length r_j. Proposition 2 implies that if

$$|K'_j| < r_j, \tag{17}$$

where K'_j is the set of element types remaining in K_j after a check of the condition (16), then there no alternatives of length r_j of subsystem j exists which satisfy the condition (14). Thus, by applying the procedure Ψ_2 to eliminate the uncompetitive alternatives, we obtain a set V^t, such that $V^t \in V$, which is further investigated for the possible presence of optimal alternatives. The procedure Ψ_2 for the analysis of alternatives $v_{jk(r_j)}$ of length r_j of subsystem j may be formalized as follows.

Procedure Ψ_2

Step 0. Set $K_{j,r_j}^{(0)} = K_j$, $r_j \in R_j$, $v = 0$.

Step 1. Find the tolerance $d_{ij}^{(v-1)r_j}$, $i \in I, j \in J, r_j \in R_j^{(v-1)}$ from (10), where

$$v_{jk(r_j-1)}^{(i)\alpha_j} = \underset{k(r_j-1)\in M_{K_{j,r_j}}^{r_j-1}(v-1)}{\arg\min} \; g_{ij}(v_{jk(r_j-1)}).$$

Checking the conditions (10)–(11) and applying the formula (12), generate the following sets $K'^{(v-1)}_{j,r_j}$, $R'^{(v-1)}_j$.

Step 2. If $R'^{(v-1)}_j = \emptyset$, end.

Step 3. Compute the tolerances $d_{ij}^{(v-1)r_j}$, $r_j \in R'^{(v-1)}_j$, from (20), where

$$v_{jk(r_j-1)} = \underset{k(r_j-1)\in M_{K_{j,r_j}}^{r_j-1}(v-1)}{\arg\max} \; p_j(v_{jk(r_j-1)}).$$

By checking the conditions (16)–(17) generate the sets $K_{j,r_j}^{(v)}$, $R_j^{(v)}$.

Step 4. If $R_j^{(v)} = \emptyset$, end.

Step 5. If $K_{j,r_j}^{(v)} = K_{j,r_j}^{(v-1)}$, $R_j^{(v)} = R_j^{(v-1)}$ end. Else go to step 1, setting $v = v + 1$.

Procedure Ψ_1

Step 0. Let $V_j^{(0)} = V_j, j \in J$, $V^{(0)} = \prod_{j \in J} V_j^{(0)}$. On the set $V^{(0)}$ define the system of tolerances $\bar{Q}^{(0)} = \bigcup_{j \in J} \bar{Q}_j^{(0)}$, $\bar{Q}_j^{(0)} = Q_j^{(0)} \cup \{d_j^{(0)}\}$, $Q_j^{(0)} = \{d_{ij}^{(0)}, i \in I\}$, where the tolerances $d_{ij}^{(0)}$ and $d_j^{(0)}$ are computed from (9), (19), respectively.

Step 1. To the constraint system (7), (8), (14) with the tolerances $\bar{Q}_j^{(v-1)}, j \in J$ apply the procedure Ψ_2. Find a new tolerance system $\bar{Q}^{(v)} = \bigcup_{j \in J} \bar{Q}_j^{(v)}$, where the tolerances $d_{ij}^{(v)}$ and $d_j^{(v)}$ are computed from the analogues of (9), (18), respectively. Go to step 2.

Step 2. If $\bar{Q}^{(v)} = \bar{Q}^{(v-1)}$, end. Otherwise, since $\bar{Q}^{(v)} \geq \bar{Q}^{(v-1)}$, set $v = v + 1$ and go to step 1.

Optimization algorithm

The algorithm to solve the problem (3)–(5) is described below.

Step 0. On the set V apply the procedure Ψ_1. It generates the set $V^{(0)} = \prod_{j \in J} V_j^{(0)}$ and the tolerance system $\bar{Q}^{(0)} = \bigcup_{j \in J} \bar{Q}_j^{(0)}$. If the number of alternatives in $V^{(0)}$ is sufficiently small, then solve the problem A by direct enumeration of the set $V^{(0)}$. Else go to step 1.

Step 1. If $M_1 = \sum_{j \in J} |V_j^{(0)}| > N_1$, then set $P_{\min} = \min_{v \in V^{(0)}} P(v)$, $P_{\max} = \max_{v \in V^{(0)}} P(v)$, $\gamma = 1$, $P_*^{(\gamma)} = (P_{\min} + P_{\max})/2$ and go to step 2. N_1 is determined by the computer resources. Otherwise, use direct enumeration to eliminate from the sets $V_j^{(0)}$ those alternatives which do not satisfy the tolerances $V_j'^{(0)}$ to generate the aggregated discrete monotone programming problem $A_{(0)}$:

$$P(v) = P(p_1(v_{1\ell_1}),\ldots,p_j(v_{j\ell_j}),\ldots,p_n(v_{n\ell_n})) \rightarrow \max;$$

$$g_i(v) = \sum_{j \in J} g_{ij}(v_{j\ell_j}) \leq b_i,\ i \in I,$$

$$v = (v_{1\ell_1},\ldots,v_{j\ell_j},\ldots,v_{n\ell_n}) \in V^{(0)} = \prod_{j \in J} V_j^{(0)},$$

where $V_j'^{(0)} = \{v_{1\ell_1},\ldots,v_{j\ell_j},\ldots,v_{j\ell^*_{jn}}\}$ is a finite set of alternatives of subsystem j satisfying the constraint subsystem (7), (8), (13) with the tolerances $\bar{Q}_j^{(0)}$.

If $|V'^{(0)}| < N$, then solve the problem $A_{(0)}$ by direct enumeration of alternatives, otherwise apply the algorithm (Volkovich and Zaslavsky 1986) to solve the problem $A_{(0)}$.

If we have found an optimal alternative $v^* = (v^*_{1\ell_1},\ldots,v^*_{j\ell_j},\ldots,v^*_{n\ell_n})$ of the problem $A_{(0)}$, then choose the j-th subsystem alternative $v_{jk(r_j)} \in V_j^{(0)}$ corresponding to the component $v^*_{j\ell_j}$, in order to obtain an optimal solution of a problem A.

If the solution of the problem $A_{(0)}$ runs into computational difficulties (the parameter N_1 is fairly large), then set $\gamma = 1$, $P_*^{(\gamma)} = (P_{\min} + P_{\max})/2$, and go to step

2. If at least for one $j \in J$ we have $V'^{(0)}_j = \varnothing$ or the problem $A_{(0)}$ has no solution, then the original problem A has no solution.

Step 2. Apply the procedure Ψ_1 using the constraint (13) with $P^{(\gamma)}_*$. This generates the tolerance system $\bar{Q}^{(\gamma)}$ and the set $V^{(\gamma)} = \prod_{j \in J} V'^{(\gamma)}_j$.

The following three cases are possible:

1. $V^{(\gamma)} = \varnothing$. Go to step 3.

2. $M_1 = \sum_{j \in J} |V^{(\gamma)}_j| \leq N_1$. Using the tolerance $\bar{Q}^{(\gamma)}$ generate on the set $V^{(\gamma)}$ the aggregated problem $A_{(\gamma)}$ analogous to $A_{(0)}$ and solve it either by direct enumeration (if $|V^{(\gamma)}| \leq N$) or by the algorithm (Volkovich and Zaslavsky 1986). The solution to the problem may lead to the following sub-cases:

2.1. v' - the optimal alternative of the problem $A_{(\gamma)}$ - has been found. Check v' for optimality using the optimality tests (Theorem 2). If the optimality tests are not satisfied, then v' is an approximate solution to the problem A.

If v^* is an optimal or an acceptable approximate solution of the problem A, then end. Otherwise, set P'_{min} $P^{(\gamma+1)}_* = P_{min}$, $\gamma = \gamma + 1$, and go to step 2.

2.2. The problem $A_{(\gamma)}$ has no feasible solutions. Go to step 3.

2.3. The solution of the problem $A_{(\gamma)}$ runs into computational difficulties. Go to step 4.

Step 3. Reduce the value of $P^{(\gamma)}_*$, setting $P^{(\gamma+1)}_* = (P_{min} + P^{(\gamma)}_*)/2$, $P^{(\gamma)}_{max} = P^{(\gamma)}_*$ and go to step 2 with $\gamma = \gamma + 1$. Here $P^{(\gamma)}_{max}$ is the upper bound on the value of the reliability function in step 2.

Step 4. Increase $P^{(\gamma)}_*$, setting $P^{(\gamma+1)}_* = (P^{(\gamma)}_* + P^{(\gamma)}_{max})/2$, $\gamma = \gamma + 1$ and go to step 2. When going to step 2 with $\gamma = \gamma + 1$, we check the condition $P^{(\gamma+1)}_* - P^{(\gamma)}_* > \varepsilon_1$, where $\varepsilon_1 \geq \varepsilon_0$, ε_0 is some positive constant which is determined by the rounding errors.

If $P^{(\gamma+1)}_* - P^{(\gamma)}_* \leq \varepsilon_1$, where $\varepsilon_1 \geq \varepsilon_0$, ε_0, is sufficiently small, but $|V^{(\gamma)}| > N$ and $M_1 > N_1$, then increase the parameters N, N_1 or stop after finding an acceptable approximate solution.

The algorithmic scheme for solving two-level discrete monotone programming problem (3)–(5) and the procedure of unpromising variants exclusion are based on the Sequential Analysis and Sifting of Variants technique that is used for formalizing and solving various applications (Volkovich et al. 1992, Afanasiev et al. 1993, Zaslavsky and Ievgiienko 2010, Kadenko et al. 1999a).

The considered procedures for the analysis and elimination of variants and the algorithmic scheme implement the mechanism of the type-variety principle when it is applied in the problem of optimal redundancy. The next section discusses the application of the type-variety principal mechanism for solving very important problems of ensuring the safety of the functioning of critical systems at the stage of their life cycle - operation. During the operation of critical systems, failures of their components occur due to various external influences, ageing of elements, etc. In this regard, the tasks of scientific and technical support of CIs arise to prevent possible failures due to the timely detection of various defects and the timely replacement of elements of these systems, which is implemented during the maintenance period.

2.4 Implementation of the mechanism of the type-variety principle for revealing defects in critical systems

The functioning of critical infrastructures, their security systems and other important components during operation may be disrupted due to defects, malfunctions, ageing, exposure to aggressive environmental factors, cyber-attacks, accidents, etc. To ensure the reliable operation of such systems, it is necessary to form a set of instruments, tools, and diagnostic systems for detecting defects and malfunctions, and failures with high reliability (Kadenko et al. 1999a, Zaslavsky and Kadenko 1999b). Indicators of the effectiveness of diagnostic systems are the probability of detecting defects, their accuracy, sensitivity and resolution, cost, duration of observation and other characteristics. The reliability of the control is characterized by a stable correspondence of the test results to the actual value of the calculated value, carried out under the same conditions. The use of the most sensitive methods does not mean that the reliability of the data will be correspondingly the greatest and considering the priority of technical indicators can lead to contradictions with economic criteria, and the formation of such a set of instruments becomes inefficient from an economic point of view. This contradiction actualizes the task of forming an optimal complex of different types of methods, instruments, and technologies for diagnosing various components of critical infrastructures and their objects.

Let us consider an important optimization problem of forming a set of methods for detecting various defects, which must be considered when ensuring and improving the reliability of critical infrastructures, their safety and security systems at the operational stage. This task is important in the implementation of the maintenance phases of the systems.

Defect detection could be performed by using a variety of inspection methods (devices, measuring techniques, expert inspections, etc.). Let us consider the mathematical formulation of the problem for designing the complex of inspection methods for defect detection.

Let's assume that the finite set of possible defects $d_\ell \in D$, are given, that may be present in a variety of systems and components, security systems, where $D = \{d_\ell, \ell \in L = \{1,2,...,\ell^*\}\}$ is a set of different types of defects, and L is a set of indexes types of defects. Defect means a mismatch of the given values of parameters required for the system's safety. The defects $d_\ell \in D$ are characterized by various parameters, for example, the potential hazard indicators, the inability to receive and distinguish signal, aggregate, and transmit information etc.

Let us assume that, for monitoring and diagnostics defects $d_\ell \in D$, various methods, devices, techniques, etc., are used, which have different characteristics of defect detection probability and cost of resources in the implementation of control.

Every modification m_{jk} of j-th method characterized $p_j(m_{jk}, d_\ell)$ probability of detection of defect d_ℓ, and $g_{i\ell}(m_{jk}, d_\ell)$, $i \in I$ - resource expenses for the realization the diagnostic by method m_{jk} (time required to oversee by a group of experts, cost of the control etc.), where $I = \{1,...,i,...,i^*\}$ is a set of indexes of expenses resources parameters.

It is necessary to design a set of detection methods that allow to detection defects from D. These methods would be most effective for the process of detecting of defects $d_\ell \in D$ in the safety and protection systems.

We suggest that strategy order (technology) application of methods for the design and the choice of set of methods are fixed. Let's designate $S_\ell = \{s_\ell = \{j_1, j_2,...,j_{\ell^*}\} | j_1, j_2,...,j_{\ell^*} \in J_\ell\}$set of combinations s_ℓ index types of methods used to diagnose the defect d_ℓ. A combination s_ℓ would be interpreted as technology combined methods for the diagnosis of components and systems, which considers the specificity of methods, devices, and technologies for the implementation of effective control and security system specifics. For each combination s_ℓ we shell define a complex $v_{s_\ell} = (m_{jk})_{j \in s_\ell}$ of various types of methods, which can be used for the diagnosis and detection of the defect $d_\ell \in D$. Let us assume, that the order of application of methods is fixed (based on existing experience of priority use of different methods in the study of different types of systems) by the order of indices combination.

The following sets could be defined: $V_{s_\ell} = \otimes_{j \in s_\ell} M_{j\ell}$ is a set of complexes of methods, which determine possible variants realization of technology s_ℓ; $V_\ell = \cup_{s_\ell \in s_\ell} V_{s_\ell}$ is a set of all possible complexes, which can be used for detection of defect d_ℓ. Here $\otimes$ is Cartesian product of the sets. The multitude of complexes for detection of all set of defects $d_\ell \in D$ is defined as follows: $V = \otimes_{\ell \in L} V_\ell$.

The probability of defects detections by the complex methods defines through the probability of defects detection parameters of by separate methods, which are part of the technology s_ℓ. Objective evaluation of the state of security systems can be significantly hampered by the effect of interference of elements on each other, which is amplified in the process of long-term operation. To increase the reliability of control in these conditions it is advisable to use the principle of redundancy control and for evaluation of the same parameter used in controlling various physical methods to investigate the nature of the components of safety systems. The use of the combination of different methods, and their consequent application only increase the probability of defects detection. The probability of defect detection d_ℓ by complete v_{s_ℓ} with a usage of a fixed set of methods in technology s_ℓ is calculated as follows:

$$P(v_{s_\ell}) = 1 - \prod_{j \in J_\ell}(1 - p_j(m_{jk}, d_\ell)). \tag{18}$$

The formula (18) gives the bottom estimation for the probability of defect detection by a complex of methods v_{s_ℓ}.

Integrated usage of the methods in the diagnosis aimed at identifying the actual state of the component, increases reliability, and efficiency and reduces risk with safety systems. Due to aging of components of security systems, increasing demands for defect detection using various techniques and control systems. Different defects D that occurs during the operation may have varying importance to provides interference with each other or is insignificant even after the prolonged operation of safety systems. Consider the following formal statement of optimization problems.

Problem 1. When implementing control all defects are to be detected.

The given condition could be interpreted as follows. The set of defects D could be considered, as a complex system, which consists of ℓ^* subsystems (subsystems, in this case, are defects). Thought a parameter of the probability of non-failure operation of the consecutive system (when the refusal even in one of the subsystems could lead to failure of the system as a whole) it is possible to consider as a parameter of the probability of detection of defects set by the complex $v = (v_{s_1}, v_{s_2}, \ldots, v_{s_{\ell^*}})$, which is defined as follows: $P_D(v) = \prod_{\ell \in L} P(v_{s_\ell})$, where $P(v_{s_\ell})$ is defined by the formula (18).

Problem 2. In the realization of diagnosis, some unimportant (non-essential) defects are considered not to be detected.

Defects of set D could have various importance and therefore it is possible to speak about the inadmissibility of opportunity of detection or not detection of essential and unessential defects, and they cannot result in dangerous consequences, the probable price of elements refusal is insignificant for serviceability the protection system.

This task is interpreted in terms of determining the reliability of complex systems with (monotone) structure (Barlow and Proschan 1975).

Let us x_ℓ, $l \in L$, is Boolean variable that is

$$x = \begin{cases} 0, \text{ than missing of defect } d_\ell \text{ can be accepted,} \\ 1, \text{ than missing of defect } d_\ell \text{ can not be accepted.} \end{cases}$$

Then $x = (x_1,...,x_\ell,...,x_{\ell^*})$ is a vector of possible combination of defect revealing. X is a set of all possible combinations $x \in X$. For example, a combination $x = (1,...,1,...,1)$ is not admitted of any not revealed defect and in a combination $x = (0,1,....,1)$ "no detection" of one insignificant defect d_1 is admitted.

Let us $\omega(x)$ Boolean indicator function that is:

$$\omega(x) = \begin{cases} 0, \text{ combination } x \text{ is not allowed to control object,} \\ 1, \text{ combination } x \text{ is allowed to control object.} \end{cases}$$

For example, if a combination $x = (1,...,1,0_\ell,1,...,1)$, and $\omega(x) = 1$, in this case for the safety system is not allowed the identification of the defect d_ℓ. It is necessary to generate such a variant of methods complex $v = (v_{s_\ell}) \in \otimes_{\ell \in L} V_\ell$, which has a maximum efficiency of revealing defects set D in view of the given constraints on resources b_i, $i \in I = \{1,....,i^*\}$. The mathematical model of the formulated problem is:

maximize

$$P_D(v) = \max_{v \in V} \left\{ \sum_{x \in X} \omega(x) \prod_{\ell \in L} P(v_{s_\ell})^{x_\ell} \left(1 - P(v_{s_\ell})\right)^{1-x_\ell} \right\}, \tag{19}$$

under constraints

$$G_i(v) = G_i(g_{i1}(v_{s_1}),...,g_{i\ell}(v_{s_\ell}),...,g_{i\ell^*}(v_{s_{\ell^*}})) \le b_i, \; i \in I, \tag{20}$$

$$v = (v_{s_1}, v_{s_2},...,v_{s_{\ell^*}}) \in V = \otimes_{\ell \in L} V_\ell. \tag{21}$$

Parameters of resource expenses $G_i(v)$, $i \in I$ for using the control system which a complex of methods V are defined, for example, by total control cost of using methods that are part of a complex, by the cost of auxiliary operations accompanying to the control, by the cost of expenses connected with the charge of energy and use of premises, the cost of replacement items etc. Thus $G_i(v)$, $i \in I$ is additive function $G_i(v) = \sum_{\ell \in L} g_{i\ell}(v_{s_\ell})$, $i \in I$ and the parameters of resource expenses for a vector v_{s_ℓ} could be determined (Volkovich et al. 1992).

$$g_{i\ell}(v_{s_\ell}) = \sum_{j \in s_\ell} g_{i\ell}(m_{jk}, d_\ell), \; i \in I, \; l \in L.$$

In the general case, indicators of resource costs can be calculated more difficult to be set in tabular form, dependent on the technology s_ℓ application methods. The problems (19)–(21) are a two-level discrete monotone programming problems, and it could be solved using procedures and algorithms discussed in Chapter 1.2. Problems (3)–(5) and (19)–(21) are also of interest in a multi-criterion setting, and their solution may give other alternatives that depend on the weights of the criteria.

3. The human factor in ensuring reliability, safety and fault tolerance of critical infrastructures

3.1 Use of the human health index for reliability

In the QRAQ Project Volume 27 (Taylor 2014) devoted to the problem of quantitative risk assessment (QRA) in the requirements section for eliminating deficiencies in the analysis risks, a requirement is included on the need to implement methods of reliability analysis of a human (human factors) in QRA and security management processes, including the number of actions of the decision maker, the operator. The works Rasmussen and Taylor 1976, and Kirwan 1994 present recommendations for the analysis of the quantitative assessment of human reliability, which has been used in this work.

It is important to note that the human factor, i.e., professionalism and the reliability of the performance of various operations by employees is a very important component of the reliability of functioning critical infrastructures and man-machine systems. Impact Study of Human Factors by the BP Company (BP and Chem 2008) showed that the occurrence of accidents at the refinery and the proportion of human errors (caused by personnel) is 33%. According to official data on railway transport in Ukraine, more than 80% of accidents are caused by the human factor (Gus et al. 2008).

Consider a CI object as a two-component system consisting of technical parts and personnel (decision maker (DM)). The probability of failure-free operation of the object $P(t)$, at a separate stage of operation in point in time t_i due to reliability of the functioning of the technical part of object $P_T(t_i)$ in point in time t_i and the reliability of the DM performing various operations $P_H(t_i)$ on the object is defined as follows

$$P(t_i) = P_T(t_i) \cdot P_H(t_i).$$

For fail-safe execution of the stage at time t_i should the technical component of CI as well as personnel be functioning. Object failure probability (risk value) is calculated by the formula

$$R(t_i) = 1 - P(t_i) = 1 - P_T(t_i) \cdot P_H(t_i).$$

At the heart of the idea of improving the reliability of the functioning of personnel, the approach, which is based on the selection of special physical exercises that belong to various physical practices to increase individual health index (*Health index – HI*) is proposed Belov et al. 2001, Horbunov et al. 2019. The human health index is considered in concepts of the information paradigm of the individual index of human health *HI* as a trinity of its physical, mental, and social side and is presented as expression:

$$HI_n = F_n\left(HI^{ph}\left(I_n^{ph}\right), HI^{ps}\left(I_n^{ps}\right), HI^{so}\left(I_n^{so}\right)\right), \tag{22}$$

and is the initial state of the indicator *HI*. Function F_n parameters are exponents: $HI^{ph}(I_n^{ph})$ is a health index, which characterizes the physical state of human health, where I_n^{ph} is an impact on the physical component of human health, which have standard methods for supporting and treating staff. Respectively $HI^{ps}(I_n^{ps})$ is the psychological index, and $HI^{so}(I_n^{so})$ is the social health index, I_n^{ps} is the impact on psychological and I_n^{so} is an impact on the social components of health.

Standard methods of support and impact on human health, defining influences I_n^{ph}, I_n^{ps}, I_n^{so} are formed based on the developed programs, professional selection of specialists (DM, operators), organization of their work, recreation, special nutrition, methods of physical culture, and medical support. In the absence of crisis situations and stressful effects on decision-makers, the reliability and safety of CI operations are usually maintained on a given level.

In times of crisis, when the vast majority of the population is exposed to prolonged, significant stress on the physical, psychological and social level (pandemic, war, food and financial crises), reliability and the safety of critical facilities and CI can be violated through the human factor, due to the deterioration of the health and professional activities of the personnel (Galea et al. 2020, Panchenko et al. 2015).

The traditional medical indicators used in this case (measurement of heart rate, blood pressure) with a certain probability do not always make it possible to identify functional states that are dangerous from the maintaining the level of professional reliability point of view but not included in pathology. These conditions, in many cases, lead to the emergence of personnel errors and consequently – accidents (Gus et al. 2008).

Using additional, unified complexes of physical exercises (*USPE*), selected individually for each DM, it is possible to improve health, eliminate dysfunction in the musculoskeletal system (MSS) and improve its professional functional state (Horbunov et al. 2021a). Impact of the *USPE* complex on physical, psychological, and social health is defined as I_n^{ph}, I_n^{ps}, I_n^{so}, respectively, the index of individual health:

$$HI_{n+u} = F_{n+u}(HI^{ph}(I_n^{ph}, I_u^{ph}), HI^{ps}(I_n^{ps}, I_u^{ps}), HI^{so}(I_n^{so}, I_u^{so})).$$

Meaning of HI_{n+u} is characterized, as an indicator, the state of health of personnel which is restored by two influences: the first – is developed for work in a peaceful time, without stress and is determined by (22); the second is additional, for the time of pandemic and war, with additional complex *UCPE*. Moreover, both influences affect all components of health. Function F_{n+u} parameters are exponents: $HI^{ph}(I_n^{ph}, I_u^{ph})$ – health index that characterizes the physical state of human health under the influence of the *UCPE* complex, where I_u^{ph} – impact on the physical component of the health of DMs by standard methods, and I_u^{ph} – impact on the physical component of health due to complex *UCPE*. Respectively $HI^{ps}(I_n^{ps}, I_u^{ps})$ – index of psychological and $HI^{so}(I_n^{so}, I_n^{so})$ – index of human social health and corresponding influences.

Then the reliability and safety of the functioning of the CI, on a series of n serial time steps t_i by time t can be defined:

$$P(t) = \prod_{i=1}^{n} P_T(t_i) \cdot P_H(t_i).$$

In the diagram shown in Figure 1, the time point t_i is defined separately, which considers the possibility of implementing the *UCPE* complex in free time or at the facility during breaks between operations carried out by personnel.

In Figure 1, the indicators $P_H(t_i^H)$ - the probability of non-failure operation of a person on t_i stage of the system operation during the time t_i^H. $\Delta P(HI_{n+u})$ – the result of a positive impact of *UCPE* implementation, is determined by the increase in the index of individual health by the value $\Delta HI_{n+u} = HI_{n+u} - HI_n$, which determines the additional reliability of the functional professional state by its impact $\Delta HI_{n+u} \Rightarrow \Box P(HI_{n+u})$ and represents an addition to the value of the object reliability indicator by increasing the reliability of the human factor.

To implement the process of supporting the normal physical condition of a person, and organizing his recovery based on the results of a video analysis of his condition, video image processing software is used using the "Skeleton" technology (Horbunov and Shcherbina 2021b). According to the

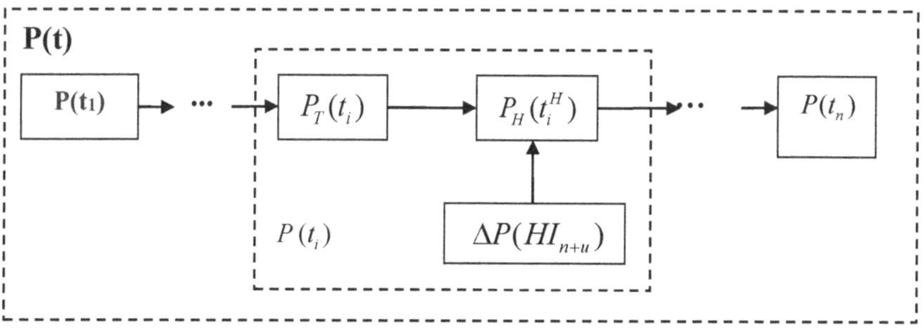

Figure 1. Scheme of serial stages in the operation of CI with the possibility of results inclusion of the *UCPE* complex impact.

video analysis of human movements, the doctor determines the elements of the musculoskeletal system that require therapeutic effects, including by performing physical exercises set *UCPE*.

Physical practice is a set of exercises that affect a certain structural part of the human body or its separate element. In this work, the problem of forming a complex of physical exercises from a variety of practices is considered. Experience of cooperation with medical institutions has shown the need to include in the therapeutic complex of exercises the following practices (Horbunov et al. 2021a):

1. The practice of "Cardio" is a set $C = \{p_{c_1}, p_{c_2},...,p_{c_\ell^*}\}$, of all possible exercises p_{c_ℓ} that affects a person's cardiovascular system to train it, where is the number of exercises in the «Cardio» practice.
2. The practice of "Flexibility" is a set $F = \{p_{f_1}, p_{f_2},...,p_{f_\ell^*}\}$, of all possible exercises p_{f_ℓ} that affect the elasticity of the musculoskeletal system;
3. The practice of "Strength" is a set $S = \{p_{s_1}, p_{s_2},...,p_{s_\ell^*}\}$ of all possible exercises p_{s_ℓ} that affect the strength qualities of the muscles;
4. The practice of "Breathing" is a set $B = \{p_{b_1}, p_{b_2},...,p_{b_\ell^*}\}$, of all possible exercises p_{b_ℓ} that affect the respiratory system;
5. The practice of "Meditation" is a set $M = \{p_{m_1}, p_{m_2},...,p_{m_\ell^*}\}$, of all possible exercises p_{m_ℓ} that affects the central nervous system;
6. The practice of "Neuro-cognition" is a set $N = \{p_{n_1}, p_{n_2},...,p_{n_\ell^*}\}$, of all possible exercises that p_{n_ℓ} affects the structures of the brain;
7. The practice of "Attention" - this is a set $A = \{p_{a_1}, p_{a_2},...,p_{a_\ell^*}\}$ of all possible exercises p_{a_ℓ} develops concentration and memory.
8. The practice of "Pain" is a set $P = \{p_{p_1}, p_{p_2},...,p_{p_\ell^*}\}$, of all possible exercises p_{p_ℓ} and those that affect trigger points in the muscles;

Table 1 shows the types of physical practices and their controlled parameters, which are used in the analysis and monitoring of health status based on the technology of video images analysis of human movements.

On the basis of a finite set of all types of physical practices $E = C \cup F \cup S \cup B \cup M \cup N \cup A \cup P$, various variants of complexes *UCPE* are formed using certain physical practices and exercises, combinations of various types of practices and exercises in them that affect the human body, taking into account its functional state, the consequences of being under stress, the presence of motor pathology and various indicators and limits. The influence of the complexes *UCPE* is presented in Figure 2 in the form of a diagram of the influence on the components of health: physical, psychological, and social.

Table 1. Exercise parameters of various types of physical practices that are calculated using video analysis methods.

Type of physical practice	**Physical practice**	**Controlled parameters of physical practice**							
		1	**2**	**3**	**4**	**5**	**6**	**7**	**8**
1. CARDIO (C)	Walking	+	+	+	--	--	+	+	*
	Nordic walking	+	+	+	--	*	+	+	*
	Gym	+	+	+	+	*	+	+	*
	Swimming	+	--	+	--	*	--	--	*
2. FLEXIBILITY (F)	Exercises in the hall on the mat	--	+	+	--	*	+	+	*
	Swedish wall	+	+	+	+	*	+	+	*
	Yoga	-	+	+	--	*	+	+	*
	Usha	--	+	+	--	*	+	+	+
3. STRENGTH (S)	Exercises with dumbbells	+	--	+	+	*	+	+	+
	Exercises with projectiles	+	--	+	+	*	*	*	*
	Exercises with a rubber band	+	--	+	+	*	+	+	+
4. BREATHING (B)	Strelnikova's system	+	--	+	+	*	*	*	*
	Buteyko system	--	--	+	--	+	*	*	*
	Testing system	--	--	--	--	--	+	--	--
5. MEDITATION (M)	Complex "Lotus"	--	--	+	*	*	*	*	*
	Calligraphy	--	+	+	*	*	+	*	+
6. NEURO-COGNITIVE (N)	Finger exercises	--	+	--	+	*	*	*	+
	"Key" exercises	--	+	--	+	*	*	*	*
	Synchronization exercises	--	+	+	+	*	+	*	+
7. ATTENTION (A)	Complex "Awakening"	--	+	+	--	*	+	+	+
8. PAIN (P)	PIR complex	--	+	+	--	*	+	+	+
	"ROLL" complex	--	+	+	+	*	*	*	*

Table 1 shows the following indicators: 1 - intensity, 2 - movement pattern, 3 - temporary load, 4 - number of iterations, 5 - displacement of test points of control, 6 - ratio of body parts, 7 - the volume of the movement, 8 - attention. Control parameters of physical practice: "+" – fixed, "--" – not fixed, "*" – not used.

The result of the complex impact I_{UCPE} obtained after *UCPE* can be expressed as HI_{n+u}:

$$I_{UCPE} => HI_{n+u} = F_{n+u}(HI^{Ph}_{n+u}, HI^{Ps}_{n+u}, HI^{So}_{n+u}).$$

Components of health presented through individual indices HI^{Ph}_{n+u}, HI^{Ps}_{n+u}, HI^{So}_{n+u}, which are defined through functions F^{ph}_{n+u}, F^{ps}_{n+u}, F^{so}_{n+u}, as an example

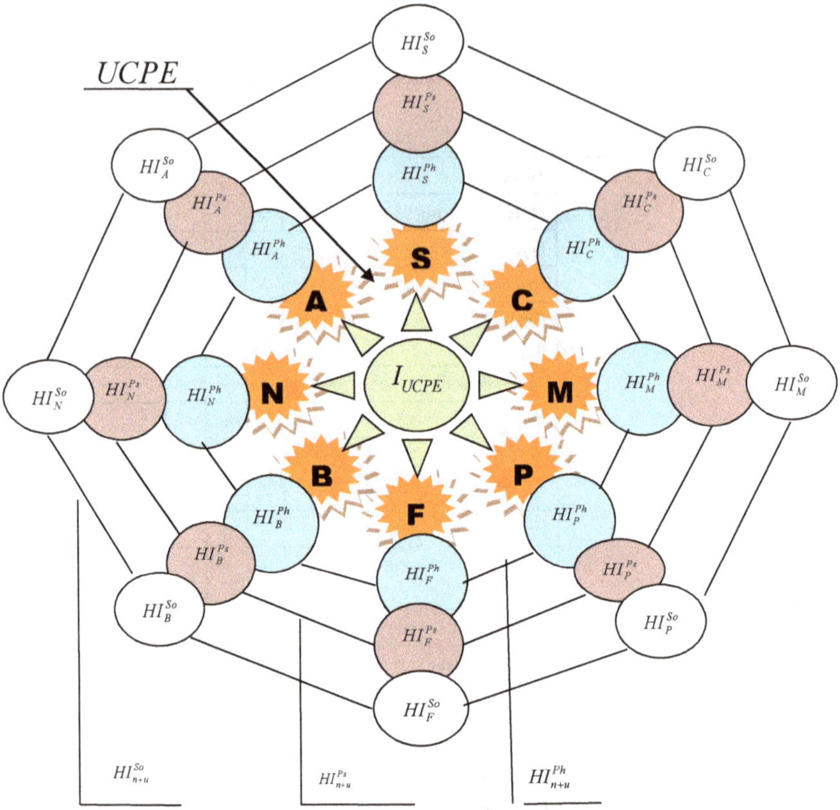

Figure 2. The impact of the *UCPE* on the integral indicator of health HI_{n+u}.

$HI^{Ph}_{n+u} = F^{Ph}_{n+u}(HI^{Ph}_C, HI^{Ph}_F, HI^{Ph}_S, HI^{Ph}_B, HI^{Ph}_M, HI^{Ph}_N, HI^{Ph}_A, HI^{Ph}_P)$, and which take into account the total impact of exercises of the complex *UCPE*, which can be built on the basis of a set of all practices $E = C \cup F \cup S \cup B \cup M \cup N \cup A \cup P$.

Calculating the health index HI_{n+u} after applying a set of exercises *UCPE*, we obtain the value of the health effect: ΔHI_{n+u}: $\Delta HI_{n+u} = HI_{n+u} - HI_n$.

The formed set of exercises in *UCPE* is intended for the purpose of its use by personnel independently, both in their free time and when performing professional activities at the workplace with the minimal participation of medical personnel. The initial data for the implementation of the process of supporting and restoring the health of personnel are the parameters of human movements, which are obtained using an intelligent software package (Horbunov and Shcherbina 2021b). Based on the video analysis data, individual sets of exercises *UCPE* are formed. The intelligent software package implements the function of monitoring the health status of CI personnel and informing responsible persons about various undesirable changes that may affect the health of personnel.

3.2 Formation of a complex of exercises UCPE using the mechanism of the principle of type-variety

As already noted, when choosing complex physical exercises *UCPE*, methods of computer video analysis of posture and elementary movements (tilts, turns of various parts of the body, and others) are used. Analysis of the parameters of video image tests allows us to identify some dysfunctions in the MSS, in muscles and ligaments, which manifest themselves in shortening the length of muscles and/or in increasing their tone (muscles or part of the muscles become dense and lose their ability to relaxation and stretching) (Travell et al. 2018).

Such conditions are not pathological MSS but are functional in nature. The danger of such dysfunctions lies in the fact that a person gets used to them and a certain stereotype is developed in movement, walking, and posture, which creates conditions for an increase in dysfunctions by the type of chain reaction (it develops especially actively in a state of stress).

If this process is not interrupted in time, then organic lesions develop in many elements of the MSS, which are expressed in the occurrence of chronic pain sensations and the appearance of specific pain points - trigger points (Travell et al. 2018). The danger of trigger points is that with a certain movement a person will experience sharp, unbearable pain. Therefore, the presence of MSS elements with dysfunctions in a person is a negative factor that reduces his functional professional reliability.

In connection with the importance of eliminating dysfunctions in the MSS elements in a person, we will consider an optimization problem, the purpose of which is to form a set of physical exercises *UCPE* that would optimally contribute to the elimination of existing dysfunctions and create conditions for blocking their development.

One of the types of exercises capable of solving this problem are exercises of physical practice "Flexibility", exercises that stretch certain elements of a person's MSS. Important requirements to *UCPE* in flexibility are the following restrictions and requirements:

- time limit for the duration of the lesson, since the time for completing the complex *UCPE* at different stages of life is different and limited;
- take into account the priorities of the impact of the complex *UCPE* on the elements of the MSS;
- the ability to add in *UCPE* new exercises based on the basic one, purposefully supplementing it. For example, if there are dysfunctions (shortenings) in the elements of the MSS, then it is necessary to select, on the basis of the basic complex *UCPE*, a new complex in which there are

additional influences (stretch marks) acting on the necessary elements of the MSS;

- in the absence of dysfunctions in the MSS, it is necessary to achieve a uniform effect on its elements.
- exclude or limit exposure to MSS elements with contraindications for various other reasons.

To provide a solution to the problem of restoring the dysfunction of "muscle shortening" as a well-known practice, we use the type of physical practice "Flexibility", which is presented in Table 1, and select the physical practice in it - "Yoga". The task of forming a complex *UCPE* using the practice of "Flexibility" is formulated as follows.

Assume that as a result of the analysis of a video image of human movements, experts have identified a finite set of elements $U = \{u_n, n \in N\}$, $N = \{1,2,...,n,...,n^*\}$MSS, for which it is necessary to use a certain set of additional physical exercises built on the basis of the well-known exercises of the "Flexibility" practice in order to support the normal functioning of the DM and eliminate the possibility of pain.

Let $q_n(u_n)$ the probability of pain in the element u_n, and $p_n(u_n) = 1 - q_n(u_n)$ the probability of the absence of pain. During the functioning of personnel in a normal environment, outside of crisis situations, the values of the indicators $p_n(u_n)$, $n \in N$ for the elements will be approximately the same (the principle of equal strength [12]) and the specified standard health requirement p^* for personnel is fulfilled, that is, $p_n(u_n) \geq p^*$, $n \in N$. During a crisis, this inequality is not met, while $p_{n_1}(u_{n_1}) \neq p_{n_2}(u_{n_2})$, n_1, $n_2 \in N$, and increases the likelihood of staff illness.

Let all elements of the set $U = \{u_n, n \in N\}$ be equivalent in terms of applying certain sets of exercises to them to implement the recovery process, built based on the practices presented in Table 1. Define $F_Y = \{y_1,...,y_j,...,y_m\}$ - the set of Yoga practice exercises, $j \in J_Y = \{1,2,...,j,...,m\}$ which can be used to restore all elements of the set U. When formalizing the problem of forming a set of exercises *UCPE* for restoring elements in a set U, we will use formalization, as in setting problem (3)–(5), since this problem is largely invariant to the original problem.

So, the analogue of the set U in problem (3)–(5) is the set of subsystems, and different types of exercises $y_j \in F_Y$ are those elements that are used to build different types of subsystems. The difference between this task and task (3)–(5) is that physical exercises $y_j \in F_Y$ are simultaneously used to restore different elements $u_n \in U$, which means they are used in different subsystems in the task (3)–(5).

The number of possible types of physical practices and elements in them $y_j \in F_Y$ that can be used for recovery is usually much greater than the number of elements $u_n \in U$, which leads to the combinatorial problem of choosing from the set of possible types of practices (exercises) such a minimum set of practices that will ensure the restoration of all elements of the set $u_n \in U$, $n \in N$. The specificity of the recovery task is also that some exercises $y_j \in F_Y$ from physical practice can be repeated. Therefore, the question arises of determining the volume of the load when performing different exercises $y_j \in F_Y$ for each element $u_n \in U$, that is, fixing quantitative indicators, such as service time, the amount of load - the number of repetitions of exercises when implementing the decision maker of physical practices from the set E. Conditions of uniformity of load formation by physical exercises for different elements $u_n \in U$ can act as additional restrictions of the task.

When choosing possible physical exercises $y_j \in F_Y$ for their implementation on a set of elements $u_n \in U$, all possible combinations of exercises are considered, just as the set of combinations $M_{K_j}^{r_j} = \{k_{\ell_j}(r_j) = \{k_1^{\ell_j},\ldots,k_p^{\ell_j},\ldots,k_{r_j}^{\ell_j}\}\}$, $k_p^{\ell_j} \in K_j$, (1) is formed, which is considered when formalizing the optimization problem (3)–(5). When forming a set of combinations of physical exercises, the wishes of the person who will perform them, the suggestions of an expert are considered in order to purposefully form the procedures for undergoing recovery, the availability of various devices for the implementation of exercises.

Suppose that for each element $u_n \in U$, $n \in N$, sets of exercises $F_Y(u_n) = \{y_{j1}^{un}, y_{j2}^{un},\ldots, y_{jk_n}^{un}\}$, $J_Y^n = \{j_1, j_2,\ldots,j_{k_n}\}$, $F_Y(u_n) \subseteq F_Y$, $n \in N$, $F_Y = \bigcup_{n \in N} F_Y(u_n)$, $J_Y = \bigcup_{n \in N} J_Y^n$, , are defined, which can be used to restore the element $u_n \in U$, $n \in N$. Moreover, the same exercises $y_j \in F_Y$ can affect different elements $u_i \in U$, $i \in N$.

Consider the sub-variant $x_{u_n} = (x_{j1}^n,\ldots,x_{j\ell}^n,\ldots,x_{jk_n}^n) = (x_j^n)$, $j \in J_Y^n$, with Boolean components

$$x_{j\ell}^n = \begin{cases} 1, \text{ for element } u_n \text{ the exercise } y_{j\ell} \text{ is involved,} \\ 0, \text{ otherwise.} \end{cases}$$

which is formed from a set of exercises $F_Y(u_n)$ to build a possible set of physical exercises for the element $u_n \in U$.

The sub-variant $x_{u_n} = (x_{j1}^n,\ldots,x_{j\ell}^n,\ldots,x_{jk_n}^n) \in X_{u_n} = \prod_{j \in J_Y^n} B_j$, $B_j = \{0,1\}, j \in J_Y$ is a certain analogue of the variant $v_{jk(r_j)} = (u_{jk_1},\ldots,u_{jk_p},\ldots,u_{jkr_j})$, which is defined in (2) when formalizing problems (3)–(5). Define τ_{jn} - the time quantum during which the exercise $y_j \in F_Y(u_n)$ is used to restore the element $u_n \in U$. Then

$$\tau_n(x_{u_n}) = \sum_{j\in J_Y^n} \tau_{jn} \cdot x_j^n, n \in N$$

- the total time of action of exercises $y_j \in F_Y(u_n)$ on the element $u_n \in U$, which are fixed in the sub-variant x_{u_n}. Let the effect of the exercise $y_j \in F_Y(u_n)$ over time τ_{jn} lead to an improvement in the state of the element $u_n \in U$ and an increase in its reliability index by the value Δp_{jn}, $\Delta p_{jn} \in (0,1)$. Then, to the value of the reliability indicator $p_n(u_n)$, it is necessary to add the values Δp_{jn}, $j \in J_Y^n$ obtained as a result of performing the exercises indicated in the corresponding sub-option x_{u_n}. This additive acts as a parallel reserve and is calculated as follows:

$$p_n(u_n, x_{u_n}) = 1-(1-p_n(u_n))\prod_{j\in J_Y^n}(1-\Delta p_{jn})^{x_j^n} = 1-q_n(u_n)\prod_{j\in J_Y^n}(\Delta q_{jn})^{x_j^n},$$

$$x_{u_n} = (x_j^n), j \in J_Y^n.$$

The indicator $p_n(u_n, x_{u_n})$ can act as a maximizable criterion, which leads to the minimization of the value $q_n(u_n, x_{u_n})$, i.e., a decrease in pain. Then the indicator $\Delta P(HI_{n+u})$ presented in Figure 1 will also improve. And it is possible to determine the magnitude of the healing effect from the complex *UCPE*, which is formed on the basis of one practice of "Flexibility" on the physical component of health: $\Delta HI_{n+u} = HI_{F,n+u}^{p\square} - HI_{F,n}^{p\square}$.

When forming variants of physical exercises x_{u_n} for elements $u_n \in U$, the following restrictions are considered from the subsystem (23)–(25) (similarly to the subsystems of restrictions (4)–(5)).

$$p_n(u_n, x_{un}) \geq p_n^*, \tag{23}$$

$$\tau_n^{\min} \leq \sum_{j\in J_Y^n} \tau_{jn} \cdot x_j^n \leq \tau_n^{\max}, \tag{24}$$

$$x_{un} = (x_{j1}^n,\ldots,x_{j\ell}^n,\ldots,x_{jkn}^n) \in X_{un}, \tag{25}$$

As a result of applying the procedures for analyzing and screening out options Ψ_1 and Ψ_2, and executing the algorithm according to the general scheme of the heterogeneity mechanism, the following two-level optimization Problem A is explicitly formed and solved, with explicitly given sets of possible options for exercise complexes $x_{u_n}^A \in X_{u_n}^A$ for each element $u_n \in U$.

Problem A.

$$R(x^A) = R(x_{u_1}^A,\ldots, x_{u_1}^A,\ldots, x_{u_1}^A) \rightarrow max(min);$$

under restrictions

$$g_k \le G_k(x^A) = G_k(x^A_{u_1},\dots,x^A_{u_1},\dots,x^A_{u_1}) \le g_k, k \in K = \{1,\dots,k,\dots,k^*\},$$
$$x^A = (x^A_{u_1},\dots,x^A_{u_n},\dots,x^A_{u_{n^*}}) \in X = \prod_{n\in N} X^A_{u_n}.$$

When forming an aggregated Problem A, various additional criteria are considered, for example, time minimization

$$\sum \sum_{j\in J^n_Y.} \tau_{jn} \cdot x^n_j \to \min,$$

the condition for the need to influence all elements $u_n \in U$.

$$\sum_{n\in N} \sum_{j\in J^n_Y} x^n_j = n^*,$$

fulfilment of the requirement to coordinate the application of a group of exercises to several elements, and others. The search $x^A_{opt} \in X$ for the optimal solution to the problem of the Problem A, which is a set of exercises *UCPE* for restoring elements in the set U, is carried out by the algorithm of Sequential Analysis and Sifting of Variants (Volkovich et al. 1992).

4. Conclusion

The type-variety principle has been demonstrated in solving the problems of ensuring the reliability, security, and resilience of CIs for a technical component and considering into account the human factor. Attention is paid to the formulation of the reliability optimization problem with type-variety elements with a demonstration of procedures and an algorithm for finding solutions that implement the type-variety mechanism. The problem of constructing optimal complexes of non-destructive testing methods for detecting defects in such systems is considered. When implementing scientific and technical support for CIs at the stages of their life cycle during crises, pandemics, and hostilities, it is necessary to pay special attention to the human factor - the personnel who are busy ensuring their reliable functioning.

From the standpoint of the type-variety principle, the influence of the human factor has been investigated and ways to improve the reliability of human functioning, when performing operations in such systems, by performing a formed heterogeneous complex of physical exercises *UCPE*, have been indicated. It has been shown how the elements of the human MSS identified by video analysis $u_n \in U$ can be used to improve the health and professional reliability of critical systems personnel using a complex of various types *UCPE*. The research results are implemented while ensuring the reliability and safety of CIs.

References

Afanasiev, V.G., V.I. Verhoturov, V.A. Zaslavsky and other. 1993. Making design of reliability satellite for communications. pp. 103–202. *In*: Reshetnev, M.F. (ed.). SSE "RASKO", Tomsk, Russia.

Ashby, W.R. 1956. An Introduction to Cybernetics, Chapman & Hall Ltd, London.

Barlow, R.E. and F. Proschan. 1975. Statistical Theory of Reliability and Life Testing Probability Models Holt, Rinehart and Winston, Inc., Boston.

Belov, V.M., A.B. Kotova and L.M. Kozak. 2001. A human: Information space of health. pp. 189–243. *In*: Gritsenko, V.I. (ed.). Bioecomedicine: A Single Information Space: Ecology - Health in the XXI Century. Nauk. Dumka, Kyiv, Ukraine.

Biryukov, D.S. and S.I. Kondratov. 2015. Green paper on the protection of critical infrastructure in Ukraine. pp. 1–176. *In*: Sukhodolia, O.M. (ed.). Proceedings of International Expert Meetings. NISS, Kiev, Ukraine.

BP and I Chem E. 2008. Hazards of Oil Refining Distillation Units, I Chem E, London.

Galea, S., R.M. Merchant and N. Lurie. 2020, 04. The mental health consequences of COVID-19 and physical distancing. The need for prevention and early intervention. JAMA Intern. Medicine, 180(6): 817–818. DOI: 10.1001.

Gus, V., M. Kutnik and V. Krot. 2008. Analysis of the state of traffic safety on the railways of Ukraine in 2007. Bulletim KNAHU, 59: 174–177.

Harris, B. 2004. Mathematical methods in combating terrorism. Risk Analysis, 24(4): 985–987.

Horbunov, O., Osadchyy, E. and V. Tereshchenko. 2019, 11. Some issues of the creation of a conceptual model of the information system for management of the health of military servants. World Science, 11(51): 40–44. DOI: 10.31435.

Horbunov, O. and P. Shcherbina. 2021b. Skeleton construction algorithms for rehabilitation systems. pp. 196–197. *In*: Egorov, B.V. (ed.). The Proceedings of the XIV International Scientific and Practical Conference, Information Technologies and Automation – 2021, Odessa, Ukraine. ONAFT.

Horbunov, O.A., Osadchyy, E.A., Klyatskiy, Y.V. and V.M. Tereshchenko. 2021a. Peculiarities of building rehabilitation programs for military personnel and combatants in sanatorium conditions with the use of information technologies. pp. 61–63. *In*: Babov, K.D. (ed.). The Proceedings of Scientific and Practical Conference, Scientific and Practical Conference with International Participation "Sanatorium-resort Treatment and Rehabilitation: Modern Development Trends", Odessa, Ukraine. Polygraph.

[IIASA] International Institute for Applied Systems Analysis. 2016. Systems Analysis for Sustainable Development. Options, Laxenburg, 31pp.

Kadenko, I., V. Zaslavskii and N. Sakhno. 1999a. Application of the complex NDT approach for inspection of NPP power system. *In*: International Symposium on Nondestructive Testing Contribution to the Infrastructure Safety Systems in the 21-st Century, Torres, Brazil, RS.

Kirwan, B. 1994. A Guide to Practical Human Reliability Assessment. Taylor and Francis, London.

Lewis, T.G. 2015. Critical Infrastructure Protection in Homeland Security: Defending a Networked Nation, 2nd Edition, Wiley, Hoboken.

Norkin, V.I., A.A. Gaivoronski, V.A. Zaslavskyi and P.S. Knopov. 2018. Optimal resource allocation models for protecting critical infrastructure. Cybernetics and Systems Analysis, 54(5): 13–26.

Ottosson, S. 2013. Fronline Innovation Management, Tervix, ABC-Tryck AB, Gothenburg.

Panchenko, O., L. Panchenko and N. Zaitseva. 2015. Social high-stress disorders of civilian population during military engagement. Science and Education, 1: 136–140.

Podkovka, O.I., M.Y. Makarchuk, N.B. Filimonova, O.S. Knyr, I.V. Pampuha and O.A. Horbunov. 2021. Neural correlates of simple sensomotor reaction as markers of information processing speed in military veterans. Fiziol. Zh. 67(6): 3–12.

Rasmussen, J. and J.R. Taylor. 1976. Notes on Human Factors Problems in Process Plant Reliability and Safety Prediction, Riso National Laboratory, Riso.

Tagarev, T. and N. Pavlov. 2007. Planning measures and capabilities for protection of critical infrastructures. Information & Security, 22: 38–47.

Taylor, J.R. 2014. The QRAQ Project Volume 27 Risk Based Safety Management and Living Risk Analysis – Closing the Gaps, ITSA, Allerod.

The Information Technology Contribution to the Building of a Safe Regional Environment. 1998. *In*: Zaslavskii, V. and MGen E. Greindl (eds.). Proceeding and Abstracts AFCEA Europe Kiev Seminar, Kiev, Ukraine. Taras Shevchenko University of Kyiv.

Travell, Simons and Simons. 2018. Myofascial Pain and Dysfunction: The Trigger Point Manual 3rd Edition, Wolters Kluwer, Alphen an den Rijn.

Tsai, S.T., M.K. Lu, S. San and C.H. Tsai. 2020. The neurologic manifestations of coronavirus disease 2019 pandemic: A systemic review. Front Neurol., 11: 1–7.

Ushakov, I.A. 1969. Methods of Solution of Simplest Optimal Redundancy Problems under Constraints, Sovetskoe Radio, Moscow.

Ushakov, I.A. 2006. Counter-terrorism: Protection resources allocation, part 2. Reliability: Theory and Applications, 2.

Volkovich, V.L. and V.A. Zaslavsky. 1986. Algorithm to solve reliability optimization problem for complex systems using type-variety redundant elements in subsystems. Kibernetika, 5: 54–61, 81.

Volkovich, V.L., A.F. Voloshin, V.A. Zaslavsky and I.A. Ushakov. 1992. Models and algorithms for reliability optimization of complex systems, Naukova Dumka, Kiev.

Zaslavsky, V.A. 2001. Risk and safety of natural and man-made systems. pp. 189–191. *In*: Durdynets, V.V., Yu.I. Saenko and Yu.O. Privalov (eds.). Social Risks and Social Safety in the Minds of Natural and Man-made Supernatural Situations and Catastrophes. Stylos, Kyiv, Ukraine.

Zaslavsky, V.A. 2006b. The type-variety principle and features of study of complex systems with the high cost of failure. Bulletin of the University of Kiev, Series: Physics and Mathematics, 1: 136–147.

Zaslavsky, V.A. 2011. Systems approach for risk management in regional systems. pp. 93–105. *In*: The International Conference "Research &Development as the Basis for Innovation in creating the competitive region". Podgorica, Montenegro, Montenegrin Academy of Sciences and Arts, 107.

Zaslavsky, V. 2014. Type-variety principle (diversity) as the base of innovation process and qualitative changes. pp. 57–60. *In*: Andersen, N. and A. Bourmistrov (eds.). Social Development through International Academic Cooperation. University of Nordland, Bodo, Norway.

Zaslavskyi, V. 2017. Type-variety principle and the algorithm of strategic planning of diversified portfolio of electricity generation sources. pp. 474–485. *In*: Zaslavskyi, V. and M. Pasichna (eds.). Advances in Intelligent Systems and Computing. Springer, Cham, Switzerland.

Zaslavskyi, V. and M. Pasichna. 2019a. System approach towards the creation of secure and resilient information technologies in the energy sector. Information & Security, 43(3): 318–330.

Zaslavskyi, V. and M. Pasichna. 2019b. Optimization techniques for modelling energy generation portfolios in Ukraine and the EU: Comparative analysis. pp 545–555. *In*: Zamojski, W., J. Mazurkiewicz, J. Sugier, T. Walkowiak and J. Kacprzyk (eds.). Contemporary Complex

Systems and their Dependability. DepCoS-RELCOMEX 2018. Advances in Intelligent Systems and Computing. Springer, Cham, Switzerland.

Zaslavsky, V.A. and A.A. Strizak. 2006a. Credit card fraud detection using self-organizing maps. Information & Security, 18: 48–63.

Zaslavsky, V.A. and I.N. Kadenko. 1999b. The role and place of non-destructive testing methods to ensure the reliability and durability of complex systems with a high price of failure. Non-destructive Testing, 1: 5–22.

Zaslavsky, V.A. and V.V. Ievgiienko. 2010. Risk analyses and redundancy for protection of critical infrastructure, monographs of system dependability. pp. 161–173. *In*: Mazurkiewicz, J., J. Sugier, T. Walkowiak and W. Zamojski (eds.). Oficyna Widawnicza Politechniki Wroclawskiej, Wroclaw, Poland.

CHAPTER 12

Informational Extended Games and Decision-Making Processes at Risk and Uncertainty

Hâncu Boris

1. Introduction

Information issues are very important for mathematical modeling decisions making problems in situations of risk and conflict. The problems with access to information generate conditions of perfect, imperfect, complete, incomplete, uncertain or asymmetric information. For the mathematical modeling of decision-making problems in situations of risk, uncertainty and informational impact a new class of games is studied. Different ways of solving games with complete and perfect information (or over the set of informational extended strategies) are studied. In the paragraphs below is a theoretical and practical analysis of how to solve games with informational extended strategies.

The Nash and Bayes-Nash solutions for informational extended games are discussed. We propose to use the new methodology for solving the complete and perfect information game. To solve games of this type we construct the incomplete and imperfect information game generated by the informational extended strategies. Then we construct an associated Bayesian game with non-informational extended strategies and determine Bayes-Nash equilibria in the incomplete information games.

Moldova State University, Republic of Moldova.
Email: boris.hancu@gmail.com

2. Information flows and the normal form of games in information extended strategies

2.1 *Normal forms of the one–way informational extended game*

One of the main problems in the game theory is to define (or to construct) the normal form (or strategic form) of the game using the "verbal" description of the game. The normal form allows beginning the more detailed studies of all approaches to solving the game. The strategic form or normal form of the game consists of a number of players, a set of strategies for each of the players, and a payoff function that associates a payoff to each player with a choice of strategies by each player.

Let

$$\Gamma = \langle I; X_i, i \in I; H_i: X \to \mathbb{R} \rangle$$

be the strategic form or normal form of the static games, where $I = \{1,2,...,n\}$ is the set of the players, X_i is a set of available alternative of the player $i \in I$, $H_i: X_i \to R$ is the payoff function of the player $i \in I$ and $X = \prod_{i \in I} X_i$ is the set of strategy profiles for the game. We will define the informational extension of the game Γ, generated by a one-way directional informational flow (Novac 2008), denoted by "$j \to i$", which meaning is: the player i, and only him, knows exactly what value of the strategy will be chosen by the player j. In the shot, we named these games as *"one–way $j \to i$ informational extended game"* or *"$j \to i$ games"*. In this case the game is done in the following way: the player i independently will choose the "actions program" as a response to the non-extended informational strategies chosen by the player j. The strategy of the player i will be defined by taking in account of all accessible pieces of information about the chosen strategy of the player j. Namely, the informational extended strategy will be constructed by the following deduction: "if the strategy x_j of the player j belongs to the specified domain then the player i will choose the specified strategy x_i", etc. On the other hand, the informational distended strategy of the player i is a response function of the player i from the chosen strategy of the player j. The normal form of the one-way $j \to i$ informational extended game will be

$$\Gamma(j \to i) = \langle I, \Theta_i, X_{[-i]}, \{\tilde{H}_p\}_{p \in I} \rangle,$$

where the set of the informational extended strategies of the player i is $\Theta_i = \{\theta_i^k : X_j \to X_i^k,\ k = 1,..., \chi_i\}$, X_i^k is the action set strategy for the player i, for all informational extended strategy θ_i^k define the set $X(\theta_i^k) = \{(x_i^k, x_j, x_{-[i,j]}):$

$x_i^k = \theta_i^k(x_j)$, $\forall x_j \in X_j$, $\forall x_{-[i,j]} \in X_{-[-ij]}\}$ of the strategy profiles for the non informational extended game Γ generated by the θ_i^k strategy and the payoff functions

$$\tilde{H}_p(\theta_i^k, x_j, x_{-[i,j]}) = H_p(x_i^k, x_j, x_{-[i,j]}),\ (x_i^k, x_j, x_{-[i,j]}) \in X(\theta_i^k).$$

2.2 *Normal forms of the double-way informational extended game*

We say that the player i is of the "$j \rightarrow i$ informational type" and as well as for the player j respectively if at any time the player i knows the precise value of the strategy which will be chosen by the player j. Now we construct the normal form of the informational extension of the game generated by a double-way informational flow (Hancu 2011), denoted by $j \leftrightarrow i$. For these games, we assume that, from the informational point of view, the player i is of the "$j \rightarrow i$ informational type" and simultaneously, the player j is of the "$j \rightarrow i$ informational type". It means that at any time the player i knows exactly the value of the strategy chosen by the player j, as well as, simultaneously, the player j knows exactly the value of the strategy chosen by player i. It should be mentioned that the players don't know the informational type of each other. In other words, the players don't know the informational extended strategies of each other and from this point of view, we can consider that game is an incomplete information structure. In this case, the game could be done in the following way: the player i independently will choose an "actions program" as a respond to the non extended informational strategies chosen by the player j and simultaneously the player j independently will choose an "actions program" as a response to the non-extended informational strategies chosen by the player i.

The action sets strategy of the payer i (correspondingly of the player j) noted by X_i^α (correspondingly X_j^β), represent the sets of the non-informational extended strategy chosen by the player i (correspondingly by the player j) in the capacity of the response to the every chosen strategy by the player j (correspondingly by the player i). As mentioned above, the informational extended strategy of the player is the detailed agenda of the chosen value of the strategies from the action sets strategy. These agenda can be defined using the functions of the following type $\theta_i^\alpha\!: X_j \rightarrow X_i^\alpha$ that $\forall x_j \in X_j$, $\theta_i^\alpha(x_j) \in X_i^\alpha$ (correspondingly $\theta_j^\beta\!: X_i \rightarrow X_j^\beta$ that $\forall x_i \in X_i$, $\theta_j^\beta(x_i) \in X_j^\beta$. We denote by

$$\Theta_i = \{\theta_i^\alpha : X_j \rightarrow X_i^\alpha,\ \alpha = 1, \ldots, \chi_i\}$$

and correspondingly

$$\Theta_j = \{\theta_j^\beta : X_i \rightarrow X_j^\beta, \beta = 1,..., \chi_j\}$$

the sets of the informational extended strategies of the player i and correspondingly j, in the $i \leftrightarrow j$ informational extended game. The $(\theta_i^\alpha \in \Theta_i, \theta_j^\beta \in \Theta_j, x_{[-ij]} \in X_{[-ij]})$ denotes the strategy profile in the double-way informational extended game.

Our goal is to determine the values of the payoff functions for the players so that the concrete value of the payoff will be determined using the payoffs of the players in the non-informational extended game (parent game) Γ. This problem is more difficult than in the one-way informational extended game. Let player i and j choose the informational extended strategy $\theta_i^\alpha \in \Theta_i$ and $\theta_j^\beta \in \Theta_j$ respectively. Denote by $X(\theta_i^\alpha, \theta_j^\beta) \subseteq X$ the set of the strategy profiles of the players in the game Γ "generated" by the informational extended strategy θ_i^α and θ_j^β. It is easy to show that

$$X(\theta_i^\alpha, \theta_j^\beta) = (gr\theta_i^\alpha \bigcup gr\theta_j^\beta) \bigcup X_{[-ij]}.$$

We mention that there exists the strategies θ_i^α and θ_j^β such that $X(\theta_i^\alpha, \theta_j^\beta) = \emptyset$, i.e., not for any couple $(\theta_i^\alpha, \theta_j^\beta)$ we can construct the carryout strategy profile in the non-informational extended game. Let us assume that the players want maximize their payoffs. Then we define payoff functions of the player as following

$$\mathcal{H}_p(\theta_i^\alpha, \theta_j^\beta, x_{[-ij]}) = \begin{cases} \max\limits_{(x_i, x_j) \in [gr\theta_i^\alpha \cap gr\theta_j^\beta]} H_p(x_i, x_j, x_{[-ij]}) & \text{if } X(\theta_i^\alpha, \theta_j^\beta) \neq \emptyset, \\ -\infty & \text{if } X(\theta_i^\alpha, \theta_j^\beta) \neq \emptyset. \end{cases}$$

Finally the normal form of the double-way i ↔ j informational extended game will be

$$\Gamma(\mathrm{i} \leftrightarrow \mathrm{j}) = \langle \mathrm{I}, \Theta_\mathrm{i}, \Theta_\mathrm{j}, \{\mathrm{X}\}_{\mathrm{p} \in \mathrm{I} \setminus \{\mathrm{i},\mathrm{j}\}}, \{\mathcal{H}_p\}_{\mathrm{p} \in \mathrm{I}} \rangle.$$

We denote the set of all strategies profiles that satisfied these definitions by $NE[\Gamma(i \leftrightarrow j)]$.

3. Properties of the Nash equilibrium profiles in the informational extended games

In this paragraph we study the properties of the strategy profiles of the non-informational extended game which can be generated by the Nash equilibrium profiles of the informational extended games $\Gamma(i \rightarrow j)$, $\Gamma(j \rightarrow i)$ and $\Gamma(i \leftrightarrow j)$.

Theorem 3.1. *The following relations are true:*

$$NE[\Gamma] \subseteq \bigcup_{\substack{(\theta_i^k,\, x_{[-j]}) \in NE[\Gamma(j\to i)] \\ k=1,\dots,\chi_i}} X(\theta_i^k) \subseteq \bigcup_{\substack{(\theta_i^\alpha,\, \theta_j^\beta,\, x_{[-ij]}) \in NE[\Gamma(j\to i)] \\ \alpha=1,\dots,\chi_i,\ \beta=1,\dots,\chi_j}} X(\theta_i^\alpha, \theta_j^\beta).$$

Proof. It can easily be proved that there are $|X_j|$ informational extended strategies $\hat{\theta}_i^\alpha \in \Theta_i$, $\alpha = \overline{1,|X_j|}$ of the player i, respectively $|X_i|$ informational extended strategies $\hat{\theta}_j^\beta \in \Theta_j$, $\beta = \overline{1,|X_i|}$ of the player j, such that $X(\hat{\theta}_i^\alpha, \theta_j^\beta) = X(\hat{\theta}_i^\alpha)$ for all $\theta_j^\beta \in \Theta_j$, respectively $(\theta_i^\alpha, \hat{\theta}_j^\beta) = X(\hat{\theta}_j^\beta)$ for all $\theta_i^\alpha \in \Theta_i$. Let $\tilde{x} \in NE[\Gamma]$ then $H_p(\tilde{x}) \geq H_p(x_p, \tilde{x}_{[-p]})$ for all $x_p \in X_p$ and for all $p \in I$. We consider the following informational extended strategy $\hat{\theta}_i^\alpha(x_j) = \tilde{x}_i$ for all $x_j \in X_j$, and then $H_i(\tilde{x}_i, \tilde{x}_{[-i]}) \geq H_i(x_i, \tilde{x}_{[-i]})\ \forall x_i \in X_i$. So in this case we have have $\tilde{x} \in X(\hat{\theta}_i^\alpha)$ since $X(\hat{\theta}_i^\alpha) = gr\hat{\theta}_i^\alpha \cup X_{[-ij]}$. Now we have to prove that $(\hat{\theta}_i^\alpha, \tilde{x}_{[-i]}) \in NE[\Gamma(j\to i)]$, that is $H_i(\hat{\theta}_i^\alpha(\tilde{x}_j), \tilde{x}_j, \tilde{x}_{-[i,j]}) = H_i(\tilde{x}_i, \tilde{x}_j, \tilde{x}_{-[i,j]}) \geq H_i(\hat{\theta}_i^\alpha(\tilde{x}_j), \tilde{x}_j, \tilde{x}_{-[i,j]})$ for all $\alpha = \overline{1,\chi_i}$, $H_j(\tilde{x}_i, \tilde{x}_j, \tilde{x}_{-[i,j]}) \geq H_j(\tilde{x}_i, \tilde{x}_j, \tilde{x}_{-[i,j]})\ \forall x_j \in X_j$, $H_p(\tilde{x}_i, \tilde{x}_j, \tilde{x}_{-[i,j]}) \geq H_p(\tilde{x}_i, \tilde{x}_j, x_p, \tilde{x}_{-[i,jp]})\ \forall x_p \in X_p$ for all $p \in I \backslash \{i, j\}$. But these inequalities are true since $\tilde{x} \in NE[\Gamma]$. So we proof that there is the strategy profile $(\hat{\theta}_i^\alpha, \tilde{x}_{[-i]}) \in NE[\Gamma(j\to i)]$, respectively $(\hat{\theta}_j^\beta, \tilde{x}_{[-i]}) \in NE[\Gamma(j\to i)]$ such that $\tilde{x} \in X(\hat{\theta}_i^\alpha)$ respectively $\tilde{x} \in X(\hat{\theta}_j^\beta)$. Consider the informational extended strategy of player $i(\hat{\theta}_i^\alpha(x_j) = \tilde{x}_i\ \forall x_j \in X_j$, and for player j the informational extended strategy of player $\hat{\theta}_j^\beta(x_i) = \tilde{x}_j\ \forall x_i \in X_i$, and then $H_i(x_i, \tilde{x}_{[-i]}) \geq H_i(x_i, \tilde{x}_{[-p]})\ \forall x_i \in X_i$ and $H_j(\tilde{x}_j, \tilde{x}_{[-j]}) \geq H_j(x_j, \tilde{x}_{[-j]})\ \forall x_i \in X_i$. Based on the above we obtained that $\tilde{x} \in X(\hat{\theta}_i^\alpha) \cap X(\hat{\theta}_i^\beta)$. Since $gr\hat{\theta}_i^\alpha \cap gr\hat{\theta}_i^\beta = (\tilde{x}_i, \tilde{x}_j)$ and $\tilde{x} \in NE[\Gamma]$ we can prove that $\max\limits_{(x_i, x_j) \in [gr\tilde{\theta}_i^\alpha \cap gr\tilde{\theta}_j^\beta]} H_i(x_i, x_j, \tilde{x}_{[-ij]}) \geq \max\limits_{(x_i, \tilde{x}_j) \in [gr\theta_i^\alpha \cap gr\tilde{\theta}_j^\beta]} H_i(x_i, \tilde{x}_j, \tilde{x}_{[-ij]})$, $\max\limits_{(\tilde{x}_i, \tilde{x}_j) \in [gr\tilde{\theta}_i^\alpha \cap gr\tilde{\theta}_j^\beta]} H_j(\tilde{x}_i, \tilde{x}_j, \tilde{x}_{[-ij]}) \geq \max\limits_{(\tilde{x}_i, x_j) \in [gr\tilde{\theta}_i^\alpha \cap gr\theta_j^\beta]} H_j(\tilde{x}_i, x_j, \tilde{x}_{[-ij]})$,

$$\max_{(\tilde{x}_i, \tilde{x}_j) \in [gr\tilde{\theta}_i^\alpha \cap gr\tilde{\theta}_j^\beta]} H_p(\tilde{x}_i, \tilde{x}_j, \tilde{x}_{[-ij]}) \geq \max_{(\tilde{x}_i, \tilde{x}_j) \in [gr\tilde{\theta}_i^\alpha \cap gr\tilde{\theta}_j^\beta]} H_p(\tilde{x}_i, \tilde{x}_j, x_p, \tilde{x}_{[-i,jp]})\ \forall p \in I \backslash \{i,j\}.$$

So we prove that $(\hat{\theta}_i^\alpha, \hat{\theta}_j^\beta, \tilde{x}_{[-ij]}) \in NE[\Gamma(i \leftrightarrow j)]$. Now we have that

$$NE[\Gamma] \subseteq \bigcup_{\substack{(\theta_i^k,\, x_{[-j]}) \in NE[\Gamma(j\to i)] \\ k=1,\dots,\chi_i}} X(\theta_i^k) \subseteq \bigcup_{\substack{(\theta_i^\alpha,\, \theta_j^\beta,\, x_{[-ij]}) \in NE[\Gamma(j\to i)] \\ \alpha=1,\dots,\chi_i,\ \beta=1,\dots,\chi_j}} X(\theta_i^\alpha, \theta_j^\beta).$$

Finaly it is easy to see that under the above

$$\bigcup_{\substack{(\theta_i^k,\, x_{[-j]}) \in NE[\Gamma(j\to i)] \\ k=1,\dots,\chi_i}} X(\theta_i^k) \subseteq \bigcup_{\substack{(\theta_i^\alpha,\, \theta_j^\beta,\, x_{[-ij]}) \in NE[\Gamma(i\to j)] \\ \alpha=1,\dots,\chi_i,\ \beta=1,\dots,\chi_j}} X(\theta_i^\alpha, \theta_j^\beta)$$

which proves the theorem. ■

This theorem shows that the set of the equilibrium profiles in the non-informational extended games is contained in the set of non-informational extended strategy profiles generated by the Nash equilibrium profiles in the informational extended game. Moreover, the set of non-informational extended strategy profiles generating by the Nash equilibrium profiles in the game Γ(j → i) or Γ(i → j) is contained in the set of non-informational extended strategy profiles generating by the Nash equilibrium profiles in the game $\Gamma(i \to j)$.

4. Informational non-extended game generated by the informational extended strategies

In this paragraph we study the case when the informational strategies of the players have already been chosen and so appears the necessity to study the informational non-extended game generated by the chosen informational extended strategies (Hancu 2012). These games differ in: (a) the sets of the strategies that are the subsets of the sets of strategies in the initial non-extended informational game; (b) how the payoff functions of the players will be constructed.
Let the payoff functions of the players be defined as $\tilde{H}_p:\prod_{p\in I}X_p \to R$ where for all $x_i \in X_i$, $xj \in X_j$, $x_{[-ij]} \in X_{[-ij]}$,

$$\tilde{H}_p (x_i, x_j, x_{[-ij]}) \equiv H_p (\theta_i (x_j), \theta_j (x_i), x_{[-ij]}).$$

The game with the following normal form $\Gamma(\theta_i, \theta_j) = \langle I, \{X_p\}_{p\in I}, \{\tilde{H}_p\}_{p\in I}\rangle\rangle$ will be called informational non-extended game generated by the informational extended strategies θ_i and θ_j. The game $\Gamma(\theta_i, \theta_j)$ is played as follows: independently and simultaneously each player $p \in I$ chooses the informational non-extended strategy $x_p \in X_p$, after that the players i and j calculate the value of the informational extended strategies $\theta_i(x_j)$ and $\theta_j(x_i)$ after that each player calculates the payoff values $H_p(\theta_i(x_j),\theta_j(x_i),x_{[-ij]})$, and with this the game is finished. To all strategy profiles $(x_i, x_j, x_{[-ij]})$ in the game Γ the following realization $(\theta_i(x_j), \theta_j (x_i), x_{[-ij]})$ in terms of the informational extended strategies will correspond. We denote by $NE[\Gamma(\theta_i, \theta_j)]$ the set of Nash equilibrium profiles of the game $\Gamma(\theta_i, \theta_j)$.

Theorem 4.1. *Let the game $\Gamma(\theta_i, \theta_j)$ satisfy the following conditions:*

1. *the X_p is a non-empty compact and convex subset of the finite-dimensional Euclidean space for all $p \in I$;*
2. *the functions θ_i (correspondingly θ_j) are continuous on X_j (correspondingly on X_i) and the functions H_p are continuous on X for all $p \in I$;*

3. *the functions* θ_i *(correspondingly* θ_j*) are quasi-concave on* X_j *(correspondingly on* X_i*), the functions* H_p *are quasi-concave on* X_p, $p \in I \setminus \{i,j\}$ *and monotonically increasing on* $X_i \times X_p$.

Then $NE[\Gamma(\theta_i, \theta_j)] \neq \emptyset$.

Proof. We define the following correspondence (point-to-set mapping) $Br: X \rightarrow X$ such that

$$Br(x) = (Br_1(x_{[-1]}),\ldots,Br_i(x_{[-i]}),\ldots,Br_n(x_{[-n]}))$$

then if $x^* \in Br(x^*)$, then $x_i^* \in Br_i(x^*_{[-i]})$ for all $i \in I$ and hence $x^* \in NE$. According to the Tikhonov's theorem: $X = \prod_{p \in I} X_p$ is compact, and according to conditions (1) and (2) for all $x_{[-i]}$ the set $Br_i(x_{[-i]})$ is non-empty. According to condition (3) $Br_i(x_{[-i]})$is also convex because $\tilde{H}_i$ is quasi-concave on X_i. Hence the set $Br(x)$ is nonempty convex and compact for all $x \in X$. The mapping Br has a closed graph because each function $\tilde{H}_p$ is continuous on X for all $p \in I$. Hence by Kakutani's theorem, the set-valued mapping has a fixed point. As we have noted, any fixed point is a Nash equilibrium. ■

4.1 *Informational extended strategies in the duopol games*

In this paragraph, we study *Cournot's Duopol* games generated by the informational extended strategies. There are two firms operating in a limited market. Market production is $p(Q) = \begin{cases} a - Q & \text{for } Q < a \\ 0 & \text{for } Q \geq a \end{cases}$ where $Q = q_1 + q_2$ for two firms, $a > 0$ represents the volume of products of type q_1 and q_2 that is in the market, p represents the price of the good. Both firms will receive profits derived from a simultaneous decision made by both on how much to produce, and also based on their cost functions: $C_i(q_i) = cq_i + C_i$ where $c \leq a$ represents the marginal cost (consumption for both firms) and $C_i \leq a$ the fixed cost. Both firms simultaneously choose the quantity of the goods that they offer on the market.

We will model the activity described in the example using game theory. Since both firms simultaneously choose the quantity of good that they offer on the market and do not cooperate, then we will use the non-cooperative static game. The normal form of these games is $\Gamma = \langle I = \{1,2\}, X = [0, N_1], Y = [0.N_2], H_1, H_2 \rangle$ where I is the set of players: firm 1 and firm 2; X, Y are the sets of strategies (the x and y represents the volume of products of type q_1 and q_2 that is in the market) of the players, $H_1(x,y) = x(a - x - y - c) - C_1$, $H_2(x,y) = y(a - x - y - c) - C_2$, represents the payoffs of the player 1 and 2. The Nash equilibrium profile is $(x^*, y^*) = \left(\frac{a-c}{3}, \frac{a-c}{3}\right)$ and firms will receive the following profits $H_i(x,y) = \frac{(a-c)}{3} - C_i$ if $x + y < a$ that is if $a > 2c$.

From the property of the equilibrium profile, we obtain that none of the firms is unilaterally convenient to offer on the market a volume of goods other than the value $\frac{a-c}{3}$, otherwise the firm will lose income.

We analyze different forms of informational extended strategies.

(A) In this case, the sets of strategies of the player *1*, respectively the player *2*, will be $\Theta_1 = \{\theta_1 : Y \to X$ such that

$$\forall y \in Y,\ \theta_1(y) = \underset{x \in X}{\operatorname{argmax}} H_1(x, y)\},$$

respective $\Theta_2 = \{\theta_2 : X \to Y$ such that

$$\forall x \in X,\ \theta_2(x) = \underset{y \in Y}{\operatorname{argmax}} H_2(x, y)\}.$$

Players do not know which function θ_1 or θ_2 will be chosen by the partners (but this is not significant because the sets Θ_i contain one element). We determine the functions $\theta_1(y)$ si $\theta_2(x)$. The functions $H_1(x,y)$ and $H_2(x,y)$ are concave, then using the necessary conditions of the functions extreme we obtane that $\underset{x \in X}{\operatorname{argmax}} H_1(x, y)\} = x \in X : x = \left\{\frac{a-y-c}{2}\right\}$ and $\underset{y \in Y}{\operatorname{argmax}} H_2(x, y)\} = y \in Y : y = \left\{\frac{a-x-c}{2}\right\}$. Then $\theta_1(y) = \frac{a-y-c}{2}$ such that $\forall y \in Y, \frac{a-y-c}{2} \in X$ (but no $\theta_1(y) = a - 2x - c$) and $\theta_2(x) = \frac{a-x-c}{2}$ such that $\forall x \in X, \frac{a-x-c}{2} \in Y$ (but no $\theta_2(y) = a - 2y - c$). So, $\tilde{H}_1(\theta_1(y), \theta_2(x)) = \frac{1}{2}[y(a-y-x-c) + x(a-c)] - C_1$. Similarly we will get $\tilde{H}_2(\theta_1(y), \theta_2(x)) = \frac{1}{2}[x(a-y-x-c) + y(a-c)] - C_2$. Then Nash equilibrium profile is $(x^*, y^*) = (a-c, a-c)$, which does not coincide with the situation in the classical case, and each company's payoffs will be

$$\tilde{H}_i(\theta_1(y^*), \theta_2(x^*)) = \frac{1}{2}(a-c)\,[(c-a) + (a-c)^2] - C_i.$$

So if every firm knows what the volume of goods is going to launch on the market, and each of them takes advantage of this, choose its strategies based on the functions of type $\theta_1(y) = \underset{x \in X}{\operatorname{argmax}} H_1(x, y)$, respectively $\theta_2(y) = \underset{y \in Y}{\operatorname{argmax}} H_2(x, y)$, then the volume of goods released on the market, based on Nash equilibrium profile, is higher than when firms do not possess that information, and the payoffs is much higher.

(B) Analise the case when "firm *1* knows what volume of goods firm *2* will be launched on market, while firm *2* does not know what volume of goods will launch firm *1*". Consider the case when the firm 1 informational extended strategy is a function $\theta_1(y) = \underset{x \in X}{\operatorname{argmax}} H_1(x,y)$ $\forall y \in Y$, but on the firm *2* is the function $\theta_2(x) = y\ \forall x \in X$. The payoffs will be

$\tilde{H}_1(\theta_1(y), \theta_2(x)) = y\left(\frac{a-y-c}{2}\right) - C_1$ and $\tilde{H}_2(\theta_1(y), \theta_2(x)) = y\left(\frac{a-y-c}{2}\right) - C_2$. In this case, the player's *1* payoff does not depend on the x strategy. We determine equilibrium profiles for this game. This is where the notion of equilibrium is already lost. Then the solution will be determined accordingly. Player 2 determines y^* that realizes $\max_{y \in Y} \left\{y\left(\frac{a-y-c}{2}\right)\right\}$, and so $y^* = \frac{a-c}{2}$. Then $x^* = \theta_1(y^*) = \theta_1\left(\frac{a-c}{2}\right) = \frac{a-c}{4}$. We note that $\left(\frac{a-c}{4}, \frac{a-c}{2}\right)$ profile coincides with the Stachelberg equilibrium profile in which player 2 makes the first move.

5. Converting the two person game with informational extended strategies to bayesian game

Consider the two person informational extended game with the following sets of the informational extended strategies

$$\Theta_1 = \{\theta_1^j : Y \to X | \forall y \in Y, \theta_1^j(y) \in X, j = \overline{1, m_1}\}$$

of the player and

$$\Theta_2 = \{\theta_1^k : X \to Y | \forall x \in X, \theta_2^k(x) \in Y, k = \overline{1, m_2}\}.$$

The payoff functions of the players are defined as follows: for all $\theta_1 \in \Theta_1$, $\theta_2 \in \Theta_2$, $\mathcal{H}_i(\theta_1, \theta_2) = H_i(\theta_1(y), \theta_2(x))$ for all $x \in X$, $y \in Y$. The game is played as follows: independently and simultaneously each player $i \in I = \{1,2\}$ chooses the informational extended strategy $\theta_1 \in \Theta_1$ and $\theta_2 \in \Theta_2$ (and players do not know what kind of the informational extended strategy will be chosen by each other's), after that the players and calculate the value of the payoff values $H_i(\theta_1(y), \theta_2(x))$ and with this the game is finished.

Remark 5.1. *The game described above will be denoted by Game(1 $\leftrightarrow$ 2) and is the game with* ***incomplete information*** *because players do not know what kind of the informational extended strategy $\theta_1 \in \Theta_1$ will be chosen by the player 1 (for example) and so the player 1 generates the uncertainty of the player 2 about the complete structure of the payoff function $H_2(\theta_1(y), \theta_2(x))$* ***in the game with non-informational extended strategies.*** *So the players do not know exactly the structure of your payoff functions and the game is in the incomplete information.*

Denote also by

$$\tilde{X}_j = \{\tilde{x}_j \in X : \tilde{x}_j = \theta_1^j(y), \forall y \in Y\}$$

and

$$\tilde{Y}_k = \{\tilde{y}_k \in Y: \tilde{y}_k = \theta_2^k(x), \forall x \in X\}$$

the set of all ranges of the informational extended strategy θ_1^j of player 1 and θ_2^k of player 2. The sets $\tilde{X}_j$ and $\tilde{Y}_k$ are the sets of informational non-extended strategies generated by the informational extended strategies of player 1 and 2 respectively. According to Harsanyi–Selten (Harsanyi and Selten 1998) principle we can reduce the analysis of a game with incomplete information to the analysis of a game with complete (but imperfect) information, which is fully accessible to the usual analytical tools of game theory. So to solve the game *Game*(1 ↔ 2) we must do the following step-intervals:

1. Construct the Bayesian game

$$\Gamma_B = \langle I = \{1,2\}, S_1(\Delta_1), S_2(\Delta_2), \Delta_1, \Delta_2, p, q, \tilde{H}_1, \tilde{H}_1 \rangle$$

that corresponds (is associated) to the game *Game*(1 ↔ 2). The normal form must consist of the following.

- A set of possible actions for player 1 is X, a for player 2 is Y.
- Because players do not know what kind of informational extended strategy will be chosen by the other player, then the uncertainty of player 1 about their own payoff function structure is generated by player 2 selected informational extended strategy and respectively, the uncertainty of player 2 about the own payoff function structure is generated by the player 1 selected informational extended strategy. The set of types for player 1 (player 2) is $\Delta_1 = \{\delta_1^j, j \in J_1\}$($\Delta_2 = \{\delta_2^k, k \in J_2\}$). In other words, player 1 (player 2) is of the type δ_1^j (of the type δ_2^k) if he generates to the player 2 (to the player 1) payoff structure uncertainty, selecting the $\theta_1^j \in \Theta_1$ ($\theta_2^k \in \Theta_2$) informational extended strategy.
- The probability function p: $\Delta_1 \rightarrow \Omega(\Delta_2)$ of the player 1, respectively $q: \Delta_2 \rightarrow \Omega(\Delta_1)$ of the player 2, means the following: if the player 1 (player 2) chooses the informational extended strategy θ_1^j (strategy θ_2^k), then he believes that the player 2 (player 1) with the $p(\delta_2^k/\delta_1^j) = \dfrac{p(\delta_2^k \cap \delta_1^j)}{p(\delta_1^j)}$ (respectively $p(\delta_1^j/\delta_2^k) = \dfrac{p(\delta_2^k \cap \delta_1^j)}{p(\delta_1^j)}$) chooses the informational extended strategy θ_2^k (strategy θ_1^j).
- The set of the strategies of the players are the set of all range of the informational extended strategies of the players

$$S_1(\Delta_1) = \{\tilde{x}_j \in X: \tilde{x}_j = \theta_1^j(y), \forall y \in Y, \forall j \in J_1\} \equiv \{\tilde{X}_j, j = 1,..., m_1\},$$

and

$$S_2(\Delta_2) = \{\tilde{y}_k \in Y : \tilde{y}_k = \theta_1^k(x), \forall x \in X, \forall k \in J_2\} \equiv \{\tilde{Y}_k, k = 1,..., m_2\},$$

So if the player 1, for example, is of the type j, i.e., he chooses the informational extended strategy θ_1^j, then the set of the informational non extended strategies, generated by θ_1^j is $\tilde{X}_j$.

- The payoff functions of the player is defined as following

$$\tilde{H}_1 : S_1(\Delta_1) \times S_2(\Delta_2) \times \Delta_1 \times \Delta_2 \to R,\ \tilde{H}_2 : S_1(\Delta_1) \times S_2(\Delta_2) \times \Delta_1 \times \Delta_2 \to R.$$

More exact, for all fixed $s_1 \in S_1(\Delta_1)$ and $s_2 \in S_2(\Delta_2)$,

$$\tilde{H}_1(s_1(\cdot), s_2(\cdot), \delta_1^j, \delta_2^k) = H_1(\tilde{x}_j, \tilde{y}_k, \delta_1^j, \delta_2^k),$$

and

$$\tilde{H}_2(s_1(\cdot), s_2(\cdot), \delta_1^j, \delta_2^k) = H_2(\tilde{x}_j, \tilde{y}_k, \delta_1^j, \delta_2^k)$$

for all $\tilde{x}_j \in \tilde{X}_j$, $\tilde{y}_k \in \tilde{Y}_k$, $j = \overline{1, m_2}$, $k = \overline{1, m_2}$.

2. For game construct the Selten-Harsanyi game

$$\Gamma_B^* = \langle J, \{R_j\}_{j \in J}, \{U_j\}_{j \in Jj} \rangle$$

with complete and imperfect information on the sets of non-informational extended strategies. Denote by $J = \{j = (i, \delta_i^j), i = 1,2, j = 1,..., m_1 + m_2\}$ the set of type-players that is equal to the sets of all informational extended strategies of the players $J = J_1 \cup J_2$. The strategy of the type-player j is $r_j = \begin{cases} \tilde{x}_j \in \tilde{X}_j & j \in J_1, \\ \tilde{y}_j \in \tilde{Y}_j & j \in J_2 \end{cases}$ and means the following: if player is of type $j = (i, \delta_i^j)$ (i.e., player $i, i = 1,2$, chooses the informational extended strategy θ_i^j), then the strategy will be equal to value of the informational extended strategy $x^j = \theta_i^j(y)$ for a fixed value of the non-extended strategy $y \in Y$. The sets of pure strategies of the players will be $R_j = \begin{cases} \tilde{X}_j & j \in J_1, \\ \tilde{Y}_j & j \in J_2. \end{cases}$, and $R = \prod_{j=1}^{m1+m2} R_j$. For all type-player $j = (1, \delta_1^j), j \in J_1$, payoff function U_j: $\tilde{X}_j \times ({}_{k \in J_2} \times \tilde{Y}_k)$ is defined as following

$$U_j(r_j, \{r_k\}_{k \in J_2}) = U_j(\tilde{x}_j, \{\tilde{y}_k\}_{k \in J_2}) = \sum_{k \in J_2} p(\delta_2^k | \delta_1^j)\, H_1(\tilde{x}_j, \tilde{y}_k),\ \forall \tilde{x}_j \in \tilde{X}_j, \tilde{y}_k \in \tilde{Y}_k.$$

In the similar mode for all type-player $j = (2, \delta_2^j), j \in J_2$ payoff function U_j: $({}_{k \in J_2} \times \tilde{X}_k) \times \tilde{Y}_j$ is defined as follows

$$U_j(\{r_k\}_{k \in J_2}, r_j) = U_j(\{\tilde{x}_k\}_{k \in J_1}, \tilde{y}_j) = \sum_{k \in J_1} q(\delta_1^k | \delta_2^j)\, H_2(\tilde{x}_k, \tilde{y}_j),\ \forall \tilde{x}_k \in \tilde{X}_k, \tilde{y}_j \in \tilde{Y}_j.$$

These utility functions have the following meaning. If, for example, player 1, had chosen information extended strategy θ_1^j, which also means he has a type-player $j = (1,\delta_1^j)$, and with the probability $p(\delta_2^k \mid \delta_1^j)$ he assumes that the player 2 will choose the information extended strategy δ_2^k, i.e., as we have the type player $k = (2,\delta_2^k)$, for all $k \in J_2$, then for all information not extended strategy $x \in X$ and $y \in Y$, average value of the payoff will be

$$\sum_{k\in J_2} p(\delta_2^k|\delta_1^j)\, H_1(\tilde{x}_j, \tilde{y}_k) \equiv \sum_{k\in J_2} p(\delta_2^k|\delta_1^j)\, H_1(\theta_1^j(y), \theta_2^k(x))$$

3. Determine Nash equilibrium profiles in the game Γ_B^* that is the Bayes-Nash equilibrium in the game Γ_B.
4. As the solution of the game Game($1 \leftrightarrow 2$) we will consider the non-informational extended strategy profile (x^*, y^*) which is generated by the Nash strategy profile in the game Γ_B^*.

Denote by $BE[\Gamma_B]$ the set of all Bayes-Nash strategies profile of the game Γ_B. Really, solving the game Γ_B is difficult, because it is transformed in the two levels dynamic game. On the first level player Nature chooses the informational extended strategies of the players, for example, (θ_1^j, θ_2^k), and on the second level each player chooses the informational non-extended strategies from the set $\tilde{X}_j$ (player 1) and from the set $\tilde{Y}_j$ (player 2). The games $\Gamma(\theta_1^j, \theta_2^k)$ are the subgames in the dynamic game defined above. Here we do not investigate a such method to solve informational extended games.

The game Γ_B^* is played as follows: for all fixed probabilities $p(\delta_2^k|\delta_1^j)$ and $q(\delta_1^k|\delta_2^j)$, independently and simultaneously each type-player $j = (i,\delta_i^j)$ chooses the strategy $r_j \in R_j$, after that each player calculates the payoff using the functions $U_j(r_j, \{r_k\}_{k\in J_2})$ or $U_j(\{r_k\}_{k\in J_1}, r_j)$ and whereupon the game is finished. In other words, because strategies defined by the informational non-extended strategies from sets X and y, for all $y \in Y$ (respectively for all $x \in X$) type-player $j = (1,\delta_1^j)$ (respectively type-player $k = (2,\delta_2^k)$) chooses the strategy $\tilde{x}_j = \theta_1^j(y)$ (respectively $\tilde{y}_k = \theta_2^k(x)$), calculates the payoff values using the functions $U_j(\tilde{x}_j, \{\tilde{y}_k\}_{k\in J_2})$ (respectively $(U_j(\{(\tilde{x}_k\}_{k\in J_2}, \tilde{y}_k)))$. and with this the game is finished.

We introduce the following definition.

Definition 5.1. *Strategy profile* $r^* = (r_1^*,\ldots,r_j^*,\ldots r_{|J|}^*)$ *is the Nash equilibrium in the game* Γ_B^* *if an only if the following conditions are fulfilled*:

$$\begin{cases} U_j(r_j^*, \{r_k^*\}_{k\in J_2}) \geq U_j(r_j, \{r_k^*\}_{k\in J_2}) \text{ for all } j \in J_1, \\ U_j(\{r_k^*\}_{k\in J_1}, r_j^*) \geq U_j(\{r_k^*\}_{k\in J_1}, r_j) \text{ for all } j \in J_2. \end{cases}$$

We get for all fixed probabilities $p(\cdot),q(\cdot)$ t that strategy profile $(\{\tilde{x}_j^*\}_{l\in J_1},\{\tilde{y}_k^*\}_{k\in J_2})$ is Nash equilibrium for the game $\Gamma_B^* = \langle J,\{R_j\}_{j\in J},\{U_j\}_{j\in J}\rangle$ if and only if the $|J_1| + |J_2|$ conditions are fulfilled:

$$\begin{cases} \sum_{k\in J_2} p(\delta_2^k|\delta_1^j)\, H_1(\tilde{x}_j^*, \tilde{y}_k^*) \geq \sum_{k\in J_2} p(\delta_2^k|\delta_1^j)\, H_1(\tilde{x}_j^*, \tilde{y}_k^*) \text{ for all } \tilde{x}_j \in \tilde{X}_j, j \in J_1, \\ \sum_{k\in J_1} p(\delta_1^j|\delta_2^k)\, H_1(\tilde{x}_j^*, \tilde{y}_k^*) \geq \sum_{k\in J_2} p(\delta_1^j|\delta_2^k)\, H_2(\tilde{x}_j^*, \tilde{y}_k^*) \text{ for all } \tilde{y}_k \in \tilde{Y}_j, k \in J_2, \end{cases}$$

Denote by $NE[\Gamma_B^*]$ the set of all Nash equilibrium strategies profile in the game Γ_B^*. The relation between the Nash equilibrium in the Harsanyi game Γ_B^* and the equilibrium at the Bayesian game Γ_B was given by Harsanyi theorem: *The set of Nash equilibria of the game Γ_B^* is identical to the set of Bayesian equilibria of the game Γ_B*

Let strategy profile $(\{\tilde{x}_j^*\}_{j\in J_1},\{\tilde{y}_k^*\}_{k\in J_2}) \in NE[\Gamma_B^*]$ then we introduce the following definition.

Definition 5.2. *For all fixed probabilities $p(\cdot)$, $q(\cdot)$ strategy profile $(x^*, y^*) \equiv (x^* (p), y^*(q))$ $x^* \in X$, $y^* \in Y$, for which the following conditions*

$$\begin{cases} \tilde{x}_j^* = \theta_1^j\,(y^*)\ \forall j \in J_1 \\ \tilde{y}_k^* = \theta_2^k\,(x^*)\ \forall k \in J_2 \end{cases}$$

are fulfilled, is called the Bayes-Nash equilibrium profile in non-informational extended strategies of the game Γ generated by the $(\{\tilde{x}_j^*\}_{j\in J_1},\{\tilde{y}_k^*\}_{k\in J_2}) \in NE[\Gamma_B^*]$.

Denote by $BN[G(1\leftrightarrow2)]$ the set of all Bayes-Nash equilibrium profiles in the game $G(1\leftrightarrow2)$. So, such as solutions of the informational extended games $Game(1 \leftrightarrow 2)$ we consider the informational non-extended Bayes-Nash equilibrium profiles $(x^*, y^*) \equiv (x^* (p), y^* (q))$ for which the relations from Definition 5.2 are fulfilled for all fixed probabilities $p(\cdot)$ and $q(\cdot)$ of the beliefs about the choice of the informational extended strategy of the other player.

For all $j = 1,2$, $k = 1,2$, denote by ϑ_1^j (respectively ϑ_2^j) an inverse function of the θ_1^j (respectively θ_2^k). If informational extended strategies $\theta_1^j(y)$ and $\theta_2^k(x)$ for all $j = 1,2$, $k = 1,2$ are bijective, then there are the inverse functions ϑ_1^j and ϑ_2^k for all $j = 1,2$, $k = 1,2$ and, so, we have that for all $(\{\tilde{x}_j^*\}_{j\in J_1},\{\tilde{y}_k^*\}_{k\in J_2}) \in NE[\Gamma_B^*]$ there is (x^*, y^*) such that $y^* = \vartheta_1^j\,(\tilde{x}_j^*)\ \forall j \in J_1$ and $x^* = \vartheta(\tilde{y}_j^*)\ \forall k \in J_2$. We can prove the following theorem.

Theorem 5.1. *Let the game satisfy the following conditions:*

1. *X and Y are non-empty compact and convex subsets of the finite-dimensional Euclidean space;*
2. *the functions $\theta_1^j, \forall j \in J_1$, and $\theta_2^k, \forall k \in J_2$, are continuous on Y (respectively on X) and the functions H_1, H_2 are continuous on $X \times Y$;*

3. *the functions* $\theta_1^j, \forall j \in J_1$, *(respectively* $\theta_2^k, \forall k \in J_2$*), are quasi-concave on* Y *(respectively on* X*), the functions* H_1 *(respectively* H_2*) are quasi-concave on* X *(respectively* Y*) and monotonically increasing on* $X \times Y$.

Then $BN[G(1 \leftrightarrow 2)] \neq \emptyset$.

Proof. Another, and some times more convenient way of defining Nash equilibrium in the game Γ_B^* is via the best response correspondences that are defined as following. For all $j \in J_1$, $Br_j: \prod_{k \in J_2} \tilde{Y}_k \rightarrow 2^{\tilde{X}_j}$, $Br_j(\{\tilde{y}_k\}_{k \in J_2}) = \{\tilde{x}_j \in \tilde{X}_j: U_j(\tilde{x}_j, \{\tilde{y}_k\}_{k \in J_2}) \geq U_j(\tilde{x}_j', \{\tilde{y}_k\}_{k \in J_2})\ \forall\ \tilde{x}_j' \in \tilde{X}_j\}$. For all $j \in J_2$, $Br_j: \prod_{k \in J_1} \tilde{X}_k \rightarrow 2^{\tilde{Y}_j}$, $Br_j(\{\tilde{x}_k\}_{k \in J_1}) = \{\tilde{y}_j \in \tilde{Y}_j: U_j(\{\tilde{x}_k\}_{k \in J_1}, \tilde{y}_j) \geq U_j(\{\tilde{x}_k\}_{k \in J_1}, \tilde{y}_j')\ \forall\ \tilde{y}_j' \in \tilde{Y}_j\}$. We define the following point-to-set mapping $Br(\{\tilde{x}_k\}_{k \in J_1}, \{\tilde{y}_k\}_{k \in J_2}) = (\{Br_j(\{\tilde{y}_k\}_{k \in J_2}\}_{j \in J_1}, \{Br_j(\{\tilde{x}_k\}_{k \in J_1})\}_{j \in J_2})$ and if $(\{\tilde{x}_j^*\}_{l \in J_1}, \{\tilde{y}_k^*\}_{k \in J_2}) \in Br(\{\tilde{x}_j^*\}_{l \in J_1}, \{\tilde{y}_k^*\}_{k \in J_2})$, then we have that $(\{\tilde{x}_j^*\}_{j \in J_1}, \{\tilde{y}_k^*\}_{k \in J_2}) \in NE[\Gamma_B^*]$. Here $\tilde{x}_j = \theta_1^j(y)$, $\forall j \in J_1$ and $\tilde{y}_k = \theta_2^k(x)$, $\forall j \in J_2$.

Denote by $\tilde{X} = \prod_{k \in J_1} \tilde{X}_k$ by $\tilde{Y} = \prod_{k \in J_2} \tilde{Y}_k$ and by grBr the graph of the point-to-set mapping Br. Since $\tilde{x}_j = \theta_1^j(y)$ $\forall j \in J_1$ and $\tilde{y}_k = \theta_2^k(x)$ $\forall k \in J_2$, then, according to condition 2), the sets $\tilde{X}_k$ $\forall k \in J_1$ and $\tilde{Y}_k$ $\forall k \in J_2$ are compacts and, according to the Tikhonov's theorem the sets $\tilde{X}$ and $\tilde{Y}$ are a non-empty compact and convex subsets of the Euclidean finite-dimensional space. For all $\{\tilde{y}_k\}_{k \in J_2}$ and $\{\tilde{x}_k\}_{k \in J_1}$ the sets $Br_j(\{\tilde{y}_k\}_{k \in J_2})$ and $Br_j(\{\tilde{x}_k\}_{k \in J_1})$ are non-empty because of conditions 1) and 2). According to condition 3), $Br_j(\{\tilde{y}_k\}_{k \in J_2})$ and $Br_j(\{\tilde{x}_k\}_{k \in J_1})$ are also convex sets. Hence the set $Br(\{\tilde{x}_k\}_{k \in J_1}, \{\tilde{y}_k\}_{k \in J_2})$ is a nonempty convex and compact for all $\tilde{x}_k \in \tilde{X}_k$, $k \in J_1$ and $\tilde{y}_k \in \tilde{Y}_k$, $k \in J_1$. According to condition 2) the mapping Br has a closed graph. Hence by Kakutani's theorem, the set-valued mapping has a fixed point. As we have noted, any fixed point is a Nash equilibrium. From the continuity of the functions θ_1^j, $\forall j \in J_1$ and θ_2^k, $\forall j \in J_2$ it results that there exist inverse functions θ_1^j and θ_2^k for all $j = 1,2$, $k = 1,2$ and so, we have that for all $(\{\tilde{x}_l^*\}_{l \in J_1}, \{\tilde{y}_k^*\}_{k \in J_2}) \in NE[\Gamma_B^*]$ there is (x^*, y^*) such that $y^* = \vartheta_1^j(\tilde{x}_j^*)$ $\forall j \in J_1$ and $x^* = \vartheta_1^k(\tilde{y}_j^*)$ $\forall j \in J_2$. The theorem is completely proved. ■

The following example illustrates the above context.

Example 5.1. *Consider the two persons game in the complete and perfect information, for which which* $X = [0,1]$, $Y = [0,1]$, *are the sets of strategies and* $H_1(x,y) = \frac{3}{2}xy - x^2$, $H_2(x,y) = \frac{3}{2}xy - y^2$ *the payoff functions of the player. Solve this game using the described above Harsanyi's principle.*

Solution. In the capacity of the informational extended strategies we will use the $\theta_1^1(y) = \underset{x \in X}{\operatorname{argmax}} H_1(x, y) = \frac{3}{2}y$, $\theta_1^2(y) = y^2$, $\theta_2^1(x) = \underset{y \in Y}{\operatorname{argmax}} H_2(x, y) = \frac{3}{4}x$, $\theta_2^2(x) = x^2$. If players will use these strategies then the payoff functions will be $H_1(x,y) = \frac{3}{2}\left(\frac{3}{4}y\right)\underbrace{\left(\frac{3}{4}x\right)} - \left(\frac{3}{2}y\right)^2 = \frac{27}{32}xy - \frac{9}{16}y^2$ and

$H_2(x,y) = \frac{3}{2}\underbrace{\left(\frac{3}{4}y\right)}\left(\frac{3}{4}x\right) - \left(\frac{3}{2}x\right)^2 = \frac{27}{32}xy - \frac{9}{16}x^2$. But because player 1, for example, will not know that player 2 as information extended strategy choose exactly $\theta_2(x) = \frac{3}{2}x$, then he will not know its payoff function So the informational extended strategies generate uncertainty of the payoff functions, such that we already have a incomplete information game.

Construct the Bayesian game Γ_B associated to the initial informational extended game.

(a) The actions sets are $X = [0,1]$ and $Y = [0,1]$.

(b) The set of the type for player 1 is $\Delta_1 = \{\delta_1^1, \delta_1^2\}$ and for player 2 is $\Delta_2 = \{\delta_2^1, \delta_2^2\}$ that means the following: player 1 (respectively 2) is of δ_1^1 type (respectively δ_2^1) if he choose the informational extended strategy $\theta_1^1(y) = \frac{3}{2}y$ (respectively $\theta_2^1(x) = \frac{3}{4}x$) and of the δ_1^2 type (respectively δ_2^2) if he choose the informational extended strategy $\theta_1^2(y) = y^2$ (respectively $\theta_2^2(x) = x^2$). Type-players will be denoted by $J_1 = \{1,2\}$, $J_2 = \{1,2\}$.

(c) If the player 2 is of the type δ_2^k, then he supposes with the probability $p(\delta_1^j/\delta_2^k) = p(\delta_1^j) = \begin{cases} p \text{ for } j = 1 \\ 1 - p \text{ for } j = 2 \end{cases}$ that the player 1 is of the type δ_1^j and, respectively, if the player 1 is of the type δ_1^j, then he suppose with the probability $p(\delta_2^k/\delta_1^j) = p(\delta_2^k) = \begin{cases} q \text{ for } k = 1 \\ 1 - q \text{ for } k = 2 \end{cases}$ that the player 2 is of the type δ_2^k.

(d) The strategies sets of the players are the following $S_1(\delta_1^1) \equiv \tilde{X}_1 = \{\tilde{x}_1 \in [0,1]: \tilde{x}_1 = \frac{3}{4}y, \forall y \in [0,1]\} = \left[0, \frac{3}{4}\right] \subseteq X$, $S_1(\delta_1^2) \equiv \tilde{X}_2 = \{\tilde{x}_2 \in [0,1]: \tilde{x}_2 = y^2, \forall y \in [0,1]\} = [0,1]$, $S_2(\delta_2^1) \equiv \tilde{Y}_1 = \{\tilde{y}_1 \in [0,1]: \tilde{Y}_1 = \frac{3}{4}x, \forall x \in [0,1]\} = \left[0, \frac{3}{4}\right] \subseteq Y$, $S_2(\delta_2^2) \equiv \tilde{Y}_2 = \{\tilde{Y}_2 \in [0,1]: \tilde{y}_2 = x^2, \forall x \in [0,1]\} = [0,1]$.

(e) The payoff functions of the player 1 is $\tilde{H}_1(\tilde{x}_j, \tilde{y}_k, \delta_1^j, \delta_1^k) \equiv H_1(\tilde{x}_j, \tilde{y}_k)$ and of the player 2 is $\tilde{H}_2(\tilde{x}_j, \tilde{y}_k, \delta_1^j, \delta_1^k) \equiv H_2(\theta_1^j(y), \theta_2^k(x))$. Finally

$$\tilde{H}_1(\tilde{x}_j, \tilde{y}_k, \delta_1^j, \delta_1^k) = \begin{cases} \frac{3}{2}\tilde{x}_1\tilde{y}_1 - (\tilde{x}_1)^2, & j = 1, k = 1, \\ \frac{3}{2}\tilde{x}_2\tilde{y}_1 - (\tilde{x}_2)^2, & j = 2, k = 1, \\ \frac{3}{2}\tilde{x}_1\tilde{y}_2 - (\tilde{x}_1)^2, & j = 1, k = 2, \\ \frac{3}{2}\tilde{x}_2\tilde{y}_2 - (\tilde{x}_2)^2, & j = 2, k = 2, \end{cases}$$

and

$$\tilde{H}_2(\tilde{x}_j, \tilde{y}_k, \delta_1^j, \delta_1^k) = \begin{cases} \frac{3}{2}\tilde{x}_1\tilde{y}_1 - (\tilde{y}_1)^2, \ j = 1, k = 1, \\ \frac{3}{2}\tilde{x}_2\tilde{y}_1 - (\tilde{y}_2)^2, \ j = 2, k = 1, \\ \frac{3}{2}\tilde{x}_1\tilde{y}_2 - (\tilde{y}_1)^2, \ j = 1, k = 2, \\ \frac{3}{2}\tilde{x}_2\tilde{y}_2 - (\tilde{y}_2)^2, \ j = 2, k = 2, \end{cases}$$

Thus the Bayesian game is $\Gamma_B = \langle I = \{1,2\}, \mathcal{S}_1(\Delta_1), \mathcal{S}_2(\Delta_2), \Delta_1, \Delta_2, p, q, \tilde{H}_1, \tilde{H}_2 \rangle$.

Now we can construct the game Γ_B^* in complete and imperfect informations, associated to the Bayesian game recently constructed. The set of the type-players is $J = J_1 \cup J_2$, where $J_1 = \{j = (1, \theta_1^j) | j = 1,2\} = \{1,2\}$ and $J_2 = \{k = (2, \theta_2^k) | k = 1,2\} = \{3,4\}$. Finally, $J = \{1,2,3,4\}$. The set of the strategy of the type-player $j \in J$ *is* $R_j = \begin{cases} \tilde{X}_j & j \in J_1, \\ \tilde{Y}_j & j \in J_2. \end{cases}$ and the strategy of the type-player $j \in J$ *is* $r_j = \begin{cases} \tilde{x}_j \in \tilde{X}_j & j \in J_1, \\ \tilde{y} \in \tilde{Y}_j & j \in J_2 \end{cases}$. So, for type player $J = 1$ the strategy set is $R_1 = \tilde{X}_1 = \left[0, \frac{3}{4}\right]$, for type-player $j = 2$ the strategy set is $R_2 = \tilde{X}_2 = [0,1]$, for type-player $j = 3$ (or $k = 1$) $R_3 = \tilde{Y}_1 = \left[0, \frac{3}{4}\right]$ and for type-player $j = 4$, the strategy set is $R_4 = \tilde{Y}_2 = [0,1]$. According to these, the strategy profile in the game Γ_B^* is $r = (r_1, r_2, r_3, r_4) = (\tilde{x}_1, \tilde{x}_2, \tilde{y}_1, \tilde{y}_2)$, where $\tilde{x}_1 \in \left[0, \frac{3}{4}\right]$, $\tilde{x}_2 \in [0,1]$, $\tilde{y}_1 \in \left[0, \frac{3}{4}\right]$ and $\tilde{y}_2 \in [0,1]$. Payoff functions of the type-players are defined as following

$$U_1(\tilde{x}_1,\tilde{y}_1,\tilde{y}_2,q) = qH_1(\tilde{x}_1,\tilde{y}_1) + (1-q)H_1(\tilde{x}_1,\tilde{y}_2) = -(\tilde{x}_1)^2 + \frac{3}{2}\tilde{x}_1(q\tilde{y}_1 + (1-q)\tilde{y}_2),$$

$$U_2(\tilde{x}_2,\tilde{y}_1,\tilde{y}_2,q) = qH_1(\tilde{x}_2,\tilde{y}_1) + (1-q)H_1(\tilde{x}_2,\tilde{y}_2) = -(\tilde{x}_2)^2 + \frac{3}{2}\tilde{x}_2(q\tilde{y}_1 + (1-q)\tilde{y}_2),$$

$$U_3(\tilde{x}_1,\tilde{x}_2,\tilde{y}_1,p) = pH_2(\tilde{x}_1,\tilde{y}_1) + (1-p)H_2(\tilde{x}_2,\tilde{y}_1) = -(\tilde{y}_1)^2 + \frac{3}{2}\tilde{y}_1(p\tilde{x}_1 + (1-p)\tilde{x}_2),$$

$$U_4(\tilde{x}_1,\tilde{y}_2,\tilde{y}_2,p) = pH_2(\tilde{x}_1,\tilde{y}_2) + (1-p)H_2(\tilde{x}_2,\tilde{y}_2) = -(\tilde{y}_2)^2 + \frac{3}{2}\tilde{y}_2(p\tilde{x}_1 + (1-p)\tilde{x}_2).$$

Thus we have obtained the following strategic game

$$\Gamma_B^* = \langle J = \{1,2,3,4\}, R_j, U_j \rangle.$$

Now, we can determine the equilibrium profile. Strategy profiles $(\tilde{x}_1^*, \tilde{x}_2^*, \tilde{y}_1^*, \tilde{y}_2^*) \in NE(\Gamma_B^*)$ using the "best response approach

$$\begin{cases} \tilde{x}_1^* \in Br_1(\tilde{y}_1^*, \tilde{y}_2^*, q), \\ \tilde{x}_2^* \in Br_2(\tilde{y}_1^*, \tilde{y}_2^*, q), \\ \tilde{y}_1^* \in Br_3(\tilde{x}_1^*, \tilde{x}_2^*, p), \\ \tilde{y}_2^* \in Br_4(\tilde{x}_1^*, \tilde{x}_2^*, p), \end{cases}$$

where
$Br_1(\tilde{y}_1^*, \tilde{y}_2^*, q) = Arg \max_{\tilde{x}_1 \in \tilde{X}_1} U_1(\tilde{x}_1, \tilde{y}_1^*, \tilde{y}_2^*, q)$, $Br_2(\tilde{y}_1^*, \tilde{y}_2^*, q) = Arg \max_{\tilde{x}_2 \in \tilde{X}_2} U_2(\tilde{x}_2, \tilde{y}_1^*, \tilde{y}_2^*, q)$,
$Br_3(\tilde{x}_1^*, \tilde{x}_2^*, p) = Arg \max_{\tilde{y}_1 \in \tilde{Y}_1} U_3(\tilde{x}_1^*, \tilde{x}_2^*, \tilde{y}_1, p)$, $Br_4(\tilde{x}_1^*, \tilde{x}_2^*, p) = Arg \max_{\tilde{y}_2 \in \tilde{Y}_2} U_4(\tilde{x}_1^*, \tilde{x}_1^*, \tilde{y}_2, p)$.
This is equivalent to the following system

$$\begin{cases} \tilde{x}_1^* = \frac{3}{4}[q\tilde{y}_1^* + (1-q)\tilde{y}_2^*] \in \left[0, \frac{3}{4}\right], \\ \tilde{x}_2^* = \frac{3}{4}[q\tilde{y}_1^* + (1-q)\tilde{y}_2^*] \in [0,1], \\ \tilde{y}_1^* = \frac{3}{4}[p\tilde{x}_1^* + (1-q)\tilde{x}_2^*] \in \left[0, \frac{3}{4}\right], \\ \tilde{y}_2^* = \frac{3}{4}[p\tilde{x}_1^* + (1-p)\tilde{x}_2^*] \in [0,1]. \end{cases}$$

In the particular case, if player 1 chooses the informational extended strategy $\theta_1^1(y) = \frac{3}{4}y$ and assume with probability q that player 2 chooses the informational extended strategy $\theta_2^1(x) = \frac{3}{4}x$, and with probability $1 - q$ the informational extended strategy $\theta_2^2(x) = x^2$, and, respectively, if player 2 chooses the informational extended strategy $\theta_2^1(x) = \frac{3}{4}x$ and assume with probability p that player 1 chooses the informational extended strategy $\theta_1^1(y) = \frac{3}{4}y$ and with probability $1 - p$ the informational extended strategy $\theta_1^2(y) = y^2$, then the informational non extended Bayes-Nash equilibrium profiles $(x^*(q), y^*(p))$ is calculated from the following system

$$\begin{cases} \theta_1^1(y) = \frac{3}{4}[q\theta_2^1(x) + (1-q)\theta_2^2(x)], \\ \theta_2^1(x) = \frac{3}{4}[p\theta_1^1(y) + (1-p)\theta_1^2(y)]. \end{cases}$$

So we have the system

$$\begin{cases} \frac{3}{4}y = \frac{3}{4}\left[q\frac{3}{4}x + (1-q)x^2\right], \\ \frac{3}{4}x = \frac{3}{4}\left[p\frac{3}{4}y + (1-q)\frac{3}{4}x\right]. \end{cases}$$

Finally, all solutions of this example are described in the following Table 1.

Table 1

Players type	Informational extended strategies	Solutions from system
$(1,1) = [(1, \theta_1^1), (2, \theta_2^1)]$	$(\theta_1^1(y), \theta_2^1(y)) = \left(\frac{3}{4}y, \frac{3}{4}x\right)$	$\begin{cases} y = \left[q\frac{3}{4}x + (1-q)x^2\right] \\ x = \left[p\frac{3}{4}y + (1-q)\frac{3}{4}x\right] \end{cases}$
$(2,1) = [(2, \theta_1^2), (2, \theta_2^1)]$	$(\theta_1^2(y), \theta_2^1(y)) = \left(y^2, \frac{3}{4}x\right)$	$\begin{cases} y^2 = \frac{3}{4}\left[q\frac{3}{4}x + (1-q)x^2\right] \\ x = \left[p\frac{3}{4}y + (1-q)\frac{3}{4}x\right] \end{cases}$
$(1,2) = [(1, \theta_1^1), (2, \theta_2^2)]$	$(\theta_1^1(y), \theta_2^2(y)) = \left(\frac{3}{4}y, x^2\right)$	$\begin{cases} y = \left[q\frac{3}{4}x + (1-q)x^2\right] \\ x = \left[p\frac{3}{4}y + (1-q)\frac{3}{4}x\right] \end{cases}$
$(2,2) = [(2, \theta_1^2), (2, \theta_2^2)]$	$(\theta_1^2(y), \theta_2^2(y)) = (y^2, x^2)$	$\begin{cases} y^2 = \frac{3}{4}\left[q\frac{3}{4}x + (1-q)x^2\right] \\ x^2 = \frac{3}{4}\left[p\frac{3}{4}y + (1-q)y^2\right] \end{cases}$

With this we finished solving the example.

References

Hancu, B. 2011. Informational extended games generated by the one and double way directional informational flow. Studia Universitatis, Seria Ştiinţe exacte şi economice, 7(47): 32–43.

Hancu, B. 2012. Solving the games generated by the informational extended strategies of the players. Buletinul Academiei de Stiinte a Republicii Moldova. Matematica, 3(70): 53–62.

Harsanyi, J.C. and R. Selten. 1998. A General Theory of Equilibrium Selection in Games. MA: MIT-Press.

Novac, L. 2008. Conditions of single Nash equilibrium existence in the informational extended two-matrix games. ROMAI Journal, 4(1): 193–201.

CHAPTER 13

Energy Production and Storage Investments and Operation Planning Involving Variable Renewable Energy Sources

A Two-stage Stochastic Optimization Model with Rolling Time Horizon and Random Stopping Time

Yuri Ermoliev, *Nadejda Komendantova* and *Tatiana Ermolieva**

1. Introduction

European Commission adopted the new Strategic Energy Technology (SET) Plan (See European Commission 2016). The SET Plan aims, among other goals, to contribute to achieving the research and innovation objectives of the EU (European Union) to become the global leader in renewable energy; facilitate consumer participation and accelerate the progress to a smart energy system; develop and reinforce energy efficient systems. Industries and companies recognize that increasing the efficiency of energy use through implementing alternative decentralized methods of production and operation with renewable energy, energy conservation and saving technologies can increase profits and energy security.

International, Institute for Applied Systems Analysis, Austria.
* Corresponding author: ermol@iiasa.ac.at

Decentralizing energy system enables renewable energy to be generated and stored close to loads/demands. Decentralizing the system encompasses a wide variety of different renewable technologies such as wind and solar power, geothermal, and biomass, etc., into the system (See Oree et al. 2017, Vezzoli et al. 2018, Weinberg et al. 1991). Relevant stakeholders, i.e., transmission system operators and distribution system operators, manufacturers and research establishments, and infrastructure managers now have more freedom in decisions to control energy sources and demand, with incentives to invest in renewable energy technologies and the possibility of participating in the energy market. Yet, this can lead to new systemic risks and potentially irreversible problems if the decisions are inappropriate for the situation that occurs.

The main challenge associated with renewable energy sources is their stochastic and dynamic characteristic. Variable renewable energy sources such as wind and solar power are non-dispatchable due to their fluctuating nature, as opposed to controllable and relatively constant gas, oil, coal, and nuclear energy sources. The shift to more renewable energy, especially the variable ones such as wind and solar, raises a concern about the so-called "Power System Flexibility (PSF)", i.e., "the extent to which a power system can modify electricity production or consumption in response to variability, expected or otherwise" (Babatunde et al. 2020). Flexibility can therefore refer to the capability to change the power supply-demand balance of the whole system or its particular units (e.g., a power plant or a factory). Load balancing is among the main services a power system must perform flexibly.

The PSF requirement, the decentralization and the variable characteristics of renewable energy generation and transmission require specific modeling approaches and performance indicators enabling to manage of stochastic supply-demand imbalances to avoid or minimize the likelihood of systemic failures associated with, e.g., periods of unstable energy generation or/and high demands. Essential is to consider both traditional and renewable (dispatchable and non-dispatchable) energy technologies along with traditional energy storage and alternative possibilities, e.g., hydrogen production, to cover variations of the residual load curve.

The scarcity of natural resources such as land and water can become an important limiting factor for renewable energy and green hydrogen production systems deployments (See European Commission 2016, Gao et al. 2018, International Energy Agency 2018). Land is required both for windmills and solar panels installation. While water demand in energy sector can be reduces through the use of renewable resources, green hydrogen production requires vast amounts of water. Often, areas suitable for wind and solar power production lack sufficient water and land resources (See Ermolieva et al. 2016, Ermoliev et al. 2022, Ermoliev et al. 2021, International Energy Agency 2018). The competition for water and land increases systemic vulnerability in the interdependent energy-land-water-environmental (ELWE) systems exacerbated by exogenous shocks from weather events and climate change. Inconsistent policies across the interacting ELWE sectors can further increase the likelihood of vital supply-demand disequilibrium in the ELWE networks leading to systemic failures similar to recent electricity deficits and price surges in US due to disintegrated electricity production and transmission based on renewables. Disruptions in ELWE networks can affect food, energy, water, and environmental security at different levels with possible global spillovers.

Therefore, our proposed energy planning methodology is a part of a broader EFWE nexus approach presented in Ermolieva et al. 2016, Ermoliev et al. 2022, Ermoliev et al. 2021. The nexus approaches consider the dynamic interlinkages between economic and natural systems, namely, energy, agricultural, water sectors and natural resources. Coherent ELWE solutions can be revealed through the use of the nexus approach, rather than conventional approaches that often overlook the interdependencies and joint resource constraints. The integrated EFWE decision support system (DSS) enables to develop of robust systemic regulations for disintegrated distributed EFWE systems in the presence of risks and uncertainties of various kinds relying on robust distributed models' linkage and optimization methods (See Ermoliev et al. 2021).

In what follows, we present the energy planning model, which is an important building block of the EFWE DSS. The model incorporates the main stages of energy flows from resources to demands: energy extraction from energy resources, primary energy conversion into secondary energy forms, transport and distribution of energy to the point of end use that results in the delivery of final energy, and the conversion at the point of end use into useful energy forms that fulfill the specific demands. Demands for useful energy products come from the main sectors of the economy: industrial, residential, transport, agricultural, water, and energy. The model accounts for the interactions and dependencies across renewable and traditional energy systems. It enables support for the design and evaluation of hybrid green energy (wind and PV)—hydrogen production-supply-demand-storage energy system accounting for potential uncertainties and risks associated with plausible production-supply-demand-storage mismatches and natural resource scarcity.

The key methodological advancement of the proposed energy planning model is that it has a random planning horizon defined by a stopping time. The stopping time represents new conditions such as a natural disaster, a market shock, the emergence of new technologies or policies, or safety constraint(s) violation, which can trigger a system's failure. The stopping time moments induce endogenous risk aversion in strategic decisions in a form of dynamic VaR-type systemic risk measures dependent on the system's structure (See Cano et al. 2016, Cano et al. 2014, Ermolieva et al. 2016, Ermoliev et al. 2022, Ermoliev et al. 2021, Ermoliev and Wets 1988, Ermoliev et al. 2012, Ermoliev and Winterfeldt 2012). Moving time horizons of the model can also be determined by the possible outcomes of stakeholders' discussions. The model builds on the advances in effective coping with uncertainty (See Ermolieva et al. 2016, Ermoliev et al. 2022, Ermoliev and Wets 1988, Gao et al. 2018, Gritsevskii and Ermoliev 1999, Gritsevskii and Nakicenovic 2000, Gritsevskii and Ermoliev 2012, O'Neill et al. 2006), which allows reducing energy consumption along with integrated management of risks, especially due to uncertainties in energy prices and loads, e.g., by employing financial instruments available for managing market risks and stochastic loads.

In the presence of deregulations and uncertainties, there is a dilemma to choose an efficient technological portfolio in real-time while pursuing long-term goals. The solution to the problem involves the so-called two-stage dynamic stochastic optimization models. Some decisions (so-called first-stage strategic decisions) regarding the deployment of new technologies are taken before uncertainties are resolved and some others (so-called second-stage operational decisions) are taken once values for uncertain parameters become known, thereby providing a trade-off between long- and short-term decisions.

Investment planning and operational optimization decisions concern the demand and supply sides of different energy sources (gas, heat, wind, solar, hydrogen, water, etc.). The demand side is affected by old and new technologies and activities including such end uses as electricity only, heating, cooling, hydrogen and fertilizers production, water desalination, smart agriculture, and energy storage and saving technologies, etc. New activities may alter peak loads whereas traditional accumulators (storage) may considerably smooth energy demand-supply processes. Green hydrogen production is being frequently considered as a "missing link" to provide additional (alternative) energy storage and to buffer volatility and intermittency of renewable energy generation. Hydrogen technologies enable the decisive step of the energy sector transformation (See European Commission 2016, International Energy Agency 2018).

Most existing energy planning models are based on a deterministic optimization approach that is unable to provide decisions robust against uncertainties and risks. The deterministic approach performs analysis of 'what-if' cases and derives scenario-specific solutions. This can lead to system failure if other than expected scenario happens. Contrary to the deterministic approach, this model allows stakeholders to analyze a portfolio of interdependent strategic long-term and operational short-term decisions, which ensures the robust performance of energy systems.

Unlike the static nature of deterministic models, the proposed stochastic model delivers solutions that are responsive to revealed information about systemic uncertainties and risks such as those related to stochastic supply, demand, prices, weather variability, and technological change, in order to adjust local or regional energy management policies in a cost-effective and risk hedging manner. Integration of the operational and strategic models under the umbrella of stochastic optimization provides an effective way to make real-time decisions consistent with the long-term strategic goals of energy system planners to guarantee secure energy provision. The solution of the problem involves adjusting operational decisions to hit long-term targets if additional information about prices, subsidies, demand, weather, and new technologies can become available in the future.

The paper is organized as follows. Section 2.1 formulates the basic deterministic energy planning model. A stochastic model is presented in Section 2.2. In particular, a basic two-stage stochastic dynamic energy sector planning model is described in Section 2.2.1. Sections 2.2.2–2.2.3. illustrate the emergence of safety constraints in the form of Value-at-Risk systemic indicators. Numerical methods are covered in sections 2.2.4–2.2.6 including the concept of a two-stage STO (stochastic optimization) model with a rolling (moving) horizon. Section 3 illustrates the benefits of STO approaches compared with deterministic counterparts using the definition of the value of stochastic solutions (VSS) or/and the value of stochastic modeling (VSM). Section 4 concludes and presents potential and current case studies involving the presented stochastic energy planning model. In particular, the model enables modeling the renewable energy system elements together with hydrogen production, which can be effectively used as energy storage and transfer media.

2. Energy planning model

Below, we discuss a rather general energy system model accounting for the available resources for energy production, potential secondary and final energy requirements/demands, and possible energy storage and transfer infrastructure. The use of storage is intended to overcome various supply-demand dis-equilibriums at different levels, in particular, in a (stand-alone off-grid) green energy production system. The model can incorporate the following five main stages of energy flows from resources to demand: energy extraction from energy resources, primary energy conversion into secondary energy forms, transport, and distribution of energy to the point of end use that results in the delivery of final energy, and the conversion at the point of end use into useful energy forms that fulfills the specific demands. Primary energy sources are coal, crude oil, gas, methanol, the solar, wind, etc. Secondary energy sources are coal, fuel oil, gas, methanol, hydrogen, electricity, ammonia, etc. Final energy products are coal, fuel oil, gas, hydrogen, ammonia, methanol, electricity, solar and wind onsite. Demands for useful energy come from the main sectors of the economy: industrial, residential, transport, agricultural, and water.

Each energy technology is characterized by united costs, efficiency, lifetime, emissions, etc. Additional sectoral (and cross-sectoral joint) constraints are imposed to capture the requirements and the limitations on natural resource use and availability and investments. The model can include the existing technologies, as well as the new zero-carbon green technologies, technologies at the beginning of implementation or even in the research stage, e.g., various renewable and carbon capture technologies. For example, the model can study the feasibility of green hydrogen and ammonia production as they have the capability to substantially impact the transition towards a zero-carbon economy. Hydrogen and ammonia can serve as effective media to store and transport energy. Both could form the basis of a new, integrated worldwide renewable energy storage and distribution solution. Depending on case study goals, the model can include further details of modeling specific energy forms and production processes, for example, green hydrogen and ammonia production.

2.1 A basic case: deterministic model

Within some time horizons, e.g., T years, the demand for electricity and other energy can be characterized by a demand profile reflecting the dynamics of demand changes over time (See Cano et al. 2016, Cano et al. 2014, Ermoliev et al. 2021, Gritsevskii and Ermoliev 1999, Gritsevskii and Nakicenovic 2000, Gritsevskii and Ermoliev 2012). For illustration purposes, let us consider only electricity demand. The demand profile within each year (time period) $t = 1,..., T$ can be adequately characterized by the demand within representative periods $j = 1,..., J$ as in Figure 1, where d_j^t, c_{ij}^t denote the energy demand and costs of technology i in period of year t, P_{ij}^t, denotes the matrix of other parameters, and y_{ij}^t are operational decisions for technology i in period j of year t. Distinguishing energy production by representative periods allows to explicitly account for seasonal, monthly, and

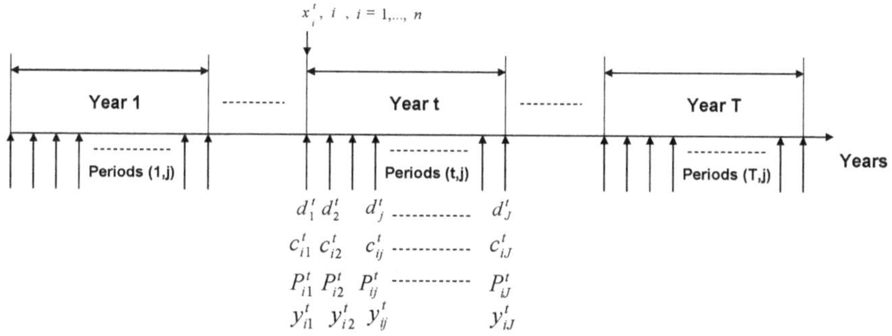

Figure 1. Temporal resolutions of the strategic planning model.

even daily variations of loads, energy prices, renewable energy resources, natural and economic resource uncertainty and variability.

The goal of the strategic model is to identify such technologies with capacities x_i^t, $i = 1,...,n$, installed at the beginning of the year (or of planning period) t that the demands d_j^t, in each representative period $j = 1,..., J$ and year t are satisfied.

Assume the planning time horizon of T years. Let x_i^t be the capacity of technology, $i = 1,..., n$ installed in year t, S_i^t – total supply by available in t, $t = 1,..., T$. Then

$$x_i^t \geq 0, i = 1,..., n, t = 1,..., T, \tag{1}$$

$$S_i^t = S_i^{t-1} + x_i^t - x_i^{t-Li} \tag{2}$$

where L_i is the lifetime of technology i, S_i^0 is the initial capacity of existent before $t = 1$.

In these equations for the simplicity of exposition, we assume that S_i^t is the energy already converted to electricity end-use to satisfy demand. For each period $j = 1: J$ within $t = 1,..., T$ we introduce the production plan decisions y_{ij}^t defining how much of technology i operates in period j to satisfy demand $d_j^t, j = 1: J$:

$$\sum_{i=1}^{n} y_{ij}^t = d_j^t, j = 1,..., J, t = 1,..., T, \tag{3}$$

$$y_{ij}^t \leq \gamma_{ij}^t S_i^t, i = 1,..., n, j = 1: J, t = 1,..., T, \tag{4}$$

$$y_{ij}^t \geq 0, \tag{5}$$

where γ_{ij}^t, $1 \leq \gamma_{ij}^t \leq 1$, may be interpreted as availability factor of technology i operating in period j in t ($\gamma_{ij}^t = 0$ for not yet existing technologies).

The deterministic dynamic strategic planning problem is to minimize the total cost

$$\sum_{t=1}^{T} (\sum_{i=1}^{n} C_i^t x_i^t + \sum_{i=1}^{n} \sum_{j=1}^{J} c_{ij}^t T_j^t y_{ij}^t), \tag{6}$$

where C_i^t—is unit investment cost for technology i in year t, c_{ij}^t—unit production cost (including fuel costs) for i in period j in t, and T_j^t is a duration of period j in t that can also be denoted as $\sum_{i,j} q_{ij}^t y_{ij}^t$.

The energy efficiency and aging processes may be introduced in equation (2) as

$$S_i^t = e_i^t S_i^{t-1} + x_i^t - x_i^{t-L_i}, \tag{7}$$

where e_i^t denotes the "depreciation" (aging) rate in t. Aging can be also modeled as a dynamic process in a detailed technology-based manner as suggested in Section 2.1.5. Investment constraint may be introduced for each $t = 1,..., T$ as

$$\sum_{i=1}^{n} C_i^t x_i^t \leq I^t.$$

2.1.1 *Energy types*

In general cases, there is a set E of available energy sources (gas, coal, solar, wind, water, biomass, hydrogen). Each technology i is characterized by its set of input-output (conversion) coefficients a_{ilk}^t, which convert the unit input energy type l into a_{ilk}^t units energy type k. In this case, i denotes different energy technology $i = 1,..., m$ to satisfy demand d_{kj}^t of energy type $k = 1,..., K$ in period $j = 1,..., n$ in $t = 1,..., T$. Define nonnegative decisions y_{ijk}^t representing technology i operating in period j of energy type k. Therefore,

$$\sum_{i,l} a_{ilk}^t y_{ijk}^t \geq d_{jk}^t \tag{8}$$

$$y_{ijk}^t \leq y_{ij}^t S_i^t. \tag{9}$$

Thus, $\sum_{i,l} a_{ilk}^t y_{ijk}^t$ is the input of energy l required to satisfy demand in energy k by production plan y_{ijk}^t for technology i. Operation costs of technology i are spread over technology life L_i. For simplification of notation, total operation cost $\sum_{i,j,l,k} q_{ijlk}^t y_{ijk}^t$ of technologies in t in all periods j includes resource costs and conversion costs from i to k.

2.1.2 *Remark 1 (Purchasing, energy efficiency and availability factors)*

Purchased electricity can be modeled as a technology, with electricity on both input and output and purchase cost on the input. Total available electricity would be a sum of electricity generated by technologies and purchased from the market. The availability factor γ^t_{ij} defines if the whole ($\gamma^t_{ij} = 1$) or only a certain portion of the capacity/output S^t_i can be used during time t and period j. This may apply, e.g., for wind-power plants and heating devices. If equation (9) is turned into equality, the portion defined by γ^t_{ij} has to be used during each time period. This may describe efficiency improvements (energy conservation). For wall insulation, which has a certain time-dependent influence on the heat demand, γ^t_{ij} is large in the winter but small in the summer.

2.1.3 Pollution constraints

Besides cost minimization and the investment constraints, there may exist environmental constraints, say, on feasible CO2 and other pollutants $l = 1,\ldots, L$ emissions:

$$\textstyle\sum_{ijl} \pi^s_{ilk}\, a^t_{ilk}\, y^t_{ijk} \leq \bar{\pi}^t_s,\ s = 1,\ldots, S,\ t = 1,\ldots, T \tag{10}$$

where π^s_{ilk} is the unit pollution, $\bar{\pi}^t_s$ is the critical level of pollution.

2.1.4 Batteries

While we have argued that the definition of technology should be general enough to accommodate all possible technologies, there is (at least) one exception–batteries and other types of energy storage. The reason is that batteries do not convert one type of energy into another, but move the energy in time (with some losses). Explicitly, storage processes can be easily represented in the operational model. In the case of a strategic model, as we assume, the load profiles are approximated by a proper number of periods j. In this case, it is possible to represent the accumulators by additional decision variables v^t_{jk}, u^t_{jk} on energy storage and its use at t and j. Equations (8) then can be modified as

$$\textstyle\sum_i a^t_{ilk}\, y^t_{ijk} = d^t_{jk} + v^t_{jk} - \mu^t_{jk}\, u^{t-1}_{jk},\ v^t_{jk} \geq 0,\ u^t_{jk} \geq 0, \tag{11}$$

$$r^t_k \geq \textstyle\sum_j v^t_{jk} \geq \sum j\, u^t_{jk}, \tag{12}$$

where r^t_k is the capacity of accumulators for energy k at time t, u^t_{jk}—loss per unit stored energy. These equations in an aggregate manner redistribute loads among periods. With a shorter time scale, it may be reasonable to introduce shifts of demands in time by substituting u^t_{jk} in (11) with u^{t-1}. In the case of the operational model, we can further specify storage processes by introducing decision variable denoting overproduction of energy at time stored/saved for production at $\tau > t$.

2.1.5 Modeling aging processes

From (7) it follows that each technology i has a fixed lifetime L_i. This does not allow for possible replacements during the technology's lifetime (either because the maintenance costs become too high or because new technology becomes available). Since the purpose of the strategic model is to decide which equipment to buy and when it is rational to do, this aspect should be modeled properly. In model (2)–(6) the driving force for the installation of new technologies is aging process, modeled by (in general random) lifetime L_i, availability factor γ_i^t, and equations (6) or (7). We shall keep in mind that the efficiency of each technology may differ, especially in the beginning (initial/learning effect) and closer to the end (diminishing marginal efficiency/aging). Ageing of technologies can be taken into account as a dynamic process by replacing (2) with

$$S_i^t = \sum_{\tau=0}^{t} e_{i,t-\tau}\, x_i^\tau,$$

where $e_{i,a}$ denotes efficiency of technology i at age a. Note that we can rewrite

$$S_i^t = S_i^{t-1} + e_{i,0}\, x_i^t + \sum_{\tau=0}^{t}(e_{i,t-\tau} - e_{i,t-\tau-1})x_i^\tau = S_i^{t-1} + e_{i,0}\, x_i^t + (e_{i,t} - e_{i,t-1})\, x_i^0 +$$

$$(e_{i,t-1} - e_{i,t-2})\, x_i^1 + \ldots + (e_{i,1} - e_{i,0})x_i^{t-1}.$$

One advantage we see in doing it this way is that we could let the maintenance cost depend on the age of the equipment: the cost of the installed technology i at time t would then be

$$\sum_{\tau=0}^{t} c_{i,t-\tau}^t\, x_i^\tau,$$

where $c_{i,a}^t$ is the cost of technology i of age a at time t, i.e., technology bought at time $t - a$:

$$c_{i,a}^t = \begin{cases} \text{purchase and maintenance cost} & \text{if } a = 0, \\ \text{maintenance cost} & \text{if } a > 0. \end{cases}$$

Note that we can again let the sum over τ let start from a negative value to include the maintenance cost of already installed equipment.

2.1.6 Replacement of installed technologies

In the deterministic model each technology i has a fixed lifetime L_i, after which it simply stops working. This assumption does not allow for replacement during the technology's lifetime (either because the maintenance costs become too high, or because a new technology becomes available). Since the main purpose of the strategic model is to decide which equipment to buy (if any—we might decide to wait for better technologies), this aspect is important and should be included in the model. The most "natural" way to model this would be adding binary variables for each i and t, denoting whether it is installed or not. The problem with this simple approach is that it can increase the solution time significantly, especially since we do not have any other integer variables (yet). Instead, we propose the following linear approximation, using new non-negative variables $z_i^{t,\tau}$ denoting the amount of technology i bought at time and decommissioned at time $\tau > t$. Obviously, this implies $\sum_{\tau > t} z_i^{t,\tau} \leq x_i^t$.

The approximation of this approach (compared to using binary variables) lies in the fact that we can decommission only parts of the technology—which, we hope, would not influence the model much. With this notation, the amount of technology i installed at time t would be

$$\sum_{\tau=0}^{t} (x_i^\tau - \sum_{a=1}^{t-\tau} z_i^{\tau,\tau+a}) = \sum_{\tau=0}^{t} (x_i^\tau - \sum_{\hat{t}=\tau+1}^{t} z_i^{\tau,\hat{t}}),$$

and the available capacity, using efficiency $e_{i,a}$ defined in (11) would become

$$S_i^t = \sum_{\tau=0}^{t} e_{i,t-\tau} (x_i^\tau - \sum_{\hat{t}=\tau+1}^{t} z_i^{\tau,\hat{t}}).$$

We can also design price $\hat{c}_{i,a}^t$ to decommission technology i aged a at time t. The cost could be negative for a price bigger than the de-installation costs. The total costs of the technology at time t, including the maintenance costs $c_{i,a}^t$ proposed in the previous section, would then become

$$\sum_{\tau=0}^{t} c_{i,t-\tau}^t (x_i^\tau - \sum_{\hat{t}=\tau+1}^{t} z_i^{\tau,\hat{t}}) + \sum_{\tau=0}^{t-1} \hat{c}_{i,t-\tau}^t z_i^{\tau,t}$$

where the first element denotes the purchase and maintenance cost and the second element the decommissioning costs. For "accounting purposes", it could be better to take the purchase costs out of $c_{i,a}^t$, which would add one more element to the above sum.

2.2 Stochastic model

Since uncertainties may seriously affect long-term decisions, the model should be formulated as a dynamic stochastic optimization problem. Among the main stochastic parameters are supply, demand, fuel prices, operational costs, and lifetime of technologies. The demands may be affected by weather conditions. It may also substantially differ by the time of the day and the day of the week.

The model can run at a rather detailed dynamic resolution (hourly, daily). Instead of considering hourly values, we can aggregate variables (demands, supply, prices, costs) into J periods representatively describing the behavior of the system within a year. In some cases, hourly values can be aggregated by month, including peak values, working days peak and peak values, and Saturday and Sunday values. Since uncertainties may seriously affect long-term decisions, the proposed model is formulated as a dynamic stochastic model. We show how the model operates in a two-stage manner with time intervals $t = 1,..., T$ implementing a rolling time horizon concept for T. Time intervals define planning intervals, for example, annual, 5- or 10-year plans, etc. The two-stage dynamic decisions suppose that so-called long-term first-stage decisions regarding investments into new energy technologies have to be taken before uncertainties are resolved and some others (so-called second-stage decisions) will be taken once values for uncertain parameters become known, thereby providing a trade-off between long- and short-term decisions.

In the presence of uncertainties, there is a dilemma to choose an efficient technological portfolio in real-time while pursuing long-term goals and continuously updating new information about uncertainties. Therefore, the solution of the problem involves the so-called two-stage dynamic stochastic optimization modeling with a rolling horizon. The model delivers solutions that are responsive to newly revealed information about systemic uncertainties and risks such as stochastic supply, demand, prices, weather variability, and technological change, in order to adjust local or regional energy structure and management policies in a cost-effective and risk hedging manner. Integration of the operational and strategic models under the umbrella of the two-stage stochastic optimization provides an effective way to make real-time decisions consistent with the long-term strategic goals of energy system planners to guarantee secure energy provision in all uncertainty scenarios. The solution to the problem involves adjusting operational decisions to hit long-term targets if additional information about prices, subsidies, demand, weather, and new technologies can become available in the future.

In the two-stage stochastic model, some decisions (so-called first-stage decisions) including investments into new energy technologies have to be taken before uncertainties are resolved and some others (so-called second-stage decisions) will be taken once values for uncertain parameters become known, thereby providing a trade-off between long- and short-term decisions.

The main goal of the dynamic stochastic two-stage strategic planning model is to find such a combination of technologies installed at the beginning of each year that the mixture of these technologies operating in each period

j would ensure safe energy "provision" plan minimizing investment costs, possibly deterministic equipment maintenance costs and stochastic operation costs which may include stochastic fuel prices. Schematically, the structure of the modeling framework is similar to Figure 1.

2.2.1 A basic two-stage stochastic dynamic optimization model

Let us consider the stochastic version only of the basic dynamic deterministic model (Section 2.1). Its generalizations can be formulated in a similar manner, but for the sake of simplicity, they are not presented in this paper. For the simplicity of notation, we consider the case when only the evolution of costs C_i^t, c_{ij}^t, and demands d_j^t is uncertain. Also, we discuss the case of uncertain lifetimes and other parameters.

Strategic first stage investment decisions x_i^t, $t = \overline{1:T}$, are made at the beginning of planning horizon $t = 1$ using a perception of stakeholders about potential future scenarios $C_i^t(\omega)$, $c_{ij}^t(\omega)$, $d_j^t(\omega)$, of costs and energy demands dependent on stochastic parameter ω. Here we use ω to denote a sequence of $\omega = (\omega_1, \omega_2, \ldots, \omega_t, \ldots, \omega_T)$ of uncertain vectors ω_t of in general independent parameters, which may affect outcomes of the strategic model, e.g., market prices, weather conditions, demands. In general, there are different components of ω_t, say, components ω_t^{dem} characterizing the variability of the demand and other components ω_t^{str}, ω_t^{ope} characterizing uncertainties associated with strategic and operational decisions. Therefore, functions $C_i(\omega)$, $c_{ij}^t(\omega)$, $d_i^t(\omega)$, depend in general only on some components of ω_t, although we indicate dependence on for simplicity of notation.

There are essential differences between the basic deterministic model (1)–(6) and its stochastic version. Because strategic decisions x_i^t depend only on t, equations (1), (2) remain the same. The second stage adaptive operational decisions y_{ij}^t are made after observing real demands and costs. They depend on observable scenario ω, i.e., $y_{ij}^t = y_{ij}^t(\omega)$. Therefore, any choice of investments decisions x_i^t, $t = \overline{1:T}$, does not yield feasible second stage solutions $y_{ij}^t(\omega)$ satisfying following equations (13)–(15) for all :

$$\textstyle\sum_{i=1}^{n} y_{ij}^t(\omega) = d_j^t(\omega), j = \overline{1:J}, t = \overline{1:T}, \tag{13}$$

$$y_{ij}^t(\omega) \le \gamma_{ij}^t S_i^t, i = \overline{1:n}, j = \overline{1:J}, t = \overline{1:T}, \tag{14}$$

$$y_{ij}^t(\omega) \ge 0, \tag{15}$$

where

$$S_i^t = S_i^{t-1} + x_i^t - x_i^{t-L_i} \tag{16}$$

$$x_i^t \geq 0,\ i = \overline{1{:}n},\ t = \overline{1{:}T}. \tag{17}$$

The feasibility of constraints (13)–(15) for any scenario ω can be guaranteed by assuming the existence of a back-stop technology with high operating costs that can also be viewed as purchasing without delay but at a high price. Without losing generality, we assume that for any period j and time t it is the same technology $i = 1$. Then the basic dynamic stochastic two-stage model is formulated as a minimization of the expected total cost's function

$$F(x) = E_\omega \sum_{t=1}^{T}\sum_{i=1}^{n}[C_i^t(\omega)x_i^t + \sum_{j=1}^{J} c_{ij}^t(\omega)T_j^t y_{ij}^t(\omega)] = \sum_{t=1}^{T}\sum_{i=1}^{n}[C_i^t x_i^t +$$

$$E_\omega \sum_{j=1}^{J} c_{ij}^t(\omega)T_j^t y_{ij}^t(\omega)] = \sum_{t=1}^{T}\sum_{i=1}^{n}[C_i^t x_i^t + E_\omega \min_{y_{ji}^t(\omega)} \sum_{j=1}^{J} c_{ij}^t(\omega)T_j^t y_{ij}^t(\omega)] \tag{18}$$

where C_i^t is expected investments cost, $C_i^t = E_\omega C_i^t(\omega)$. Because the backup technology is available in unlimited amounts, $C_1^t = 0$, $t = 1,\ldots, T$; $c_{1j}^t = C$ for all j where C is a large enough positive number. If optimal $y_{ij}^t > 0$ for some scenarios ω, then this indicated the existence of risk that installed technologies (excluding $i = 1$) are not able to satisfy demands. The following example illustrates this risk.

2.2.2 Example 1 (Back-stop technology and induced risks)

As this simple example shows, the back-stop technology $i = 1$ induces critically important safety constraints connected with a type of Value at Risk (VaR) measures of risks. Assume that $T = 1$, $J = 1$, and there are only two electricity-supplying technologies $i = 1,2$, where $i = 1$ is the back-stop technology enabling instantaneous "purchasing" electricity at a high price c_1^1; $i = 2$ is a traditional technology that may require investments with expected unit costs $c_2^1 > 0$, whereas $c_1^1 = 0$. Demand $d^1(\omega)$ depends on ω. In this case the model (13)–(17) is formulated as minimization of the function

$$C_2^1 x_2^1 + E_\omega [c_{11}^1 y_{11}^1(\omega) + c_{21}^1 y_{21}^1(\omega)]$$

subject to

$$y_{11}^1(\omega) + y_{21}^1(\omega) = d^1(\omega),$$

$$S_1 = x_2^1,$$

$$y_{21}^1 \le x_2^1,$$

$$x_2^1 \ge 0, y_{11}^1 \ge 0, y_{21}^1 \ge 0,$$

where $S_0 = 0$, $\gamma_{21}^1 = 1$. The capacity of the back-stop technology $i = 1$ is assumed to be unlimited, and $c_{11}^1 > C_2^1 > c_{21}^1$. Therefore, the back-stop technology $i = 1$ ensures the satisfaction of the constraints (14) at high operations costs $c_{11}^1 y_{11}^1$. For any strategic decision x_2^1 and scenario ω there may be two situations

$$x_2^1 \ge d^1(\omega) \text{ or } x_2^1 < d^1(\omega).$$

From the structure of costs follows, the corresponding optimal operations decisions

$$y_{21}^{1Opt}(\omega) = d^1(\omega),\ y_{11}^{1Opt}(\omega) = 0, \text{ or}$$

$$y_{21}^{1Opt}(\omega) = x_2^1,\ y_{11}^{1Opt}(\omega) = max\{0, d^1(\omega) - x_2^1\}.$$

Therefore, the minimization of function (18) subject to constraints (13)–(17) in this example is reduced to minimizing the function

$$F(x_2^1) = C_2^1 x_2^1 + E_\omega [c_{11}^1\ max\{0, d^1(\omega) - x_2^1\} + c_{21}^1\ max\{x_2^1, d^1(\omega)\}]$$

$$\text{for } x_2^1 \ge 0.$$

Consider only the case when has a continuous probability density, i.e., function $F(\cdot)$ is continuously differentiable. Otherwise, we need to use a more complicated lengthy analysis. It is clear that an optimal strategic decision x_2^1 is positive, therefore, the optimality condition $F'(x_2^1) = 0$ yields the following important equations:

$$C_2^1 - c_{11}^1\ Prob\ [d^1(\omega) \ge x_2^1] + c_{21}^1\ Prob[d^1(\omega) < x_2^1] = 0$$

that can be easily rewritten as $C_2^1 + c_{21}^1\ (C_{21}^1 - c_{11}^1)\ Prob\ [d^1(\omega) \ge x_2^1] = 0$ or

$$Prob\ [d_2^1(\omega) \ge x_2^1] = (C_2^1 + c_{21}^1)/(C_{11}^1 - c_{21}^1). \tag{19}$$

Under large enough back-stop unit costs C_{11}^1, the value $(C_2^1 + c_{21}^1)/(C_{11}^1 - c_{21}^1) < 1$, i.e., the optimal solution satisfies important safety constraints (19). Therefore, the use of back-stop technologies induces safety constraints (19) which are used in reliability theory, the insurance industry, and financial applications. The safety level (right hand of equations (19)) can be regulated by the back-stop unit cost C^1_{1t}

2.2.3 *Remark 2 (VaR and CVaR risk measures)*

In model (13)–(18) the variability of outcomes is characterized by stochastic second-stage operational decisions $y^t_{ij}(\omega)$ with unit costs $c^t_{ij}(\omega)$. Although strategic first-stage decisions x^t_i have random costs $c^t_i(\omega)$, total cost function (18) ignores their variability by using mean values $C^t_i = EC^t_i(\omega)$. If total stochastic costs $\sum^n_{i=1} C^t_i(\omega)x^t_i$ cannot well enough approximate a normal distribution, then instead of the mean value of these costs, it is advantageous to use in equation (18) the median or other quantiles, especially when these costs may be affected by extreme events. This variability can be easily modeled by additional decision variables $z_t \geq 0$ which jointly with variables x^t_i, $y^t_{i,j}(\omega)$ minimize the function

$$\sum^T_{t=1}[z_t + (1-\rho_t)^{-1} E\, max\{0, \sum C^t_i(\omega)x^t_i - z_t\} + E\sum^J_{j=1} c^t_{ij}(\omega)y^t_{ij}(\omega)] \quad (20)$$

subject to constraints (13)–(17). The minimization of function (20) includes now additional subproblems on minimization of functions with respect to $z_t \geq 0$, $t = 1, ..., T$,

$$R(z_t) = z_t + (1-\rho_t)^{-1} E\, max\{0, \sum C^t_i(\omega)x^t_i - z_t\} \quad (21)$$

subject to $z_t \geq 0$, where ρ_t is a risk factor controlling those stochastic costs $\sum^n_{i=1} C^t_i(\omega)x^t_i$ don't exceed a desirable robust safety level z_t.

If the safety vector $\rho_t < 1$, then the minimization of risk function (20) yields optimal values z^*_t satisfying the following safety constraints which in financial applications are well known as the Value at Risk indicator

$$Prob\, [\sum C^t_i(\omega)x^t_i \geq z_t] = \rho_t,\ t = 1, ..., T. \quad (22)$$

Sub-problems (21) correspond to Conditional Value at Risk (CVaR) minimization. This problem has the following interpretation. Foreseen investments costs $\sum^n_{i=1} C^t_i(\omega)x^t_i$ at time t are planned to be covered by ex-ante credit at price $\rho_t < 1$ and ex-post borrowing $max\{0, \sum^n_{i=1} C^t_i(\omega)x^t_i - z_t\}$ fore relatively greater price. This provides more flexibility (robustness) compared only with ex-ante planning of investments using mean values of perceived unit costs $EC^t_i(\omega)$. It is also important that ex-ante credit is evaluated by (22) as the quantile with desirable safety levels ρ_t, $t = 1, ..., T$, rather than mean values which are especially misleading in the case of non-normal distributions.

2.2.4 Numerical methods

The model (13)–(18) is formulated in the space of variables (x^1_i, $y^t_{ij}(\omega)$, $t = \overline{1:T}$, $i = \overline{1:n}$, $j = \overline{1:J}$, $\omega \in \Omega$), where the set of scenarios Ω may include a finite number of implicitly given scenarios, e.g., by scenario trees. This induces often extremely large size of models. A realistic practical model (13)–(18) excludes analytically tractable solutions, although the model has an important block-structure that is usually utilized for most effective numerical solutions. In particular, the second-stage submodels often have simple solutions as in Example 1. In this case, the two-stage model is reduced to optimization in the space of only strategic solutions x^t_i, $t = 1,..., T$, similar to optimization of function $F(\cdot)$ in this Example.

In a rather general case, Ω contains or can be approximated by scenarios ω^s_i, $s = \overline{1:S}$, characterized by probabilities p_s, $s = \overline{1:S}$. Then the model (13)–(18) is formulated as minimization of function

$$\sum_{s=1}^{S} p_s[\sum_{t=1}^{T}\sum_{i=1}^{n} C^t_i(\omega_s)x^t_i + \sum_{j=1}^{J} c^t_{ij}(\omega_s)T^t_j\, y^t_{ij}(\omega_s)] \tag{23}$$

subject

$$\sum_{i=1}^{n} y^t_{ij}(\omega_s) = d^t_j(\omega_s), j = \overline{1:J}, t = \overline{1:T}, s = \overline{1:S},$$

$$y^t_{ij}(\omega_s) \le \gamma^t_{ij} S^t_i, i = \overline{1:n}, j = \overline{1:J}, t = \overline{1:T}, s = \overline{1:S},$$

$$S^t_i = S^{t-1}_i + x^t_i - x^{t-Li}_i, i = \overline{1:n}, j = \overline{1:J}, t = \overline{1:T}, s = \overline{1:S},$$

$$x^t_i \ge 0, i = \overline{1:n}, t = \overline{1:T}.$$

2.2.5 Remark 3 (General model).

Example 1 shows that although objective function (18) uses mean values of total costs, $\sum_{i=1}^{n} C^t_i x^t_i$, from equations (19) follows that the optimal solution, in fact, is a quantile of the energy demand. Stochastic versions of more general Section 2 models are formulated similarly to the basic deterministic model (1)–(5). Example 1 also illustrates the main ideas involved in the formulation of general stochastic models. In general cases, the main issue is the "curse of dimensionality", i.e., the use of proper solution methods. For example, stochastic quasigradient (SGQ) methods (See Ermoliev 2009, Ermoliev 1976) allow to avoid a priori generations of a large number of scenarios that dramatically increase the size of models. Instead, SQG methods use only a single sequentially generated at each iteration jointly with adjustments of a current approximate solution.

2.2.6 Remark 4 (Two-stage models with rolling horizons)

Initial model is focused on time horizon $[1,T]$. Robust strategic solution can be written as $x^{[1,T]} = (x_i^{1,[1,T]},\ldots, x_i^{T,[1,T]}, i = \overline{1{:}n})$. Solutions $(x_i^{1,[1,T]})$, $i = \overline{1{:}n}$, are implemented at $t = 1$. This creates a basis for readjustments of scenarios $\omega^{[1,T]}$ perceived at the beginning of time horizon $[1,T]$. New set of scenarios $\omega^{[2,T]}$ are evaluated, new robust strategic solutions $(x_i^{2,[1,T]})$, $i = \overline{1{:}n}$ and so on. Thus, initially a long-term strategic trajectory $x^{[1,T]}$ is evaluated, the first time interval solutions $(x_i^{1,[1,T]})$, $i = \overline{1{:}n}$, are implemented, new data is received, new scenarios $\omega^{[2,T+1]}$ are adjusted, and so on.

3. Value of stochastic solution

In this section, we discuss the value of stochastic optimization models that is often (See, for instance, Birge 1982, Ermoliev and Wets 1988) termed as the value of stochastic solutions (VSS) although this notion has a misleading character because the two-stage models incorporate both ex-ante deterministic first stage decisions chosen before observations of uncertain parameters (events) and ex-post stochastic adaptive decisions chosen when additional information becomes available. We have to emphasize that the VSS or maybe better the VSM (Value of Stochastic Modeling) is different from the expected value of perfect information which is defined as the improvement of the objective function by learning perfect information about parameters of the true deterministic model. In other words, the advantage of using deterministic models with exact values of parameters which are the mean values of observable random variables. Contrary, the VSS considers deterministic models as an approximation of real stochastic optimization models with inherently uncertain parameters which cannot be evaluated by using real observations. The disadvantages of using deterministic approximations of stochastic models become clear from Example 1. In this example only demand $d^1(\omega)$ is uncertain because the stock of back-stop technology is unbounded. In the deterministic version of the model $d^1(\omega)$ is replaced by the mean value $d^1 = Ed^1(\omega)$. Accordingly, the stochastic strategic planning model is reduced to minimization of the cost function

$$C_2^1 x_2^1 + c_{11}^1 y_{11}^1 + c_{21}^1 y_{21}^1$$

subject to

$$y_{11}^{1} + y_{21}^{1} = d^{1},$$

$$S_{1} = x_{2}^{1},$$

$$y_{21}^{1} \leq x_{2}^{1},$$

$$x_{2}^{1} \geq 0, y_{11}^{1} \geq 0, y_{21}^{1} \geq 0,$$

That clearly has a trivial degenerated solution $x_2^1 = d^1$, $y_{11}^1 = 0$, $y_{21}^1 = d^1$. This solution assumes that the future demand is exactly known, therefore the operational decision coincides with the strategic decision. Definitely, this is an unrealistic solution because the average energy demand d_2^1 never occurs in reality. The strategic robust solution of the stochastic energy supply model may be significantly different from the mean value as it is indicated by equation (19).

The VSS is calculated by the non-negative difference

$$F(x^{*det}) - F(x^{*sto})$$

where $F(x)$ is defined by (18), x^{*det} is the optimal solution of deterministic model (1)–(6) used in the objective function of the stochastic model (18), (22), and x^{*sto} is the optimal solution of this stochastic model. Definitely, the value $F(x^{*sto})$ is smaller than $F(x^{*det})$ because the stochastic model has a richer set of feasible solutions, i.e., the deterministic solution x^{*det} is a degenerated version of x^{*sto}.

The solution x^{*det} is often combined with the sensitivity analysis of the deterministic model with respect to variations of parameters (say, demand) which have been substituted by mean values. This analysis may be rather misleading because the deterministic model is focused on one only scenario (mean value) that may never occur in reality. As (19) demonstrates, the robust solution of the stochastic model depends on the whole probability distribution, therefore variations in the mean values may be misleading especially for multimodal distributions. For example, in the case of two scenarios ± 10 with a probability 0.5, the mean value is even outside the set of feasible scenarios. In addition to the sensitivity analysis, the so-called scenario analysis is applied, i.e., a set of possible future "trajectories" of uncertain parameters is considered and for each of them, optimal solutions of the deterministic model are calculated. This generates a set of degenerated deterministic solutions without identifying a solution that is good enough (robust) with respect to all potential scenarios. Again, Example 1 nicely illustrates this: any scenario ω^s, $s = 1,\dots, N$, of deterministic model has trivial solution $x_2^1 = d^1(\omega_s)$, $y_{11}^1 = 0$, $y_{21}^1 = d^1(\omega^s)$, $s = 1,\dots, N$. The analysis ignores the essential specifics of strategic solutions which have to be made before scenarios ω^s, $s = 1,\dots, N$, become known. The strategic stochastic optimization (programming) model aims to find solutions robust with respect to all potential scenarios. Equations (19) show that solutions of such models depend on the whole probability distribution, i.e., on all scenarios.

4. Conclusions

Energy systems investment planning for sustainable energy transition is a critical component of the EU's policy. Rapid shift towards intermittent renewable energy resources such as wind and solar raises concerns about the Power System Flexibility (PSF) defined by the extent to which a power system can balance stochastic energy supply and demand relations. Also, areas suitable for wind and solar power production often lack sufficient water and land resources. The competition for water and land increases systemic vulnerability in the interdependent energy-land-water-environmental (ELWE) systems exacerbated by exogenous shocks from weather events and climate change. Therefore, our stochastic energy investments and operation planning model is a part of a broader modeling framework for the analysis of the energy-land-water-environmental (ELWE) security nexus. The model uses two-stage stochastic programming techniques. The model can include both traditional and renewable, dispatchable and non-dispatchable, energy technologies, traditional energy storage and alternative possibilities, such as, e.g., hydrogen production, to cover variations of the residual load curve.

The PSF requirement, the decentralization and the variable characteristics of renewable energy generation and transmission required specific two-stage dynamic stochastic optimization modeling techniques and robust performance indicators enabling to the management of stochastic supply-demand imbalances to minimize systemic failures due to, e.g., unstable wind or solar energy availability. The model incorporates rolling random planning time horizons bounded by stopping time moments defined either by the model's security indicators or by stochastic events of new policy introduction or market conditions requiring revisions of decisions. This allows us to analyze and model systemic impacts of potential extreme events and structural changes emerging from policy interventions and experts' and stakeholders' dialogues, which may occur at any moment of the decision-making process. In fact, model-based decision-making provides an environment that can guide the necessary experts and stakeholders' dialogues and negotiations with the model and among the stakeholders themselves, likely with different conflicting motivations and targets. Therefore, these dialogues/negotiations can provide endogenous and exogenous feedback into the decision-making process. Since the decision-making can be considered as an iterative process, the model experts/stakeholder outcomes may provide endogenous feedback and revisions of the model and the energy system structure through stopping time moments and moving time horizons.

The presented energy model allows the representation of all relevant energy subsystem components (e.g., traditional and renewable) and their interactions, dealing with both strategic and operational planning, and including technological and financial-oriented decisions. Energy storage are represented and modeled in a rather general way. For example, the excess electricity can be used for hydrogen production and fertilizer production or "exported". The stopping time moments induce endogenous risk aversion in strategic decisions in a form of dynamic VaR-type systemic risk measures dependent on the system's structure. Unlike the static nature of deterministic models, the proposed stochastic model and the DSS deliver solutions that are responsive to revealed information about systemic uncertainties and risks such as those related to stochastic supply, demand, prices, weather variability, technological change, in order to adjust local or regional energy structure and management policies in a cost-effective and risk hedging manner. Integration of the operational and strategic models under the umbrella of the two-stage stochastic optimization provides an effective way to make real-time decisions consistent with the long-term strategic goals of energy system planners to guarantee secure energy provision in all uncertain scenarios. The solution of the problem involves adjusting operational decisions to hit long-term targets if additional information about prices, subsidies, demand, weather, and new technologies can become available in the future.

The energy model has been used within the EnRiMa (Energy Efficiency and Risk Management in Public Buildings) project (Cano et al. 2016, Cano et al. 2014, EnRiMa 2015), whose overall objective is to develop a DSS for operators of energy-efficient buildings and spaces of public use. The model permitted for a decision-making process on optimal planning and operation of building infrastructure in a dialogue with stakeholders. The model was revised in Gao et al., 2018 to address the energy-water-agricultural security nexus in the case study of coal-based energy production expansion in water-scarce China regions characterized by competition among energy, agricultural, and industrial sectors for natural resources.

The energy model enables to investigate the feasibility of green energy sector developments in the presence of competition for natural resources between energy and agricultural sectors, in view of potential water and agricultural land scarcity, climate change, market risks and other inherent risks (see Ermoliev et al. 2021).

Acknowledgements

The development of the presented methodologies and models was supported by a joint project between the International Institute for Applied Systems Analysis (IIASA) and the National Academy of Sciences of Ukraine (NASU) on "Integrated robust modeling and management of food-energy-water-land use nexus for sustainable development". The work has received partial support from the National Research Foundation of Ukraine, grant No. 2020.02/0121. and project CPEA-LT-2016/10003 jointly with Norwegian University for Science and technology. The paper also contributes to EU PARATUS (CL3-2021-DRS-01-03, SEP-210784020) project on "Promoting disaster preparedness and resilience by co-developing stakeholder support tools for managing systemic risk of compounding disasters".

References

Arrow, K. and A. Fisher. 1974. Preservation, uncertainty, and irreversibility. Q. J. Econ., 88: 312–319.

Babatunde, O.M., J.L. Munda and Y. Hamam. 2020. Power system flexibility: A review. Energy Report, 6: 101–106.

Birge, J.R. 1982. The value of the stochastic solution in stochastic linear programs with fixed recourse. Math Progr., 24: 314–325.

Cano, E.L., J.M. Moguerza, T. Ermolieva and Y. Ermoliev. 2014. Energy efficiency and risk management in public buildings: Strategic model for robust planning. Comput. Manag. Sci., 11(1-2): 25–44. doi:10.1007/s10287-013-0177-3.

Cano, E.L., J.M. Moguerza, T. Ermolieva and Y. Yermoliev. 2016. A strategic decision support system framework for energy-efficient technology investments. TOP: 1–22.

Energy Efficiency and Risk Management in Public Buildings (EnRiMa).2015. EU FP7 project http://www.iiasa.ac.at/web/home/research/researchPrograms/AdvancedSystemsAnalysis/Enrima.html.

Ermoliev, Y. 1976. Methods of Stochastic Programming. Nauka, Moscow, Russia, 1976 (in Russian).

Ermoliev, Y. 2009. Two-stage stochastic programming: Stochastic quasigradient methods. pp. 3955–3959. *In*: Floudas, C.A. and P.M. Pardalos (eds.). Encyclopedia of Optimization. Springer-Verlag, New York, USA, 2009.

Ermoliev, Y. and D. von Winterfeldt. 2012. Systemic risks and security management. pp. 19–49. *In*: Ermoliev, Y., M. Makowski and Marti K. (eds.). Managing Safety of Heterogeneous Systems: Decisions under Uncertainty and Risks. Lecture Notes in Econom. Math. Systems. Springer Verlag, New York, USA, 2012; Volume 658.

Ermoliev, Y. and R. Wets (eds.). 1988. Numerical techniques of stochastic optimization. Computational Mathematics, Berlin, Springer Verlag.

Ermoliev, Y., A.G. Zagorodny, V.L. Bogdanov, T. Ermolieva, P. Havlik, P. Rovenskaya, N. Komendantova and M. Obersteiner. 2022. Robust energy-food-water-environmental security management: iterative stochastic quasigradient procedure for linkage of distributed optimization models under asymmetric information and uncertainty. Cybernetics and System Analysis, 2022 (forthcoming).

Ermoliev, Y., M. Makowski and K. Marti. 2012. Robust management of heterogeneous systems under uncertainties. pp. 1–16. *In*: Ermoliev, Y., M. Makowski and K. Marti (eds.). Managing Safety of Heterogeneous Systems. Springer-Verlag, Heidelberg, Germany (February 2012).

Ermoliev, Y., N. Komendantova and T. Ermolieva. 2021. Robust energy production and storage investments and operation planning involving variable renewable energy sources: a two-stage stochastic optimization model with stopping time and rolling time horizon. *In*: Mathematical modeling, optimization and information technologies, proceedings of web conference, 15–18 November, 2021.

Ermolieva, T., P. Havlík, Y. Ermoliev, A. Mosnier, M. Obersteiner, D. Leclère, N. Khabarov, H. Valin and W. Reuter. 2016. Integrated management of land use systems under systemic risks and security targets: A stochastic global biosphere management model. Journal of Agricultural Economics, 67(3): 584–601.

European Commission. 2016. Transforming the European Energy System through Innovation (https://op.europa.eu/en/publication-detail/-/publication/9546f4e9-d3dc-11e6-ad7c-01aa75ed71a1/language-en/format-PDF/source-93755709).

Gao, J., X. Xu, G. Cao, Y.M. Ermoliev, T.Y. Ermolieva and E.A. Rovenskaya. 2018. Optimizing regional food and energy production under limited water availability through integrated modeling. Sustainability, 10(6), 10.3390/su10061689.

Goransson, L. and F. Johnsson. 2009. Dispatch modeling of a regional power generation system–Integrating wind power. Renew. Energy, 34(4): 1040–9.

Gritsevskii, A. and N. Nakicenovic. 2000. Modeling uncertainty of induced technological change. Energy Policy, 28(13): 907–921.

Gritsevskii, A. and Y. Ermoliev. 1999. An energy model incorporating technological uncertainty, increasing returns and economic and environmental risks. In Proceedings of the International Association for Energy Economics 1999 European Conference, 30 September–1 October, Paris, France.

Gritsevskii, A. and Y. Ermoliev. 2012. Modeling technological change under increasing returns and uncertainty. pp. 109–136. *In*: Ermoliev, Y., M. Makowski and K. Marti (eds.). Managing Safety of Heterogeneous Systems. Springer-Verlag, Heidelberg, Germany (February 2012).

International Energy Agency. 2018. Status of power system transformation 2018: Advanced power plant flexibility [Internet] Paris: IEA. 2018, Available from: https://www.oecd-ilibrary.org/energy/status-of-power-system-transformation-2018_9789264302006-en.

Komendantova, N., S. Neumueller and E. Nkoana. 2021. Public attitudes, co-production and polycentric governance in energy policy.//Energy Policy. – 2021.- Vol. 153, 10.1016/j.enpol.2021.112241.

O'Neill, B., Y. Ermoliev and T. Ermolieva. 2006. Endogenous risks and learning in climate change decision analysis. pp. 283–300. *In*: Marti, K., Y. Ermoliev, M. Makowski and G. Pflug (eds.). Coping with uncertainty, modeling and policy issues. Springer, Berlin.

Oree, V., S.Z.S. Hassen and P.J. Fleming. 2017. Generation expansion planning optimisation with renewable energy integration: A review. Renew. Sustain Energy Rev., 69: 790–803.
Vezzoli, C. et al. 2018. Distributed/decentralised renewable energy systems. *In*: Designing Sustainable Energy for All. Green Energy and Technology. Springer, Cham. https://doi.org/10.1007/978-3-319-70223-0_2.
Weinberg, C., J. Iannucci and M. Reading. 1991. The distributed utility: technology, customer, and public policy changes shaping the electrical utility of tomorrow. Energy Syst. Policy, 5(4): 307–322.

CHAPTER 14

How the Market Power of Electricity Suppliers and a Carbon Price Together Affect the Restructured Electricity Markets

Zhonghua Su[a,*] and *Alexei A Gaivoronski*[a]

1. Introduction

The restructuring of electricity markets worldwide leads to more competition and less regulation, but the market power of electricity suppliers still exists here and there (Ciarreta et al. 2016). Suppliers can affect the market price of electricity by withholding the electricity generation capacity (or ramp rate (Moiseeva et al. 2015)) to get the maximum profits. In some special cases, they can even take advantage of the network structure and/or transmission constraints to achieve local monopoly (Moreno et al. 2012). Accordingly, there are some negative effects on the electricity markets, e.g., lower consumers' welfare. Furthermore, the development of electricity markets is intertwined with the development of carbon markets (Maryniak et al. 2019). All the time, the electricity industry is an important area for greenhouse gas reduction and carbon dioxide is the key greenhouse gas traded in any emissions trading scheme (Teng et al. 2017). A proper carbon price can curb greenhouse gas

[a] Norwegian University of Science and Technology, Department of Industrial Economics and Technology Management, NO-7491 Trondheim, Norway.
* Corresponding author: zhonghuasu@iot.ntnu.no

emissions and promote the development of electricity markets sustainable (Brauneis et al. 2013). However, low carbon prices exist widely (Lundgren et al. 2015), for which too many (free) allocation of carbon permits should bear some of the blame (Teixidó et al. 2019). Then, a proper carbon tax or carbon price floor has been imposed in more and more countries (Ramaker 2018). Finland is a pioneer in imposing a carbon tax in 1990 (Vehmas, 2005). Thereafter, Sweden, Norway, Slovenia, Italy, Germany and some other countries also imposed carbon taxes (Ghazouani et al. 2020). In 2013, the UK implemented a carbon price floor. Accordingly, the coal power output dropped from 13.1 TWh in 2013 to 0.97 TWh in 2019.

The trends of restructuring electricity markets and regulating carbon prices can have great effects on the electricity markets, especially when the two trends are interacting with each other (Teng et al. 2017). In this paper, we quantitatively analyze how the market power of electricity suppliers and a carbon price together affect the restructured electricity markets on generation capacity expansion, consumption, carbon emissions, and electricity price.

Many researchers have investigated the effects of market power on the electricity markets, e.g., higher prices of electricity (Amountzias et al. 2017), lower electricity consumption and lower social welfare (Nazemi and Mashayekhi 2015). Moreover, market power could lead to inefficient dispatch of electricity, e.g., more usage of carbon-intensive power generators and less usage of low-carbon power generators, especially during peak demand periods, as the study by Browne et al. (2015) indicates. Further, when the market power of suppliers goes deep into the emission permit market, the permit price tends to be lowered, thereby increasing carbon emissions (Viskovic et al. 2021). Then, to reduce market power many efforts have been made, e.g., transmission network switching (Tohidi et al. 2017), and demand shifting (Ye et al. 2018). More importantly, the overall reform of electricity markets can make a bigger impact. For example, Yin et al. (2019) investigate different scopes of Chinese electricity market reform and the results indicate that more competition (i.e., weaker market power) on the generation and distribution sides can enhance efficiency in the electricity sector.

The effects of a carbon price/tax on the electricity markets have been studied extensively. Generally, the higher a carbon price/tax is, the lower the electricity production/consumption (Ghaith and Epplin 2017, Zhao et al. 2018) and carbon emissions (Rocha et al. 2015, Zhao et al. 2018) are; the higher the carbon price/tax is, the higher the electricity price/cost (Woo et al. 2018, Wild et al. 2015) is. Besides, the comparative cost-advantage of renewable power is largely influenced by the carbon price, as Li et al. (2015) show. Moreover, Petitet et al. (2016) indicate that a high and stable carbon price can boost the development of wind power. Further, Wilson and Staffell (2018) find that a strong carbon price floor boosts the switching of coal power to gas power in Great Britain.

Very few researchers have investigated the effects of market power along with a carbon/tax price on the electricity markets. Luckily, the two factors attract the attention of Leslie (2018), due to a real-world example of the Western Australian wholesale electricity market. When a carbon tax was imposed on the electricity market in a certain market structure/power setting, the corresponding carbon emissions increased. Two years later, when the market structure/power changed and this carbon tax was repealed, the carbon emissions decreased. Leslie (2018) analyzes how the carbon tax and market power/structure affect electricity generation, carbon emissions, and the day-ahead price of electricity in a short-term period. However, neither capacity expansion nor renewable power is discussed and very few cases of a carbon tax and market power are analyzed.

Additionally, the research question and method in Oderinwale and van der Weijde (2017) are closer to this paper. They use a Nash-Cournot model to depict the oligopolistic generators and the price-taking transmission system operator. Totally 12 cases of a carbon tax and/or feed-in tariff, the number of competing generators, and the transmission condition are investigated to compare the market responses, e.g., capacity expansion, and electricity prices. The results imply that a different combination of a carbon policy and market competition degree causes a different set of market responses. However, the results are based on the assumption of Nash-Cournot market power. In practice, the degree of market power to influence the electricity price/quantity can be from Nash-Cournot market power to no market power (i.e., perfectly competitive), where some intermediate cases of market power lie.

Then, by referring to Huppmann and Egging (2014), in this paper, we construct an equilibrium model with market power parameters (Egging et al. 2010, Huppmann and Egging 2014, Su et al. 2015) that can flexibly depict the degree of market power. We implement the model on a stylized market setting based on Chinese electricity market data in 2016 and investigate thousands of cases of market power and carbon price. To the best of our knowledge, this is the first time quantitatively analysis on how the market power of electricity suppliers and a carbon price together affect the restructured electricity markets.

The remainder of the paper is organized as follows. Section 2 discusses market power. Section 32 presents the mathematical formulation of the equilibrium model. Section 43 shows the basic assumptions, the cases of market power and a carbon price, and the market setup. Section 54 analyzes the results. Section 65 concludes and proposes some future work.

2. A discussion of market power

In this paper, we define the market power of electricity suppliers as the ability to affect the market price of electricity by withholding generation capacity.

Market power can be broadly extended to the degree of market competition, the distinction of which two concepts is not always clear-cut. The degree of market competition among electricity supplies is affected by several factors, e.g., the market rules, the number of competing suppliers, the capacities, and the transmission conditions. To measure the degree of market competition, there are several indicators. For instance, the Herfindahl-Hirschman Index (HHI) is defined as the sum of the square of each market share (Hirschman 1964). The 4-firm concentration ratio is the total market share of the four biggest suppliers. Besides, the Lerner Index compares the actual market price with the associated marginal cost (Spierdijk and Zaouras 2016). However, they all are ex-post indicators based on historical data, which can not predict current and future market competition, especially during a market transition time.

Then, when making ex-ante studies of market power, researchers resort to equilibrium models. In an equilibrium model, the strategic behavior of suppliers is specified and that is pre-existing market power. The most widely used equilibrium model in electricity markets is the Nash-Cournot equilibrium model, where each independent oligopolistic supplier makes the optimal decision on quantity with a belief that the rival will not alter the quantity. Every belief affects every supplier to make the decision, resulting in the equilibrium market price, which in reverse indicates the strategic behavior of suppliers.

The market power parameter in this paper is essentially conjectural variation (CV), which is a generalization of the Nash-Cournot conjecture. Moreover, CV is a belief a supplier has on how the rival will react to its decision. In this paper, we define the market power parameter as the first-order partial derivative of the total quantity with respect to its own supplied quantity, i.e., $\frac{\partial \sum_{i' \in I, k' \in K} q_{ot}^{i'k'}}{\partial q_{ot}^{ik}}$. In Appendix B, we use a stylized model to mathematically justify how the market power parameter affects the electricity market price and quantity. Generally, as the market power parameter increases, the market price of electricity increases; the total quantity of electricity decreases. Furthermore, the market power parameter to some extent can reflect the competition degree of suppliers. That is why in this paper we use market power parameters to depict market power, though it is an approximation. Besides, a similar CV approach can be found in Ruiz et al. (2012), Yang et al. (2018), Mousavian et al. (2020) and Oliveira and Ruiz (2021).

3. The equilibrium model

By referring to Huppmann and Egging (2014), we construct an equilibrium model to depict the interactions of suppliers, consumers, and a carbon permit auctioneer (CPA) in the electricity markets, as illustrated in Figure 1. The suppliers buy carbon permits from CPA and sell electricity to the consumers. Besides, all the agents solve their own optimization problems to maximize profits, which results in an equilibrium problem.

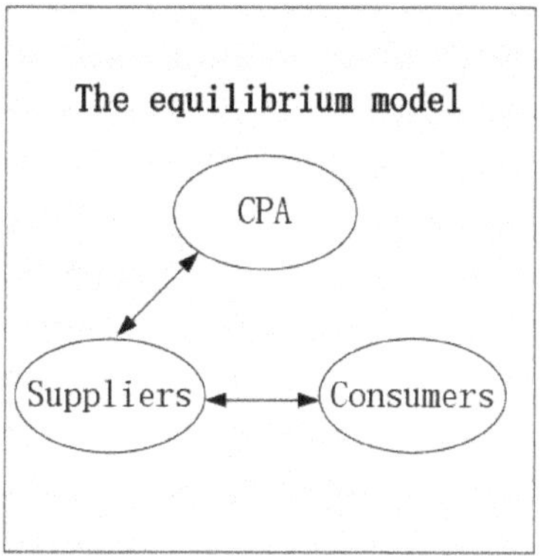

Figure 1. An illustration of the equilibrium model.

Moreover, all the agents are assumed to play a static game based on complete information, simultaneously making decisions on multi-period investments and/or operations. The suppliers are assumed to be able to exert market power to some extent to affect the electricity price. Meanwhile, the suppliers are assumed to be price takers with respect to the price of carbon permits.

Specifically, we differentiate two timescales in the model. The timescale of operation is a short-term period, e.g., hours/days/months/seasons, denoted by $t \in T$. The timescale of investment is a long-term period, e.g., years, denoted by $o \in O$. Here, O is an ordered set. Besides, we use shorthand notation $o' < o$ to denote years o' previous to years o. Likewise, $o' > o$ denotes years after years o. Due to the mathematical characterization of the model, there is no capacity expansion in the last investment period.

3.1 The agents' optimization problems

The supplier

There are several electricity suppliers in the markets. Each supplier uses one or several generation technologies $k \in K$ (e.g., wind power, coal power) to produce electricity and sell it to the consumers. To transmit electricity to the consumers, the supplier is charged of an electricity transmission fee. Further, it is assumed that all electricity generated (after production and transmission losses) is consumed by the consumers.

The supplier's profit comes from electricity sales, subtracting costs of transmission, production, investment in capacity expansion, and buying carbon permits. Moreover, the time value of profits in future periods is considered by the suppliers. Then, the supplier's problem is specified as below. (see Table 1 for the notation explanation.)

$$\max_{f_{ot}^{ik}, q_{ot}^{ik}, z_o^{ik}} \sum_{\substack{o \in O, t \in T, \\ k \in K}} d_o \{[h_o^i \Pi_{ot}(\cdot) + (1 - h_o^i) p_{ot}] q_{ot}^{ik} - e_{ot}^{ik} q_{ot}^{ik} - c_{ot}^{ik} f_{ot}^{ik} - u_{ot} r_{ot}^{ik} f_{ot}^{ik} - m_o^{ik} z_o^{ik}\} \quad (1)$$

$$\text{s.t.} \quad f_{ot}^{ik} \leq l_{ot}^{ik} (Q^{ik} + \sum_{o' < o} z_{o'}^{ik}) \qquad (\alpha_{ot}^{ik}) \qquad (2)$$

$$q_{ot}^{ik} = (1 - s_{ot}^{ik})(1 - g_{ot}^{ik}) f_{ot}^{ik} \qquad (\phi_{ot}^{ik}) \qquad (3)$$

In the objective function, we assume linear cost functions of production, transmission, investment and buying carbon permits. Moreover, we bring in the market power parameter h_o^i (Egging et al. 2010), which describes to what degree the supplier can influence the market price of electricity. Specifically, h_o^i takes value from the range [0,1]. When $h_o^i = 1$, based on the knowledge of the inverse demand curve $\prod_{ot}(\cdot)$, the supplier exerts Nash-Cournot market power to influence the market price of electricity. When $h_o^i = 0$, the supplier has no market power (i.e., a perfectly competitive market) to influence the price. Other values of h_o^i can describe intermediate cases of market power.

Equation (2) regulates that the quantity of electricity generation is no greater than the available generation capacity. It is assumed that the generation capacity to expand in the investment period o' is put into operation since the next neighboring investment period o. Besides, what is in parenthesis represents the corresponding dual variable of the constraint. Equation (3) is a mass balance constraint regulating that the quantity of electricity sold must be equal to the quantity of electricity generation after losses of production and transmission.

The carbon permit auctioneer

The carbon permit auctioneer (CPA) is an authority that allocates limited carbon permits to electricity suppliers. Here, a uniform-price auction with a price floor of carbon permits is adopted. Then, the CPA's problem can be

Table 1. Notation explanation.

Notation	Explanation
Sets:	
$i \in I$	electricity suppliers
$k \in K$	electricity generation technologies
$o, o' \in o$	long-term investment periods
$t \in T$	short-term operation periods
Parameters:	
a_{ot}	intercept of the inverse demand curve
b_{ot}	slope of the inverse demand curve
c_{ot}^{ik}	unit operation cost of generating electricity
d_o	discount factor
e_{ot}^{ik}	unit electricity transmission cost
g_{ot}^{ik}	production loss rate
h_o^i	market power parameter
l_{ot}^{ik}	availability factor of generation capacity
m_o^{ik}	unit investment cost of expanding generation capacity
Q^{ik}	existing generation capacity
r_{ot}^{ik}	carbon emissions rate of generating electricity
s_{ot}^{ik}	transmission loss rate
V_{ot}	carbon permit cap
w_{ot}	price floor of carbon permits
Variables:	
f_{ot}^{ik}	quantity of electricity generation
p_{ot}	market-clearing price of electricity
q_{ot}^{ik}	quantity of electricity sold to consumers
u_{ot}	market-clearing price of carbon permits
v_{ot}	quantity of carbon permits
z_o^{ik}	quantity of generation capacity expansion
α_{ot}^{ik}	dual variable for the constraint of available generation capacity
β_{ot}	dual variable for the constraint of carbon permit cap
ϕ_{ot}^{ik}	dual variable for the constraint of mass balance
Function:	
$\prod_{ot}(\cdot)$	inverse demand curve of the consumers

modeled through the following linear program. (see Table 1 for the notation explanation.)

$$\max_{v_{ot}} (u_{ot} - w_{ot})v_{ot} \tag{4}$$

$$\text{s.t. } v_{ot} \le V_{ot} \qquad (\beta_{ot}) \tag{5}$$

Specifically, Equation (5) regulates that the carbon permit quantity v_{ot} can not exceed the cap V_{ot}. The mathematical form of Equation (4) ensures that the market-clearing price of carbon permits u_{ot} is no lower than the price floor w_{ot}. Both the carbon permit cap and the price floor are exogenous parameters set by the policymaker. Besides, we assume that the quantity of carbon permits allocated to the suppliers is equivalent to the number of carbon emissions during electricity generation. Then, market clearing of carbon permits is specified in Equation (6).

$$v_{ot} = \sum_{i \in I, k \in K} r_{ot}^{ik} f_{ot}^{ik} \qquad (u_{ot}) \tag{6}$$

The consumer

A linear inverse demand curve is assumed to depict the relationship between the electricity demand of rational consumers and the electricity price. (see Table 1 for the notation explanation.)

$$p_{ot} = a_{ot} - b_{ot} \sum_{i \in I, k \in K} q_{ot}^{ik} \tag{7}$$

3.2 Mixed complementarity problem

After specifying each agent's optimization problem, we derive the corresponding Karush-Kuhn-Tucker (KKT) conditions for the optimal solution of each problem (Cottle et al. 1992). Concatenating all the KKT conditions and the market-clearing conditions (in Appendix A), we pose a mixed complementarity problem (MCP), to solve which we get a Nash equilibrium solution of the equilibrium problem. Since every agent's maximization problem is with a concave objective function and convex inequality constraints and/or linear equality constraints, we guarantee the existence of a global optimum solution, but not necessarily an unique solution (Oggioni et al. 2012, Gabriel et al. 2012). All the cases in the following are solved by using the PATH solver in GAMS (Brook et al. 1988, Ferris and Munson 2000).

4. Case study

The equilibrium model is implemented in a stylized market setting based on Chinese electricity market data in 2016. Wherein, we investigate 6611 cases of market power and a carbon price to find how sensitive the market responses (during years 2016–2046) are to market power and a carbon price. The interesting market responses of interest are the aggregate generation capacity expansion during 2016–2036, the aggregate electricity consumption, the aggregate carbon emissions, and the average electricity price during 2016–2046. Figure 2 shows the inputs and outputs of one-time estimation using the equilibrium model.

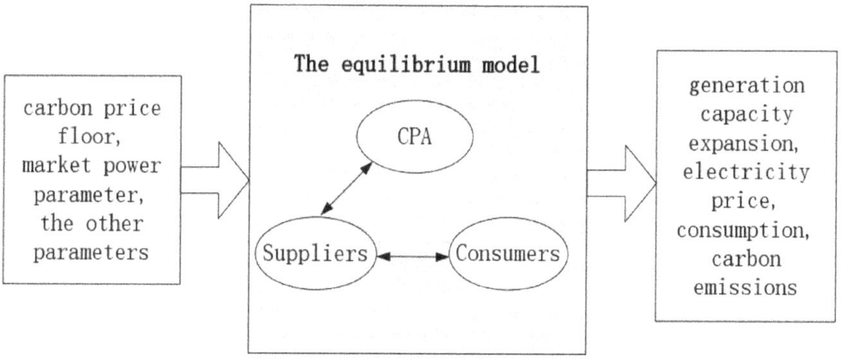

Figure 2. The inputs and outputs of one-time estimation using the equilibrium model.

4.1 Basic assumptions

From the year 2016 to 2046, every investment period is ten years, and every operation period is one hour. Besides, we assume the parameters do not change the values within each investment period. Then, by watching the electricity market dynamics during a typical hour, we roughly estimate the market responses over ten years. We assume a discount rate of 5% annually for all the agents. Moreover, we set a very large value of the carbon permit cap (indicating an infinite cap) to ignore the influence of the carbon permit cap on carbon price formation. Then, the market-clearing price of carbon permits equals the carbon price floor.

4.2 The cases of market power and a carbon price

Here, we investigate 11 cases of market power, where h_o^i takes value from a discrete set of {0, 0.1, 0.2, 0.3, 0.4, 0.5, 0.6, 0.7, 0.8, 0.9, 1.0}. In each case, it is assumed that all the suppliers have the same market power in all the periods. We use notations MP-0, MP-0.1, MP-0.2, MP-0.3, MP-0.4, MP-0.5, MP-0.6, MP-0.7, MP-0.8, MP-0.9, and MP-1.0 to represent the cases. For example, MP-0.4 denotes the case $h_o^i = 0.4$. In each case of market power, we investigate 601 cases of a carbon price, which is imposed in 2016 and increases at a rate of 5% per year. The 601 cases of a carbon price are of values from 0 to 600 Yuan/tonne with a step size of 1. Altogether, we estimate 6611 cases of market power and a carbon price.

4.3 The market setup

Our study is based on Chinese electricity market data on generation quantities and capacities of various generation technologies in 2016. We assume that in the markets there are six electricity suppliers each with a unique type of generation technology, i.e., hydropower, coal power, gas power, nuclear power, wind power, and solar power. We illustrate the economic meaning of the market setting in Appendix C by mathematical proof. Data relevant to the suppliers' investment and operation in 2016 are summarized in Table 2.

Table 2. Data relevant to the suppliers' investment and operation in 2016.

	Operation cost (thousand Yuan/GWh)	Investment cost (Yuan/kW)	Emissions rate (tonnes/GWh)	Generation capacity (GW)
hydro power	65.0055	8783.70	0	332.11
coal power	244.401	4999.95	890	983.62
gas power	459.528	3856.05	390	70.26
nuclear power	102.3975	13776.00	0	33.64
wind power	60.024	8583.60	0	148.64
solar power	104.4885	7472.70	0	77.42

Generation capacity

The existing generation capacity data in 2016 are mainly from National Energy Administration.[1] Since we omit oil power, we slightly adjust data of coal power and gas power by further referring to World Energy Outlook 2017 2017 (IEA 2017) and China Electricity Council's data.

Emissions rate

The data of emissions rate of coal power and gas power in 2016 derive from China's Thirteenth Five-Year Plan (2016–2020) on Electric Power Development. Further, we assume the rates reduce at a rate of 1% annually.

[1] See http://www.nea.gov.cn/2017-01/16/c_135986964.htm.

Operation cost

The operation cost data in 2016 come from NEA et al. (2015). We assume that the operation cost data remain the same during 2016–2046.

Investment cost

The investment cost data of hydro power, wind power, and solar power are collected from REN21 (2017) by using the exchange rate of 6.9 for USD-Yuan on 31 December 2016; The data for coal power, gas power, and nuclear power is collected from NEA et al. (2015) by using the exchange rate of 6.15 as the paper suggested. We assume that the investment costs of hydropower, coal power, gas power, and nuclear power remain the same during 2016–2036. However, the investment cost of wind/solar power is assumed to drop greatly according to BloombergNEF's prediction that wind/solar power cost falls by 58%/71% from 2018 to 2050.[2] Based on their prediction, we derive the compound annual growth rates to obtain the wind/solar investment costs in 2026 and 2036.

Availability factor

Recall that in the case study we estimate the power market dynamics for ten years by watching the dynamics during a typical hour. Then, we should cautiously choose data for the typical hour to approach the real markets. Note that the model itself ignores several features of generation technologies, e.g., the starting time of each generator, and the intermittency of renewable power. Here, we collect data of capacity factors to serve as the availability factors. The capacity factor of a certain generation technology describes its average utilization ratio during a long period, which aids to make up for the model deficiency.

In Table 3, the data for 2016 are generally derived by using the real generation capacity and quantity data from National Energy Administration,[3] National Bureau of Statistics of China[4] and IEA (2018). Data for 2020 and 2030 are derived by using the projected generation quantities and capacities in New Policies Scenario in World Energy Outlook 2017 (IEA 2017).

---

[2] See https://bnef.turtl.co/story/neo2018.

[3] See http://www.nea.gov.cn/2017-02/04/c_136030860.htm.

[4] See http://data.stats.gov.cn/easyquery.htm?cn=C01.

Table 3. Specification of availability factors.

	2016	2026	2036
hydro power	0.410194	0.356129	0.350827
coal power	0.491289	0.447733	0.431946
gas power	0.331239	0.415362	0.429542
nuclear power	0.723786	0.846036	0.86675
wind power	0.182069	0.234427	0.252375
solar power	0.097611	0.147042	0.154555

Inverse demand curve parameters

To derive the values of a_{ot} and b_{ot} in 2016, we use the real data of price elasticity of electricity demand (i.e., 0.84), the real data of electricity consumption in 2016 as the equilibrium quantity, and an assumed equilibrium electricity price. Specifically, the assumed equilibrium electricity price (i.e., 614.83 thousand Yuan/GWh) is the real average electricity price in 2016 collected from National Energy Administration. We deduce that a_{ot} =1346.770476 (Thousand Yuan/GWh) and b_{ot} = 1.177469 (Thousand Yuan/GWh^2) in 2016. Further, we assume that a_{ot} varies over time at a growth rate of 2.4% annually; b_{ot} is constant over time.

Other parameters

We get the data on the transmission loss rate and production loss rate in 2016 from the National Energy Administration. Further, we assume them to decrease at a rate of 1% annually. In the model, we don't consider any subsidy on any type of generation technology and assume the electricity transmission costs 0. This facilitates the data consistency in 2016 and avoids the equilibrium electricity price (obtained by running the model) from upwardly deviating from the assumed value due to the high costs of renewable power in 2016.

5. Result analysis

To manage the numerical magnitude, when running the model, we proportionally contract data from a ten-year scale to a one-hour scale. We don't focus on any specific number in the results. However, regarding each market response, we compare the results under all the cases of market power and a carbon price in the same picture. In this visualisation way, we can see the market trends as market power or/and carbon price varies.

In practice, before running the three-period equilibrium model, we run a one-period equilibrium model and test a lot more cases of market power to check data consistency in 2016. We find that the market power parameters in the range of $[0, 0.67]$ can support the data consistency in 2016. However, market power can change over time. Anyway, in the following, we will display the results under all the cases of market power and then discuss some market responses when the market power parameters decrease from 0.6 to a certain level.

5.1 Electricity consumption during 2016–2046

Figure 3 shows the quantity of total electricity consumption during 2016–2046. Generally, the consumption quantity has a decreasing tendency as the carbon price increases, independent of the market power. We can average the results in all the cases of market power to get the average consumption quantity with respect to the carbon price. For example, as the carbon price increases from 0 to 50, 100, 200, or 300 (Yuan per tonne), the respective consumption quantity is reduced by 0.71%, 1.97%, 4.71%, or 6.77% average. When the carbon price varies in a relatively low range, i.e., $[0, 200]$, and when the market structure tends to be perfectly competitive, i.e., the cases of MP-0, MP-0.1, and MP-0.2, to increase the carbon price has very small effects on the

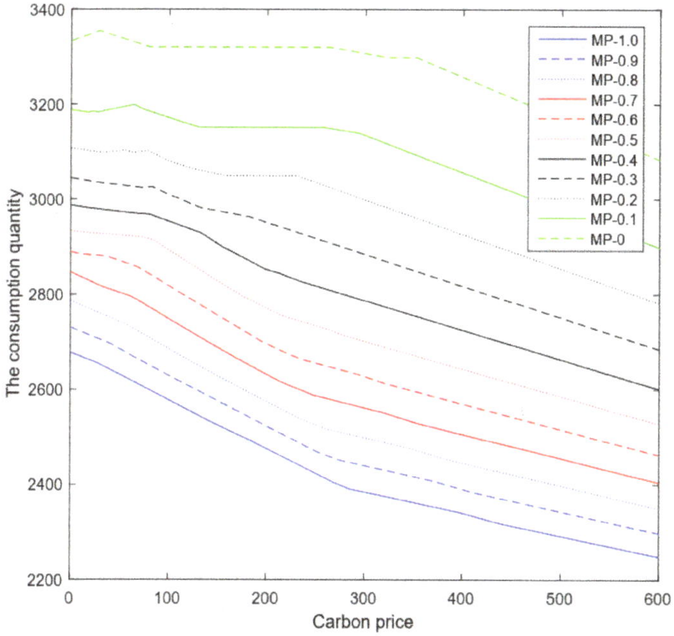

Figure 3. The quantity of total electricity consumption during 2016–2046.

total electricity consumption quantity. Moreover, the consumption quantity has an increasing tendency as the market power decreases, independent of the carbon price.

Besides, the market share of electricity generated from each technology is demonstrated in Figure 4. Generally, in each case of market power, as the carbon price increases, the market share of nuclear/hydro power increases; the market share of coal power is decreasing. Independent of the carbon price, as the market power decreases, the market share of gas/wind/solar power decreases; the share of nuclear power increases.

Moreover, Figure 5 displays the market share of electricity generated from the renewable technologies including hydro power, wind power, and solar power. Generally, there is an increasing tendency of the share of renewable power as the carbon price increases, independent of the market power. However, as the market power decreases, the share generally decreases, independent of the carbon price.

5.2 Generation capacity expansion during 2016–2036

Figure 6 displays the quantity of total generation capacity expansion during 2016–2036. Generally, a higher carbon price together with stronger market power promotes more expansion. For instance, as the carbon price increases from 0 to 50, 100, 200, or 300 (Yuan per tonne), the respective expansion quantity is increased by 3.35%, 18.10%, 57.15%, or 70.10% on average of the results in all the cases of market power. However, there is a threshold value of the carbon price in each case of market power, beyond which the value the expansion quantity doesn't vary as the carbon price further increases. Moreover, the smaller the market power is, the smaller the threshold value is.

Besides, Figure 7 displays the share of each type of generation capacity expansion quantity in the total capacity expansion quantity. The capacity expansion in nuclear power always occurs regardless of the market power. Regardless of the market power and carbon price, there is always capacity expansion in nuclear; there is no capacity expansion in coal. Also, except in the extreme case of MP-0 (i.e., a perfectly competitive market), there is always capacity expansion in hydropower; there is no capacity expansion in coal power. Generally, weaker market power promotes capacity expansion in nuclear power and hydropower. When the market power parameters start to decrease from 0.6, the capacity expansion in gas power and solar power disappears. Further, when the market power parameters are lower than 0.2, there is no capacity expansion in wind power. As the market power decreases,

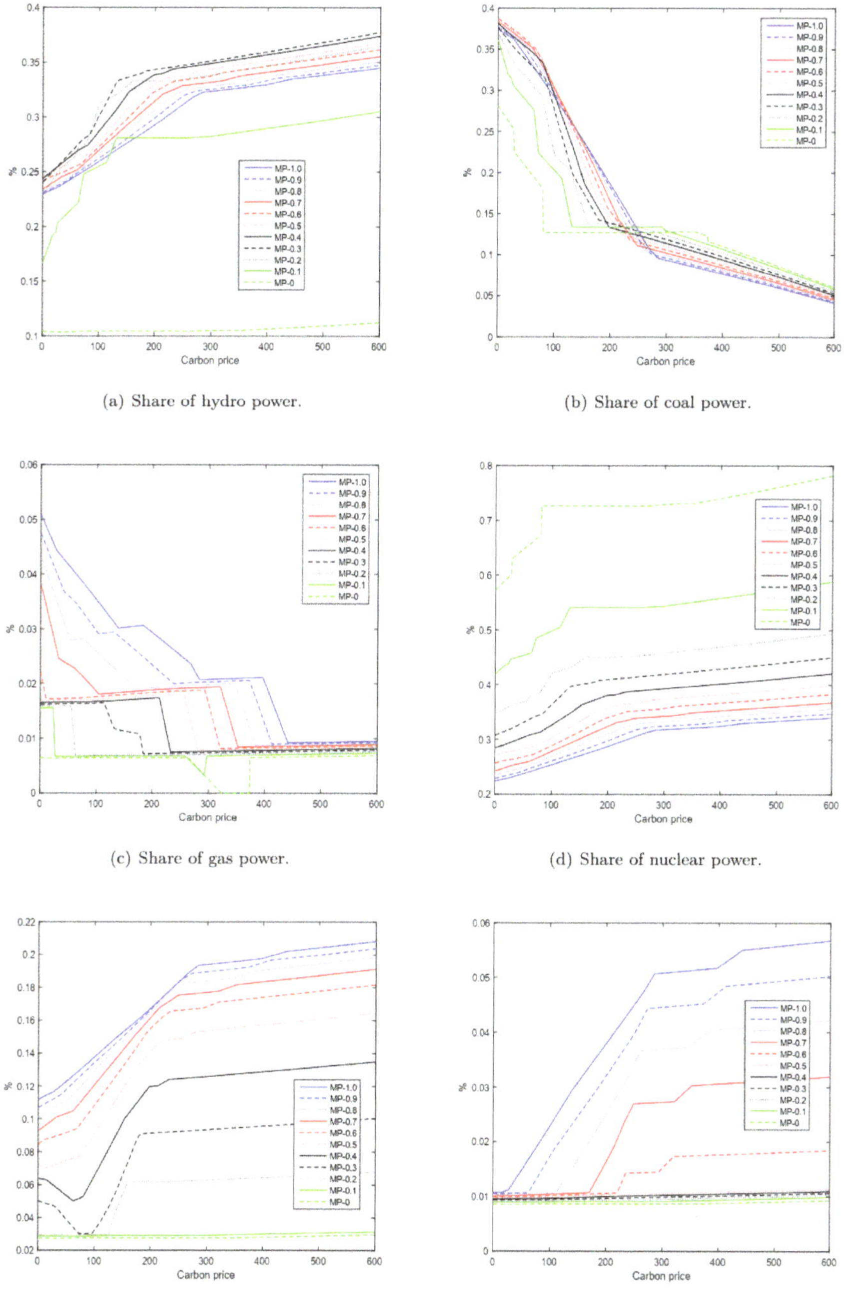

(a) Share of hydro power.

(b) Share of coal power.

(c) Share of gas power.

(d) Share of nuclear power.

(e) Share of wind power.

(f) Share of solar power.

Figure 4. The market share of electricity generated from each technology.

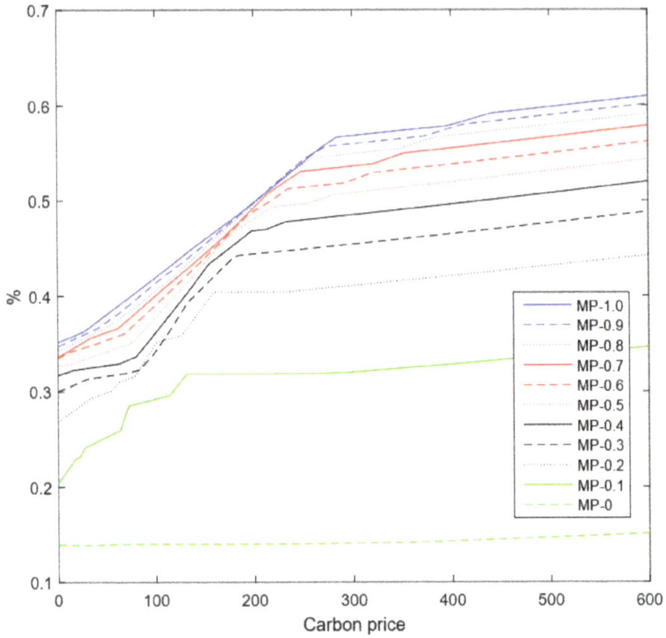

Figure 5. The market share of electricity generated from renewable technologies during 2016–2046.

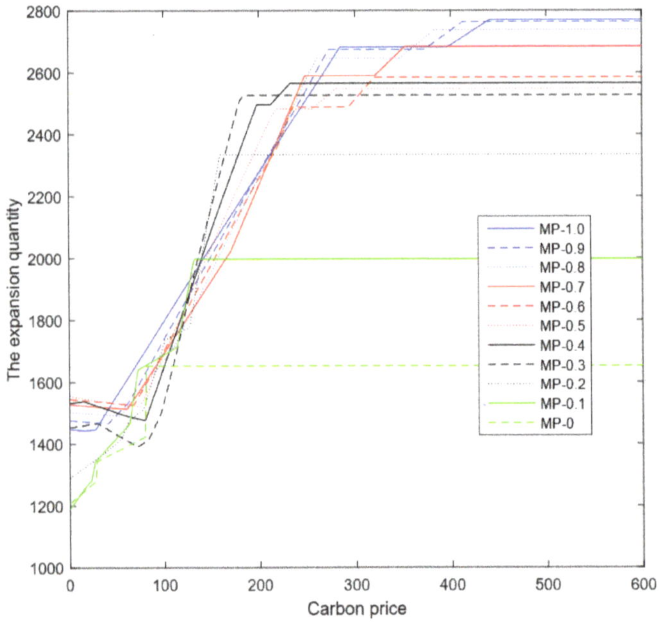

Figure 6. The quantity of total generation capacity expansion during 2016–2036.

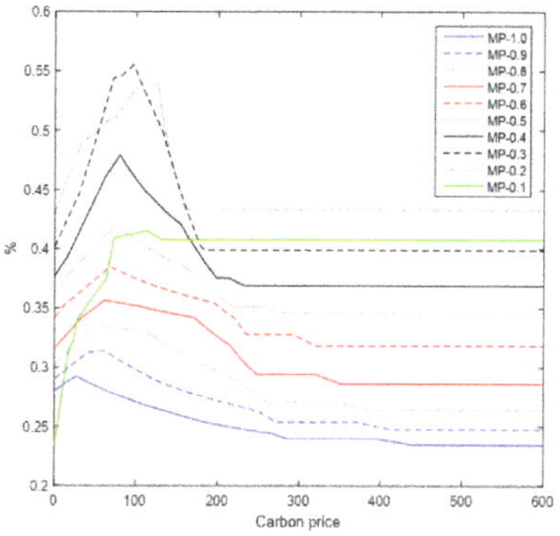

(a) Share of hydro power.

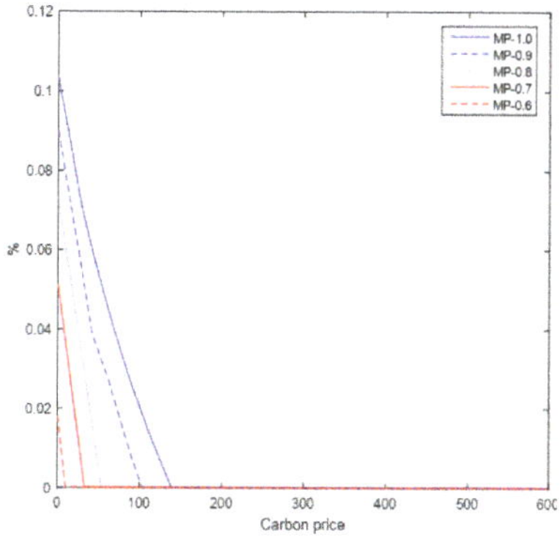

(b) Share of gas power.

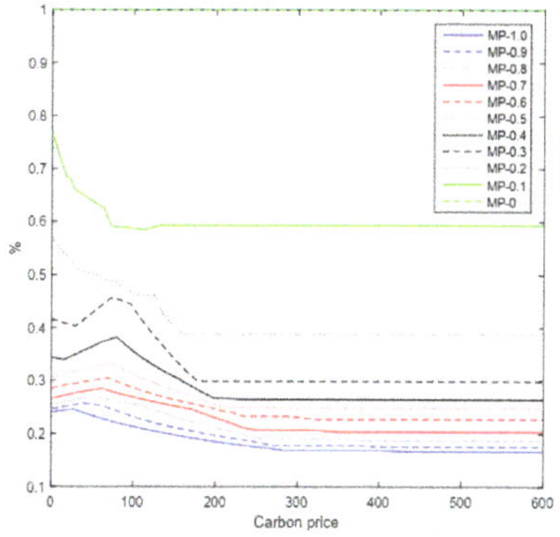

(c) Share of nuclear power.

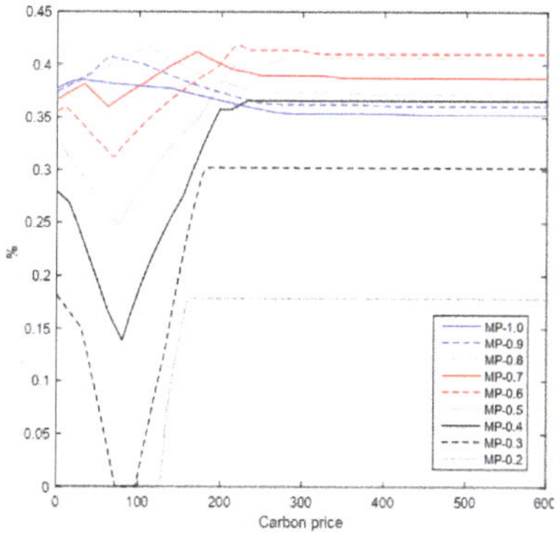

(d) Share of wind power.

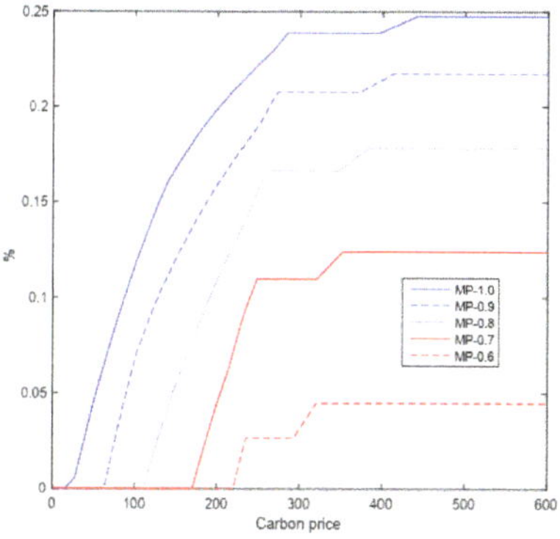

(e) Share of solar power.

Figure 7. The share of each type of generation capacity expansion quantity in the total capacity expansion quantity.

the share of gas/solar power has a decreasing trend; the share of nuclear power noticeably increases. When the market power is relatively strong, as the carbon price increases, the share of gas power capacity expansion quantity in the total capacity expansion has a decreasing trend; the share of solar power has an increasing trend. Specifically, strong market power promotes the capacity expansion in gas power, only if the carbon price is relatively low the expansion in gas power occurs when the carbon price is relatively low and the market power is relatively strong. In fact, to generate one unit quantity of electricity, gas power emits more than half the carbon dioxide of coal power. (see Table 2 for the emissions rates.) Strictly, gas power is not a carbon-friendly substitute for enough to substitute coal power. Generally, weaker market power facilitates the capacity expansion of nuclear power and hydro power. In the extreme case of MP-0, the only expansion happens to nuclear power.

Further, Figure 8 displays the share of renewable capacity expansion quantity in the total capacity expansion quantity during 2016–2036. Generally, the share decreases as the market power decreases (except for a small number of cases).

5.3 Carbon emissions quantity during 2016–2046

Figure 9 shows the total carbon dioxide emissions quantity during 2016–2046. Obviously, the emissions reduction is significant when the carbon price comes into play in the electricity markets. As the carbon price increases from 0 to 50, 100, 200, or 300 (Yuan per tonne), the respective emissions quantity is reduced by 11.81%, 28.96%, 57.44%, or 67.98% on average. Generally, the carbon price has a much greater impact on the emissions quantity than the market power. However, when the carbon price is higher than 300 Yuan per tonne, the weaker the market power is, the more emissions quantity there is.

5.4 Averagely discounted-to-present price of electricity during 2016–2046

Figure 10 shows the averagely discounted-to-present (i.e., discounted to 2016) price of electricity during 2016–2046. We assume a discount rate of 5% per year. Generally, the electricity price increases as the carbon price increases. Besides, it decreases as the market power decreases. As the carbon price increases from 0 to 50, 100, 200, or 300 (Yuan per tonne), the respective electricity price is increased by 1.90%, 4.33%, 9.69%, or 15.05% on average.

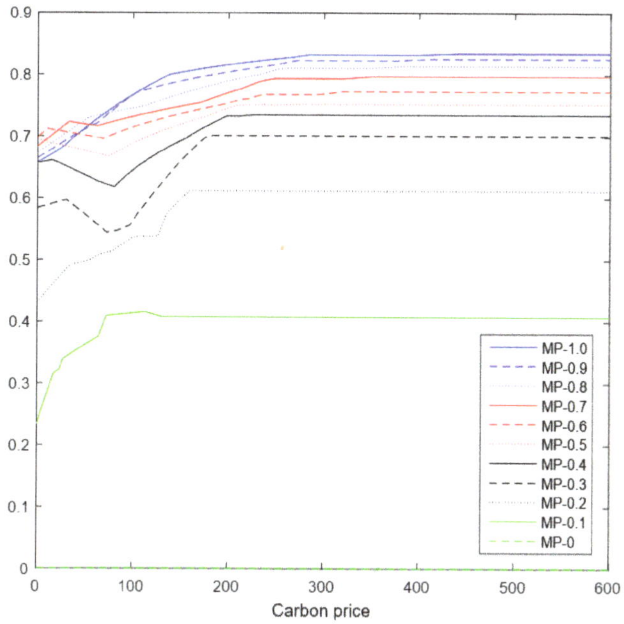

Figure 8. The share of renewable capacity expansion quantity in the total capacity expansion quantity during 2016–2036.

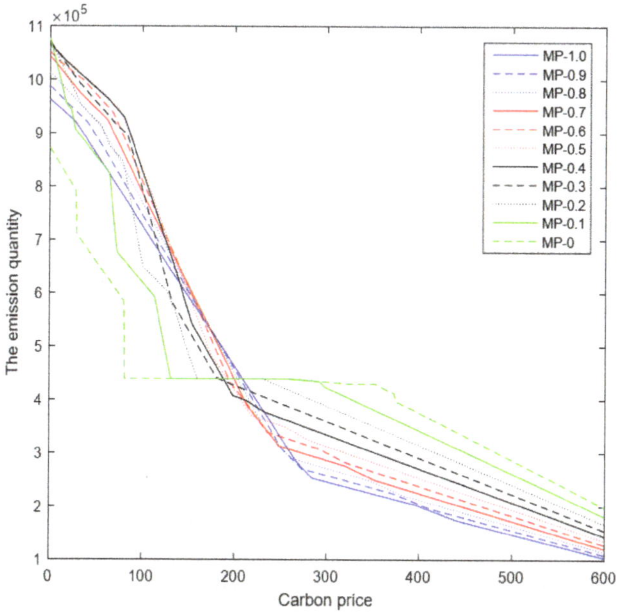

Figure 9. The expected quantity of the total emissions during 2016–2046.

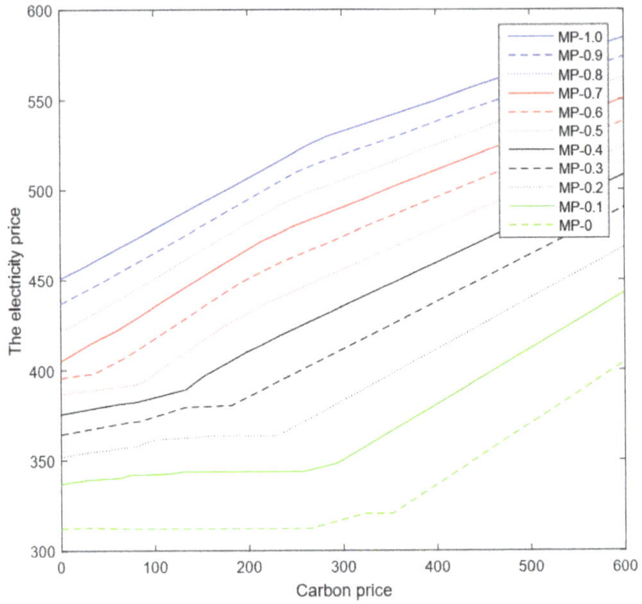

Figure 10. Averagely discounted-to-present price of electricity during 2016–2046.

When the carbon price varies in a relatively low range, i.e., [0, 200], and when the market structure tends to be perfectly competitive, i.e., the cases of MP-0, MP-0.1, and MP-0.2, to increase the carbon price has very small effects on the electricity price.

5.5 Some average data

Regarding each market response, we average over the results in all the cases of market power to get the average rates of change when the carbon price increases from 0 to a certain value. (see Table 4.) As the carbon price increases, there are decreasing tendencies of the carbon emissions and consumption, increasing tendencies of the generation capacity expansion and the electricity price.

Regarding each market response, we average over the results in the cases of a carbon price from 0 to 200 (Yuan per tonne) to get the average rates of change when the market power parameters decrease from 0.6 to a certain level. Table 5 summarizes the rates of change.

When the carbon price varies in a relatively low range (e.g., [0, 200]), decreasing the market power (from a medium level) can reduce the electricity price and increase the consumption quantity. However, it does not necessarily

Table 4. The average rates of change of the electricity price, the emissions quantity, the consumption quantity and the expansion quantity when the carbon price increases from 0 to a certain value.

	Rate of change			
Carbon price (Yuan per tonne)	**Electricity price**	**Emissions quantity**	**Consumption quantity**	**Expansion quantity**
50	1.90%	–11.81%	–0.71%	3.35%
100	4.33%	–28.96%	–1.97%	18.10%
150	7.02%	–45.97%	–3.43%	40.19%
200	9.69%	–57.44%	–4.71%	57.15%
250	12.33%	–64.43%	–5.81%	67.23%
300	15.05%	–67.98%	–6.77%	70.10%
350	17.83%	–70.78%	–7.71%	71.16%
400	20.93%	–74.04%	–8.76%	72.18%
450	24.04%	–77.16%	–9.80%	72.87%
500	27.12%	–80.17%	–10.83%	72.87%
550	30.21%	–83.19%	–11.85%	72.87%
600	33.30%	–86.20%	–12.87%	72.87%

Table 5. The average rates of change of the electricity price, the emissions quantity, the consumption quantity and the expansion quantity when the market power parameters decrease from 0.6 to a certain level.

	Rate of change			
Market power parameters	**Electricity price**	**Emissions quantity**	**Consumption quantity**	**Expansion quantity**
0.5	–3.79%	0.66%	2.33%	1.36%
0.4	–7.33%	–0.32%	4.61%	1.92%
0.3	–10.65%	–3.79%	6.89%	1.01%
0.2	–14.23%	–9.17%	9.55%	–0.38%
0.1	–18.44%	–16.78%	12.82%	–4.65%
0	–25.51%	–29.91%	18.40%	–13.97%

reduce the emissions quantity or increase the capacity expansion quantity, which depends on the degree of market power.

6. Conclusion

In this paper, we investigate how the market power of electricity suppliers and a carbon price together affect the restructured electricity markets on generation capacity expansion, consumption, carbon emissions, and electricity price. To do so, we construct an equilibrium model to depict the interactions of suppliers, consumers, and a carbon permit auctioneer. The suppliers can exert market power to influence the market price of electricity to some extent; however, they are price takers with respect to the price of carbon permits. We implement the model on a stylized market setting based on Chinese electricity market data in 2016, where various generation technologies play a role. By analyzing the results of 6611 cases of market power and a carbon price, we find how sensitive the restructured electricity market is to a carbon price or/ and market power.

1. Introducing a carbon price can greatly reduce carbon emissions quantity; obviously increase the generation capacity expansion quantity; modestly increase the electricity price; slightly reduce the consumption quantity.
2. Strong market power promotes the capacity expansion in gas power, only if the carbon price is relatively low. After all, gas power is not a carbon-friendly substitute for coal power.
3. Regardless of the market power, there is a threshold value of the carbon price, beyond which value the total generation capacity expansion quantity doesn't vary as the carbon price further increases. However, when below the threshold value, generally the expansion quantity increases rapidly as the carbon price increases. Moreover, the weaker the market power is, the smaller the threshold value is. From an economic perspective, a carbon price can drive up the electricity price, to some threshold of which there comes saturated consumption. Then, there is no incentive for more investment, thereby a threshold appearing.
4. At any carbon price, decreasing the market power can decrease the electricity price and increase the electricity consumption quantity. However, it does not necessarily reduce the emissions quantity or increase the capacity expansion quantity, which actually depends on the levels of carbon price and market power. For instance, when the carbon price is higher than 300 (Yuan per tonne), the weaker the market power is, the more emissions quantity there is.
5. Regardless of the market power, as the carbon price increases, both renewable power and nuclear power increase their market shares. Moreover, regardless of the carbon price, the weaker the market power is, the less renewable (/the more nuclear) power is in the shares of capacity expansion and consumption. From an economic perspective, the weaker the market power is, the easier the advantage of the levelized cost of each generation of technology displays. Though both the production cost and investment cost of nuclear power are high, the high utilization ratio of nuclear power makes its levelized cost the most competitive. That is why

renewable power loses the market shares of nuclear power as the market power decreases.

We expect the findings in this paper to hold true for more complicated market settings. This paper can provide some insights for policymakers to set carbon prices (floor) in consideration of the market power so as to balance among the factors of energy, environment and economy. For instance, if policymakers want to promote the investment in renewable power and do not want to decrease a lot electricity consumption, pay attention to the threshold value of a carbon price. Moreover, if policymakers want gas power to serve for energy security in the existence of a carbon price, more subsidies (e.g., in carbon capture and sequestration) should be given to gas power, especially when the electricity markets tend to be more competitive. Besides, in consideration of the market trends castcasted in this paper, e.g., electricity price variation with a carbon price or market power, some price-taking take suppliers can decide how much to produce to create more profits.

In the future, we may update the equilibrium model by involving uncertain characteristics of renewable power and electricity storage. Besides, a more complicated market setting, e.g., involving competitive fringe suppliers, could be studied to further investigate the market trends.

Acknowledgment

We are grateful to our editors and three anonymous referees for their insightful comments which helped us improve the paper considerably.

References

Amountzias, C., H. Dagdeviren and T. Patokos. 2017. Pricing decisions and market power in the UK electricity market: A VECM approach. Energy Policy, 108: 467–473.

Brauneis, A., R. Mestel and S. Palan. 2013. Inducing low-carbon investment in the electric power industry through a price floor for emissions trading. Energy Policy, 53: 190–204.

Brook, A., D. Kendrick and A. Meeraus. 1988. GAMS, a user's guide. ACM Signum Newsletter, 23(3-4): 10–11.

Browne, O., S. Poletti and D. Young. 2015. How does market power affect the impact of large scale wind investment in 'energy only' wholesale electricity markets? Energy Policy, 87: 17–27.

Ciarreta, A., S. Nasirov and C. Silva. 2016. The development of market power in the Spanish power generation sector: Perspectives after market liberalization. Energy Policy, 96: 700–710.

Cottle, R.W., J. Pang and R.E. Stone. 1992. The Linear Complementarity Problem. Society for Industrial & Applied Mathematics.

Egging, R., F. Holz and S.A. Gabriel. 2010. The World Gas Model: A multi-period mixed complementarity model for the global natural gas market. Energy, 35(10): 4016–4029.

Ferris, M.C. and T.S. Munson. 2000. Complementarity problems in GAMS and the PATH solver. Journal of Economic Dynamics and Control, 24(2): 165–188.

Gabriel, S.A., A.J. Conejo, J.D. Fuller, B.F. Hobbs and C. Ruiz. 2012. Complementarity Modeling in Energy Markets. International Series in Operations Research & Management Science. Springer.
Ghaith, A.F. and F.M. Epplin. 2017. Consequences of a carbon tax on household electricity use and cost, carbon emissions, and economics of household solar and wind. Energy Economics, 67: 159–168.
Ghazouani, A., W. Xia, M.B. Jebli and U. Shahzad. 2020. Exploring the role of carbon taxation policies on CO_2 emissions: Contextual evidence from tax implementation and non-implementation European countries. Sustainability, 12: 8680.
Hirschman, A. 1964. The paternity of an index. The American Economic Review, 54(5): 761–762.
Huppmann, D. and R. Egging. 2014. Market power, fuel substitution and infrastructure - a large-scale equilibrium model of global energy markets. Energy, 75: 483–500.
IEA. 2017. World energy outlook 2017. Paris: Organisation for Economic Cooperation and Development.
IEA. 2018. Electricity Information 2018. Paris: Organisation for Economic Cooperation and Development.
Leslie, G. 2018. Tax induced emissions? Estimating short-run emission impacts from carbon taxation under different market structures. Journal of Public Economics, 167: 220–239.
Li, Y., Z. Lukszo and M. Weijnen. 2015. The implications of CO_2 price for China's power sector decarbonization. Applied Energy, 146: 53–64.
Lundgren, T., P. Marklund, E. Samakovlis and W. Zhou. 2015. Carbon prices and incentives for technological development. Journal of Environmental Management, 150: 393–403.
Maryniak, P., S. Trück and R. Weron. 2019. Carbon pricing and electricity markets—The case of the Australian Clean Energy Bill. Energy Economics, 79: 45–58.
Moiseeva, E., M.R. Hesamzadeh and D.R. Biggar. 2015. Exercise of market power on ramp rate in wind-integrated power systems. IEEE Transactions on Power Systems, 30(3): 1614–1623.
Moreno, B., A.J. López and M.T. García-Álvarez. 2012. The electricity prices in the European Union. The role of renewable energies and regulatory electric market reforms. Energy, 48(1): 307–313.
Mousavian, S., A.J. Conejo and R. Sioshansi. 2020. Equilibria in investment and spot electricity markets: A conjectural variations approach. European Journal of Operational Research, 281(1): 129–140.
Nazemi, A. and M. Mashayekhi. 2015. Competitiveness assessment of Iran's restructured electricity market. Energy Economics, 49: 308–316.
NEA, IEA, and OECD. 2015. Projected Costs of Generating Electricity 2015. Paris: Organisation for Economic Cooperation and Development.
Oderinwale, T. and A.H. van der Weijde. 2017. Carbon taxation and feed-in tariffs: Evaluating the effect of network and market properties on policy effectiveness. Energy Systems, 8: 623–642.
Oggioni, G., Y. Smeers, E. Allevi and S. Schaible. 2012. A generalized Nash equilibrium model of market coupling in the European power system. Networks and Spatial Economics, 12(4): 503–560.
Oliveira, F.S. and C. Ruiz. 2021. Analysis of futures and spot electricity markets under risk aversion. European Journal of Operational Research, 291(3): 1132–1148.
Petitet, M., D. Finon and T. Janssen. 2016. Carbon price instead of support schemes: Wind power investments by the electricity market. Energy Journal, 37(4): 109–140.
Ramaker, A. 2018. A carbon price floor: today's solution for yesterday's problem? Renewable Energy Law and Policy Review, 9(1): 19–29.
REN21. 2017. Renewables 2017 Global Status Report. Paris: Renewable Energy Policy Network for the 21st Century Secretariat.

Rocha, P., T. Das, V. Nanduri and A. Botterud. 2015. Impact of CO_2 cap-and-trade programs on restructured power markets with generation capacity investments. International Journal of Electrical Power & Energy Systems, 71: 195 –208.

Ruiz, C., S.J. Kazempour and A.J. Conejo. 2012. Equilibria in futures and spot electricity markets. Electric Power Systems Research, 84: 1–9.

Spierdijk, L. and M. Zaouras. 2016. The Lerner index and revenue maximization. Applied Economics Letters, 24(15): 1075–1079.

Su, Z., R. Egging, D. Huppmann and A. Tomasgard. 2015. A Multi-Stage Multi-Horizon Stochastic Equilibrium Model of Multi-Fuel Energy Markets. CenSES Working paper. ISBN 978-82-93198-15-4.

Teixidó, J., S.F. Verde and F. Nicolli. 2019. The impact of the EU Emissions Trading System on low-carbon technological change: The empirical evidence. Ecological Economics, 164: 106347.

Teng, F., F. Jotzo and X. Wang. 2017. Interactions between market reform and a carbon price in China's power sector. Economics of Energy & Environmental Policy, 6(2): 39–54.

Tohidi, Y., M.R. Hesamzadeh, R. Baldick and D.R. Biggar. 2017. Transmission network switching for reducing market power cost in generation sector: A Nash-equilibrium approach. Electric Power Systems Research, 146: 71–79.

Vehmas, J. 2005. Energy-related taxation as an environmental policy tool-the Finnish experience 1990–2003. Energy Policy, 33(17): 2175–2182.

Viskovic, V., Y. Chen, A.S. Siddiqui and M. Tanaka. 2021. Economic and environmental consequences of market power in the south-east Europe regional electricity market. The Energy Journal, 42(6).

Wild, P., W. Bell and J. Foster. 2015. Impact of carbon prices on wholesale electricity prices and carbon pass-through rates in the Australian National electricity market. Energy Journal, 36: 137–153.

Wilson, I.A.G. and I. Staffell. 2018. Rapid fuel switching from coal to natural gas through effective carbon pricing. Nature Energy, 3: 365–372.

Woo, C., Y. Chen, J. Zarnikau, A. Olson, J. Moore and T. Ho. 2018. Carbon trading's impact on California's real-time electricity market prices. Energy, 159: 579–587.

Yang, L., Z. Wang, F. Deng, M. Liang and W. Lin. 2018. Bidding strategy of automatic generation control market based on conjectural variation. In 2018 International Conference on Power System Technology (POWERCON), pp. 699–702.

Ye, Y., D. Papadaskalopoulos and G. Strbac. 2018. Investigating the ability of demand shifting to mitigate electricity producers' market power. IEEE Transactions on Power Systems, 33(4): 3800–3811.

Yin, J., Q. Yan, K. Lei, T. Baležentis and D. Streimikiene. 2019. Economic and efficiency analysis of China electricity market reform using computable general equilibrium model. Sustainability, 11(2): 350.

Zhao, Y., H. Li, Y. Xiao, Y. Liu, Y. Cao, Z. Zhang et al. 2018. Scenario analysis of the carbon pricing policy in China's power sector through 2050: Based on an improved CGE model. Ecological Indicators, 85: 352–366.

Appendix

A. The Karush-Kuhn-Tucker (KKT) conditions and the market clearing conditions

KKT conditions of the suppliers

$$-d_o(-h_o^i b_{ot} q_{ot}^{ik} + a_{ot} - b_{ot} \sum_{i' \in I, k' \in K} q_{ot}^{i'k'} - e_{ot}^{ik}) + \phi_{ot}^{ik} \geq 0 \perp q_{ot}^{ik} \geq 0 \quad (8)$$

$$-d_o(-c_{ot}^{ik} - u_{ot} r_{ot}^{ik}) + \alpha_{ot}^{ik} - (1 - s_{ot}^{ik})(1 - g_{ot}^{ik})\phi_{ot}^{ik} \geq 0 \perp f_{ot}^{ik} \geq 0 \quad (9)$$

$$d_o m_o^{ik} - \sum_{o' > o, t \in T} l_{o't}^{ik} \alpha_{o't}^{ik} \geq 0 \perp z_o^{ik} \geq 0 \quad (10)$$

$$l_{ot}^{ik}(Q^{ik} + \sum_{o' < o} z_{o'}^{ik}) - f_{ot}^{ik} \geq 0 \perp \alpha_{ot}^{ik} \geq 0 \quad (11)$$

$$(1 - s_{ot}^{ik})(1 - g_{ot}^{ik}) f_{ot}^{ik} - q_{ot}^{ik} = 0 \perp \phi_o^{ik} \ (free) \quad (12)$$

KKT conditions of the carbon permit auctioneer

$$-(u_{ot} - w_{ot}) + \beta_{ot} \geq 0 \perp v_{ot} \geq 0 \quad (13)$$

$$V_{ot} - v_{ot} \geq 0 \perp \beta_{ot} \geq 0 \quad (14)$$

The market clearing conditions

$$v_{ot} - \sum_{i \in I, k \in K} r_{ot}^{ik} f_{ot}^{ik} = 0 \ \perp u_{ot} \ (free) \quad (15)$$

B. A mathematical justification of how the market power parameter affects the electricity market price and quantity

In an electricity market, there are electricity suppliers i, $i' \in I$, each using a generation technology of a unit cost x^i, to make a one-period electricity production decision q^i. A linear inverse demand curve is assumed:

$$p = a - b\sum_{i \in I} q^i \quad (16)$$

The electricity price p is assumed high enough for all the generators to run, i.e., $q^i > 0 (\forall i \in I)$. For each supplier i, it solves an optimization problem as below:

$$\max_{q^i} (pq^i - x^i q^i) \tag{17}$$

$$\text{s.t. } q^i \leq Q^i \quad (\alpha^i) \tag{18}$$

Here, α^i is the dual variable associated with the capacity constraint. The KKT conditions of supplier 's problem are:

$$\alpha^i + bq^i \frac{\partial \sum_{i' \in I} q^{i'}}{\partial q^i} + x^i - a + b\sum_{i' \in I} q^{i'} = 0 \tag{19}$$

$$Q^i - q^i \geq 0 \perp \alpha^i \geq 0 \tag{20}$$

The KKT conditions are sufficient and necessary for an optimal solution since the objective function and constraint are linear. By solving the system of KKT conditions of all the suppliers, we can obtain an equilibrium solution.

By defining a market power parameter:

$$h^i = \frac{\partial \sum_{i' \in I} q^{i'}}{\partial q^i}, \tag{21}$$

which is the conjecture of supplier i about how the total market quantity varies as its own quantity varies, we rearrange Equation (19) into:

$$bq^i h^i + \alpha^i + x^i - a + b\sum_{i' \in I} q^{i'} = 0 \tag{22}$$

Assuming $h^i \neq 0$, we divide both sides of Equation (22) by h^i and obtain:

$$\frac{1}{h^i}(bq^i h^i + \alpha^i + x^i - a + b\sum_{i' \in I} q^{i'}) = 0 \tag{23}$$

To sum the above equation over all $i \in I$, we obtain:

$$\sum_{i \in I} \frac{1}{h^i}(bq^i h^i + \alpha^i + x^i - a + b\sum_{i' \in I} q^{i'}) = 0 \tag{24}$$

To solve for the total quantity, we rearrange the terms in the above equation to obtain:

$$\sum_{i \in I} q^i = \frac{a\sum_{i \in I} \frac{1}{h^i} - \sum_{i \in I} \frac{\alpha^i + x^i}{h^i}}{b(1 + \sum_{i \in I} \frac{1}{h^i})} \tag{25}$$

Taking the first-order partial derivative of the total quantity $\sum_{i \in I} q^i$ with respect to a market power parameter h^i, we obtain:

$$\frac{\partial \sum_{i \in I} q^i}{\partial h^i} = \frac{-a + x^i + \alpha^i}{b^2 (h^i)^2 (1 + \sum_{i \in I} \frac{1}{h^i})^2} \tag{26}$$

By rearranging the terms in Equation (22), we obtain:

$$-a + x^i + \alpha^i = -b \sum_{i' \in I} q^{i'} - bq^i h^i < 0 \tag{27}$$

Substituting Equation (27) into Equation (26), we obtain:

$$\frac{\partial \sum_{i \in I} q^i}{\partial h^i} < 0 \tag{28}$$

Then,

$$\frac{\partial p}{\partial h^i} = \frac{\partial (a - b \sum_{i \in I} q^i)}{\partial h^i} = -b \frac{\partial \sum_{i \in I} q^i}{\partial h^i} > 0 \tag{29}$$

Hence, as the market power parameter increases, the market price of electricity increases; the total quantity of electricity decreases.

C. An illustration of the economic meaning of the market setting in the case study

To illustrate the meaning of the market setting in the case study of the paper, we apply two equilibrium problems, i.e., problem (EP-A) and problem (EP-B), each occurring in a different market setting.

Problem (EP-A)

In problem (EP-A), the electricity suppliers make one-period electricity production decisions in an electricity market. The market setting is that:

(1) there are electricity suppliers ($i \in I$) and consumers, all of which are price-takers and no one exerts market power;

(2) all the suppliers are symmetric/identical, each applying the same types of generation technology ($k \in K$) with the same production capacities ($Q^{ik} = Q^{i'k}, \forall i, i' \in I, \forall k \in K$), and the same production costs (x^k) to generate electricity q^{ik};

(3) the market price of electricity is $p = a - b \sum_{i \in I, k \in K} q^{ik}$ and p is assumed high enough for all the generators to run, i.e., $q^{ik} > 0 (\forall i \in I, \forall k \in K)$.

For each supplier i, it solves an optimization problem, referred to as problem (OP-A) hereafter.

$$\text{(OP – A):} \qquad \max_{q^{ik}} \sum_{k \in K} (pq^{ik} - x^k q^{ik}) \tag{30}$$

$$\text{s.t. } q^{ik} \le Q^{ik} \quad (\alpha^{ik}) \tag{31}$$

Here, α^{ik} is the dual variable associated with the capacity constraint.

Problem (EP-B)

In problem (EP-B), the electricity suppliers make one-period electricity production decisions in an electricity market. The market setting is that:

(1) there are electricity suppliers and consumers, all of which are price-takers and no one exerts market power;

(2) each supplier uses a different generation technology $k \in K$ (i.e., total $|K|$ suppliers) with the production capacity Q^k and the production cost x^k to generate electricity q^k;

(3) the market price of electricity is $p = a - b \sum_{k \in K} q^k$ and p is assumed high enough for all the generators to run, i.e., $q^k > 0 (\forall k \in K)$.

For each supplier with its particular technology k, it solves an optimization problem, referred to as problem (OP-B) hereafter.

$$\text{(OP – B):} \qquad \max_{q^{k}} \sum_{k \in K} (pq^{k} - x^k q^{k}) \tag{32}$$

$$\text{s.t. } q^{k} \le Q^{k} \quad (\alpha^{k}) \tag{33}$$

Here, α^k is the dual variable associated with the capacity constraint.

Assumptions on the two market settings

We assume several factors in the two market settings to be the same. They are:

(1) the types of generation technology (represented by the same notation $k \in K$);

(2) the production cost of each generation technology (represented by the same notation x^k);

(3) the total capacity of each power generation technology ($Q^k = |I|Q^{ik}$, $\forall i \in I, \forall k \in K$);

(4) the market price-quantity relationship.

The above assumptions provide conditions for the following theorem.

Theorem 1. *Problem (EP-A) and problem (EP-B) are equivalent in achieving the same electricity price and quantity.*

Proof. To solve the problem (EP-A), we begin with the problem (OP-A), whose KKT conditions are sufficient and necessary for an optimal solution (due to the linear objective function and constraint). Next, we concatenate the KKT conditions of all the suppliers' problems, to solve which we obtain an equilibrium solution of the problem (EP-A). Recall that in problem (EP-A), q^{ik} is assumed strictly positive. Then, the KKT conditions of the problem (OP-A) are:

$$\alpha^{ik} + x^k - a + b \sum_{i \in I, k \in K} q^{ik} = 0 \tag{34}$$

$$Q^{ik} - q^{ik} \geq 0 \quad \perp \alpha^{ik} \geq 0 \tag{35}$$

Due to the symmetric feature of the suppliers, we have the relationship of the variables:

$$q^{ik} = q^{i'k}, \ \forall i, i' \in I, \forall k \in K \tag{36}$$

$$\alpha^{ik} = \alpha^{i'k}, \ \forall i, i' \in I, \forall k \in K \tag{37}$$

Then, the KKT conditions can be recast into:

$$\alpha^{ik} + x^k - a + b \sum_{k \in K} |I| q^{ik} = 0, \tag{38}$$

$$|I| Q^{ik} - |I| q^{ik} \geq 0 \quad \perp \alpha^{ik} \geq 0 \tag{39}$$

Let's bring in new variables q^k and α^k and let

$$q^k = |I| q^{ik}, \tag{40}$$

$$\alpha^k = \alpha^{ik}. \tag{41}$$

Recall the assumption:

$$Q^k = |I| Q^{ik}. \tag{42}$$

Substituting Equation (40), Equation (41) and Equation (42) into Equation (38) and Equation (39), we obtain:

$$\alpha^k + x^k - a + b\sum_{k\in K} q^k = 0, \tag{43}$$

$$Q^k - q^k \geq 0 \quad \perp \alpha^k \geq 0. \tag{44}$$

Equation (43) and Equation (44) are the KKT conditions of the problem (OP-B). Similarly, we get the equilibrium solution of problem (EP-B), by concatenating the KKT conditions of all the suppliers' problems, i.e., problem (OP-B), and solving them.

Obviously, a solution to the problem (EP-B) can lead to a solution to the problem (EP-A) by a linear transition in Equation (40), and vice versa. The electricity quantities in the two problems lead to the same electricity prices in the two problems. Hence, problem (EP-A) and problem (EP-B) are equivalent in achieving the same electricity price and quantity. □

A discussion of the market setting in the case study of the paper

Recall the market setting in the case study of the paper: six electricity suppliers each with a unique type of generation technology, which in the case of perfectly competitive markets (i.e., MP-0) is similar to that in problem (EP-B) (e.g., the costs of transmission and buying carbon permits can be counted as the production cost in problem (EP-B)). Problem (EP-B) displays the competition among different generation technologies, but in a perfectly competitive market, it is equivalent to the problem (EP-A) where there are symmetric competing suppliers each using several generation technologies.

In either perfectly competitive markets or imperfectly competitive markets, every rational supplier with several generation technologies gives priority to generation technologies with lower average costs. This implicitly reflects the competition among different generation technologies in electricity markets. This is an ideological basis for the market setting in the case study of the paper, especially when the paper is focusing on the electricity industry trends (rather than a big supplier squeezing a small one).

CHAPTER 15

Safety of Water Resources of a River Basin

Ibraim Sh Didmanidze,[1] *Zurab N Megrelishvili*[1] and *Volodymyr A Zaslavskyi*[2,*]

1. Introduction

At the turn of the 21st century, experts from different countries are trying to answer the question: what could cause conflicts in the next century? Often they name territorial claims, racial and ethnic problems, and religious contradictions. Recently, the most probable reason for the outbreak of war is associated with the problem of access to water resources. Such a forecast is contained in the materials of the International Congress held in Tashkent in November 2018 and which considered the political and environmental problems related to water. This congress was attended by more than 50 states of the Middle East, Africa and Southeast Asia (Independent Newspaper 2001, Resolution of the International Conference of the NWO EECCA 2018). According to the UN, two million people on the earth do not have direct access to drinking water. Its sharp shortage is noted in 80 countries. Poorly treated drinking water containing life-threatening bacteria caused the death of at least one million people in a year. According to the documents of the congress, 2% of the 3% of the world's freshwater reserves, are the glaciers of the Arctic and Antarctic. The remaining 1% is sufficient to provide water for future generations unless it is polluted and used unreasonably.

[1] Batumi Shota Rustaveli State University, Georgia.
[2] Taras Shevchenko National University of Kyiv, Ukraine.
* Corresponding author: zas@unicyb.kiev.ua

2. Problems of water resources

Environmental pollution, which is the result of human activity, directly depends on the number of people. The population of the earth doubled from 1650 to 1850, i.e., for 200 years, then from 1850 to 1950, i.e., for 100 years, then from 1950 to 1975, i.e., for 25 years (Independent Newspaper 2001). The American weekly magazine "US News & World Report" has published the following data on the population of the earth and large cities (Tables 1 and 2).

In the 21st century, more than two dozen countries of the world, in particular the Arab countries, face a sharp water shortage. This data is presented in the materials of the seminar "Resources of the Middle East", which was held in Damascus. Today, the region's population has reached 300 million people who consume water with a norm of 200 litres per day per person. Under these circumstances, a sharp shortage of water is already felt. The experts participating in the seminar raised the question: what will happen in 25 years when the number of people living in the region reaches 500 million, considering that the neighboring 8 states control 87% of the water resources of rivers flowing through the territory of the Arab states?

At an international conference on water supply issues, held in 1998 in Cairo, scientists from 32 countries in Asia and Africa spoke about the need to take urgent measures to prevent a future crisis (Independent Newspaper 2001). In the mid-21st century, China may face severe water shortages. This was stated in special reports of the Chinese Academy of Sciences and a number of National institutes. The materials emphasized that only urgent measures can save the country from the crisis since almost 300 large cities are already experiencing water shortages.

Many examples can be given. But it is not only the lack of water that creates tension. Today there are more than 40 thousand dams that regulate

Table 1. Population growth in some regions of the world (million people).

Region	1983	2000	Growth %
Africa	513	851	65.9
Latin America	390	564	44.6
Asia	2730	3564	30.5
Oceania	24	29	20.8
North America	259	302	16.6
Europe	489	511	4.3
Total	4677	6130	31.1

Table 2. Population growth in some large cities (million people).

City	1983	2000	Growth %
Beijing	13.5	22.5	68.9
Mexico	17.0	27.6	62.4
Shanghai	16.6	25.9	57.0
Sao Paulo	14.0	21.5	53.6
Tokio-Yokohama	21.1	23.8	12.8
New York	18.1	19.5	7.7

the flow of rivers and affect the economic development of arid regions in the world. Dams and ponds have been built on around 200 large rivers that cross state borders. Moreover, a third of these rivers do not have international agreements which would regulate the consumption of water from these rivers.

These conflicts are probably related to the access to drinking water which are most likely related to the Middle East. The flowing rivers Tigris and Euphrates flow here and originate in the territorial region of Turkey (Figure 1), they can regulate the flow of water to the northern regions of Iraq

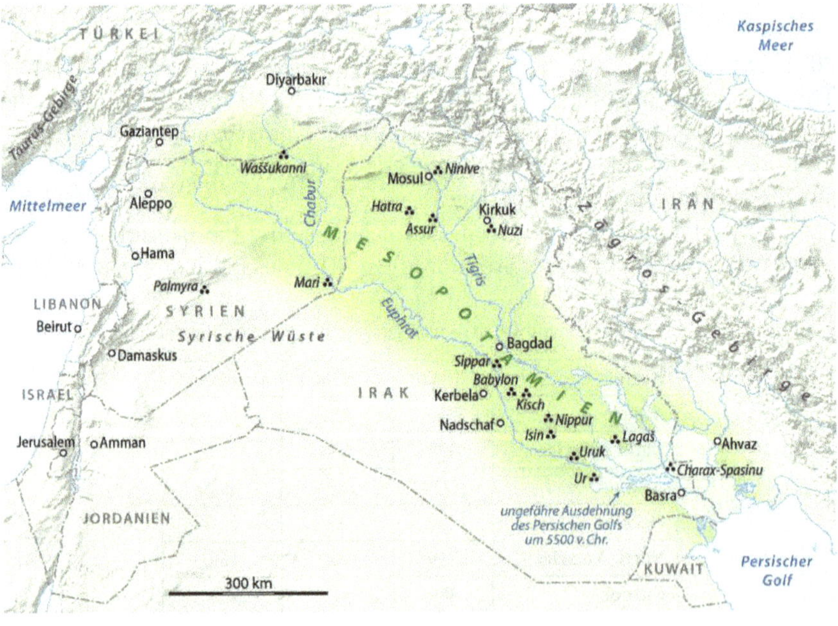

Figure 1. The countries of the Tigris and Euphrates rivers.

and Syria at their own wish. In the territory of Turkey, in the province of Anatolia, 22 dams and hydroelectric power plants were built, which limited the supply of water to Iraq and Syria by a third (Independent Newspaper 2001). Iraq and Syria viewed the construction of dams as a violation of international standards and a violation of the rights of the populations of neighboring countries to use the water of these rivers.

Western observers believe that a conflict between Ethiopia, Sudan and Egypt due to the waters of the Blue Nile is possible (Figure 2). Ethiopia, whose population has reached 60 million people, is facing a population explosion. Therefore, today the country has been claiming its rights to tributaries of the river Nile, which are necessary for watering fields and generating electricity. Egypt fears that Ethiopia, where the upper reaches of the river are located, may regulate the water, which will lead to a decrease in the flow rate of the region's full-flowing river. International experts do not count out that in the case of such a development of phenomena, Egypt will not abandon the use of military forces (Resources of surface waters of the USSR 1974).

Israel, for its part, regulates the flow of water from sources located in the upper river Jordan in the Golan Heights, where drinking water comes from (Figure 3). This is constantly a hotbed of tension. Many experts believe that Syria will not abandon the Golan Heights, while Israel will not allow anyone to control the sources of water located in this territory (Independent Newspaper 2001).

The Zambezi River flows through six countries of southern Africa - Angola, Botswana, Zambia, Mozambique, Zimbabwe and Namibia (Figure 4). There are two of the region's largest hydroelectric power plants located there - Cabora Bassa in Mozambique and Cariba on the border of Zambia and Zimbabwe. There is a plan to build several more dams in the region. Zimbabwe is going to carry water from the river to arid provinces. With this in mind, tension has arisen between Zambia and Zimbabwe.

In fact, two African countries were on the verge of starting a "Water War". The reason was the beginning of the construction of a water supply system from the Zambesi River to the second largest city in Bulawayo from Zimbabwe. The project, valued at US$ 437 million, aims to provide water to the city and nearby arid lands. Zambia and Botswana were against this project. They believe that taking such a quantity of water will significantly reduce the proportion of water resources. Zambia is particularly against the project as it believes that the project puts the implementation of the Zambian program for the development of the southern and southwestern regions of the country in doubt, also it will reduce the capacity of the Livingston

Figure 2. The countries of the river basin Nile and Horn of Africa.

Figure 3. The countries of the river basin Jordan.

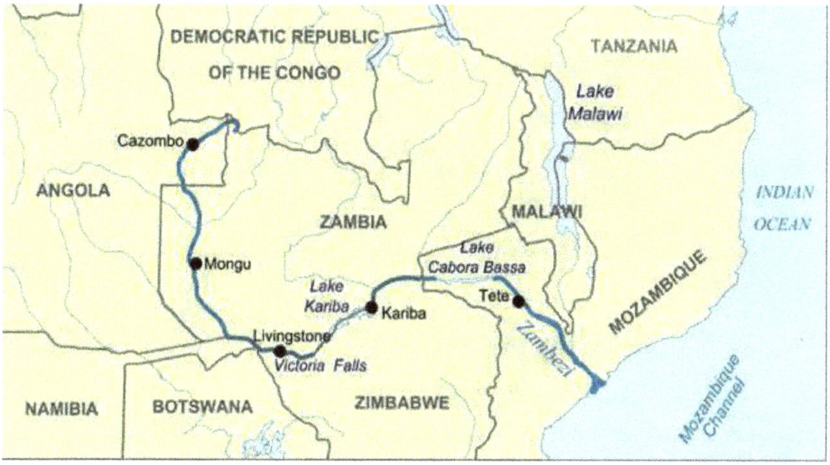

Figure 4. The countries of the river basin Zambezi.

Hydroelectric Power Station and reduce the water level at Victoria Falls, the source of tourism revenue in this part of the country (Independent Newspaper 2001).

Under the given circumstances, experts propose the creation of a regional council on the use of water. A significant step in this direction can be considered with the creation of the "Commission on Water" within the framework of the Commonwealth of Development of South Africa. Countries were recommended to adopt a law to avoid the desire of resolving the water issue by using weapons.

185 km section of the upper river Mtkvari is located in Turkey. Only 27 km of the 435 km long river Chorokhi is on the territory of Georgia (Figure 5) (The State Water Cadastre 1985). The construction of river dams in Turkish territory has already caused environmental problems for Adjarian Black Sea coast of Georgia. Dam construction has reduced the amount of river sediment discharged into the Black Sea. This creates the danger of erosion of the coastal zone of Adjara.

Industrial development in those regions of Turkey where the river Mtkvari flows can affect not only the flow of the river but also the quality of the water, which can affect the drinking water resources in Georgia. On this basis, the rational use of water resources is an urgent issue for two neighboring countries.

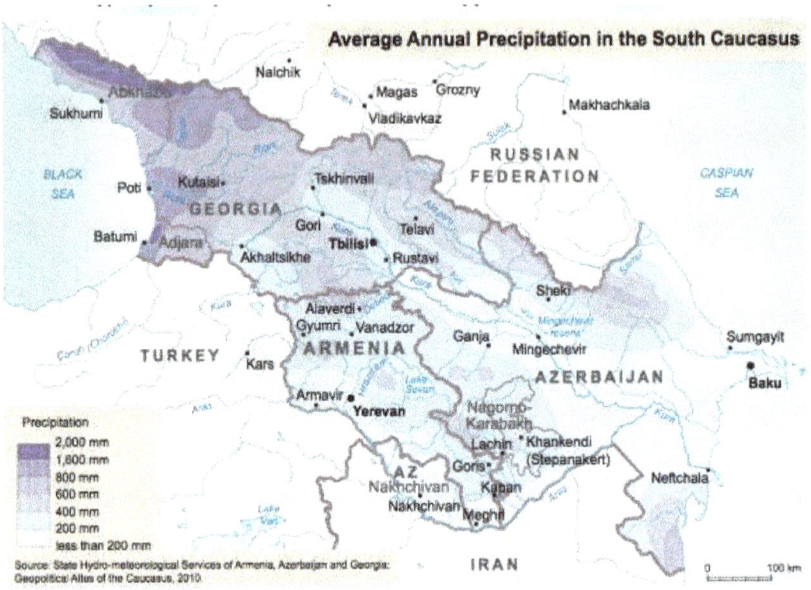

Источник: ЗОИ (2011).

Figure 5. Mtkvari and Chorokhi River Basin countries.

3. Methodology for calculating environmental damage

Not only can unreasonable decisions on the use of waters of transboundary rivers and anthropogenic pollution of the environment lead to armed conflicts, but it can also create serious environmental problems. This is due to river pollution with various substances and excess MPC during the development of the industrial potential of neighboring countries. Experts see a way out in holding two multilateral conferences and treaties, in developing international documents that will regulate equal access to drinking water resources for the entire population of the earth (Resolution of the International Conference of the NWO EECCA 2018, Kazakhstan Today 2001).

One of the issues for international discussion and development of a common approach may be the methodology for determining the damage caused to the environment proposed by us.

Determining the damage to the environment from wastewater discharges taking into account biological, chemical, and physics-chemical changes is almost impossible. Changes caused by wastewater discharges into water ponds can occur after decades when it will be impossible to correct the changes (Zaslavskyi 2005, Baglaeva et al. 2013).

The value of environmental and economic damage can be represented as the sum of various costs (Megrelishvili et al. 2014):

$$U_i = \sum_{i=1}^{n} x_i q_i \tag{1}$$

where x_i – changes *i*-like factor q_i – price evaluation of *i*-like factor.

Considering $x_i = V_i n_i$ and marking q_i through q_B, damage to the environment, caused by wastewater discharge, can be calculated with the following formula:

$$U_i = V_i n_i q_B \tag{2}$$

where V_i – is the volume of wastewater discharged into the reservoir per unit time, with the i - pollutant, m^3; n_i - required dilution ratio for reaching the maximum permissible concentration (MPC) for the first substance; q_B - economic assessment of the water resource (cost of natural water) lari/m^3.

Since the water of the reservoir may contain the i pollutant, the dilution ratio is determined depending on:

$$n_i = \frac{c_i}{c_{impc} - c_{iv}} \tag{3}$$

where c_i – is concentration, *i*-того substance in wastewater, g/m^3; c_{impc} – maximum permissible concentration of the *i*-like substance in natural water, g/m^3; c_{iv} - concentration of the *i*-like substance in the reservoir, g/m^3;

The expression $V_i n_i$ in formula (2) is the amount of natural water that is necessary to dilute the wastewater in order to achieve the MPC for the substance. Taking into account this fact, formula (2) will look as follows:

$$U_i = W_i q_B \tag{4}$$

where W_i - volume of natural water required for dilution with respect to the *i*-like substance, m^3 per unit time.

According to the current legislation of Georgia, the amount of payment for the use of water resources is different and varies from 10 to 3% depending on the market price. For thermal power plants and enterprises of irrigation systems, the value is 1% of the basic payment, for hydro power plants 0.01% of the basic payment, for municipal and agricultural water supply, the payment for 1 m^3 of water is 0.01 tetri (tetri- the hundredth part of the lari). Since the market price of 1 m^3 of natural water is 1.0 GEL, according to the minimum payment for the use of water resources is 3.0, for thermal power plants 1.0, for hydro power plants 0.01, for communal and agricultural water supply 0.01 tetri.

Table 3. Calculation of damage to the environment.

No.	Pollutant	C_{impc} mg/lt	Minimum amount of water to dilute up to C_{impc}, m^3	Payment, Lari	
				According to the proposed method	According to the legislation
1	2	3	4	5	6
1.	Chlorides	350	2857	0.29	0.2
2.	Titanium	0.1	10^7	1000	390
3.	Selenium	0.001	10^9	1000000	39000
4.	Tellurium	0.01	10^8	10000	3900
5.	Arsenic	0.05	$2*10^7$	2000	790
6.	Iron	0.5	$2*10^5$	200	78
7.	Aluminum	0.1	10^7	1000	78
8.	Beryllium	0.0002	$5*10^9$	500000	390000
9.	Surfactants*				
	minimum	10	10^5	10	390
	maximum	0.001	10^9	100000	390
10.	Oil products	0.05	$2*10^7$	2000	780
	Gasoline	0.1	10^7	1000	780
11.	Ketones**				
	minimum	2.0	$5*10^5$	50	780
	maximum	0.02	$5*10^7$	5000	780
12.	Phenols	0.001	10^9	100000	39000

* Surfactants – 4 items.

** Ketones – 11 items.

Payment (fine) for environmental pollution with a harmful substance in accordance with the current legislation of Georgia, in certain cases does not correspond to the real damage to the environment. Table 3 shows the results of calculation of damage during the discharge of one ton of pollutants, carried out according to the formulas (1)–(4) and according to the methodology of the current legislation.

In the calculation, the assumption was made that there is no pollutant in natural water. In another case, the amount of natural water for dilution and, consequently, damage to the environment will increase. The cost of natural water is a minimum of 0.01 tetri.

The obtained data shows that the damage calculated according to the proposed methodology is 1.4–250 times higher than the amount calculated

Table 4. Hydrogeological data of some rivers of Georgia.

No.	River	Long-term average consumption, m³/s		Flow speed, m/s	Length, km	Consumption m³/h	Pollution** according to the legislation
		Исток	Устье				
1.	Rioni	10.3	409.0	0.3–4.2	337.0	$1.47*10^6$	1
2.	Mtkvari*	32.6	291.0	0.2–2.0	513.0	$1.05*10^6$	2,3
3.	Inguri	45.5	192.0	0.6–5.8	213.0	691200	2
4.	Chorokhi*	–	159.0	0.7–2.5	26.0	572400	–
5.	Alazani	14.4	112.0	0.8–3.5	351.0	403200	3
6.	Tskhenistkali	23.0	90.4	0.5–1.5	176.0	325440	–
7.	Adjaris tskali	8.73	51.5	0.6–1.3	90.0	185400	–
8.	Khobi	7.43	50.5	0.5–2.5	150.0	181800	3
9.	Supsa	2.56	46.0	0.5–1.2	108.0	165600	–
10.	Natanebi	2.12	33.5	0.4–1.5	60.0	120600	3
11.	Mokva	4.50	18.1	1.0–1.5	47.0	65160	3
12.	Kintrishi	4.25	17.3	0.7–1.8	45.0	62280	3
13.	Gubis tskali	7.55	16.4	0.2–0.4	36.0	59040	1
14.	Choloki	0.66	7.03	0.3–2.5	24.0	25308	–
15.	Zhokvera	1.79	6.11	0.5–3.5	20.0	21996	1
16.	Achkva	–	1.66	0.1–1.1	19.0	5979	–

* On the territory of Georgia.
** Pollution: 1- pollution of a special level, 2 – pollution of a high level, 3 – polluted.

according to the current legislation. It should be noted that taking into account the market price of water, the amount of damage will increase significantly.

Along with this, attention must be paid to the amount of water that is needed to dilute the substance to the MPC. Most rivers (especially in the Transcaucasian republics, and particularly in Georgia) cannot provide such a quantity of water. This is clearly seen in Table 2, which is compiled according to (The State Water Cadastre 1985, Resources of surface waters of the USSR 1974).

A full-flowing river (Table 4) 680 hours will be required for the allocation of the maximum amount of dilution water (Table 3). During this time, the river water, with an average speed of 2.0 m/s, will pass 4896 km-s, which is significantly longer than its length. This means that the concentration of the harmful substance in water will be significantly higher than the maximum permissible concentration. Similar calculations are possible for any of the given rivers. If such a river crosses the state border, the problem of water quality will be the cause of a possible conflict between neighboring countries.

4. Conclusions

Many transboundary rivers flowing through the territories of several states do not have bilateral or multilateral agreement documents on the use or accessibility of these rivers. Modern equipment and technologies make it possible to significantly regulate the flow of such rivers, which can cause future conflicts without an agreement between neighboring countries. Unreasonable decisions on the use of waters of transboundary rivers and anthropogenic pollution of the environment can not only lead to armed conflicts but also create serious environmental problems. Experts see a way out in holding two multilateral conferences and agreements, as well as in developing international documents that will regulate equal access to drinking water resources for the entire population of the earth. The methodology for determining the damage to the environment caused by wastewater discharges, which may become one of the subjects for discussion in the development of international solutions for transboundary rivers, is discussed and proposed.

References

Africa http://www/otd/ru/gto/Afica/ 2001.

Baglaeva, V.O., I.M. Potravny and V. Zaslavskyi. 2013. Justification environmental infrastructure construction options, depending on the scenarios of water management ecological-economic systems: The interaction of government, business, science and society. Proceedings - Irkutsk: Publishing House of the Institute of Geography, 2013: 176–179.

Independent Newspaper. http://world/ng/ru/problem/ 2001.

Kazakhstan today http://www/gazeta/kz/ 2001.

Megrelishvili, Z., L. Turmanidze, Z. Kalandarishvili and N. Dondoladze. 2014. Probability risk model for environmental decision making. Hydroengineering, Tbilisi, 1-2(17-18)2014: 63–71.

Resolution of the International Conference of the NWO EECCA Water for Land Reclamation, Water Supply for Economic Sectors and the Environment under Climate Change. Tashkent. November 6–7, 2018.

Resources of surface waters of the USSR. Hydrographic description of rivers, lakes and reservoirs. Vol. 9. Transcaucasia and Dagestan. Edit. 1. Western Transcaucasia under edit. V. Sh. Tsomaia – L. Hydrometeoizdat, 1974.

The State Water Cadastre. Annual data on the regime and resources of land surface water. Vol. VI, Soviet Republic of Georgia – Tbilisi: 1985.

Zaslavskyi, V. 2005. Multi-stage optimization model of the problem of water management in the river basin. Scientific and Practical Journal World of Information Technologies, 2(2005): 18–22.

CHAPTER 16

Optimization Problems for Retrial Queues with Unreliable Server

Eugene A Lebedev and *Mykhailo M Sharapov**

1. Introduction

Nowadays, the interest in stochastic systems with repeated calls is due to their widespread use in various models for computers, satellites, telephones, security, telecommunication systems, etc. In such systems, the call that was refused becomes a source of repeated calls. Queueing systems with retrials are a specific class of stochastic models that describes important properties of the service process and is very useful and efficient for modeling real-life structures.

A large number of works are devoted to the classical models of systems with repeated calls (see, for instance, Falin et al. 1997, Yang et al. 1987, Artalejo et al. 2008, Nazarov et al. 2018, Lebedev et al. 2011, Gomez-Corral et al. 1999, Walrand 1998). In comparison with the classical models, the case of unreliable servers is much less commonly used (Thiruvengadam 1963, Li et al. 1997, Artalejo 1994, Wartenhorst 1995, Vinod 1985, Wang et al. 2001, Artalejo et al. 2002). In these models, the problem of calculating the basic system characteristics is in general significantly more complicated. If we consider stochastic processes describing the operation of such systems, they are Markov processes which have a countable number of states, and the infinitesimal matrix usually does not have any special properties that would

Taras Shevchenko National University of Kyiv, Faculty of Computer Science and Cybernetics.
* Corresponding author: boxus@ukr.net

help to obtain a solution in an explicit form. In this chapter, we consider one of these models with an unreliable server as well as approaches to its stationary distribution calculation.

2. Model description, statement of the problem and main result

In this chapter, we consider a system with repeated calls and one unreliable server. Thus, the system consists of one server and an orbit. If the orbit is not empty, then the orbital call flow rate $\theta > 0$ is constant, and if the orbit is empty, then this rate is zero. Let the input flow of primary calls arrive in the system, its rate $\lambda = \lambda_k > 0$ is assumed to be dependent on the number of calls at the orbit $k = 0,1,...$. If a primary (or repeated) call arrives into the system and finds the server idle, its service begins immediately. Service rate is $\mu > 0$. If the server is busy, the call is directed to the orbit. The server is assumed to be unreliable. It means that its lifetime is exponentially distributed random variable a with failure rate $\alpha > 0$. When the server breaks down, it is directed for repairing, service of a call is interrupted, and this call goes to the orbit, all new calls arriving for the service are directed to the orbit immediately. The rate of the recovery time is $\beta > 0$. The next service of an interrupted call is independent of the failed one.

We introduce a service process in the system with repeated calls and an unreliable server as two-dimensional Markov chain $Q^T(t) = (Q_1(t), Q_2(t))$. The first component of the chain $Q_1(t) \in \{0, 1, 2\}$ points out the status of the server at time t: if its state is 0, it means that the server is idle and ready to serve calls, if it is 1, then the server is busy, if it is 2, then the server in on repairing (restoring of working condition). The second component $Q_2(t) \in \{0, 1, 2...\}$ is the number of sources of repeated calls. For $(i, j) \neq (i', j')$, (i, j), $(i', j') \in S = \{0, 1, 2\} \times \{0, 1,...\}$, infinitesimal transition rates $a_{(i,j)(i',j')}$

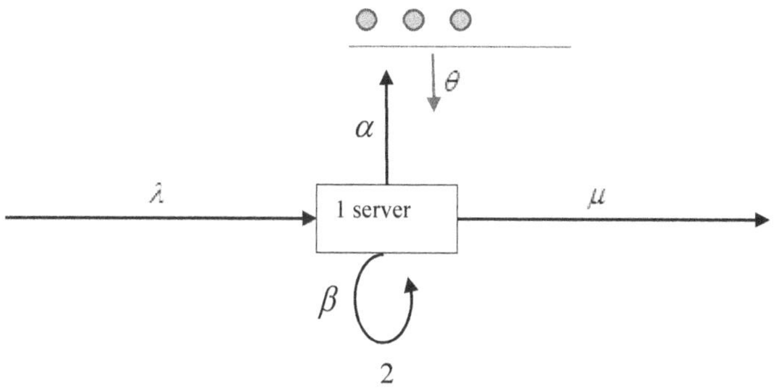

Figure 1. A system with repeated calls and one unreliable server.

of the Markov process $Q(t)$ can be expressed via the system parameters, λ_k, $k = 0, 1, \dots, \mu, \alpha, \beta, \theta$ as follows:

$$a_{(i,j),(i',j')} = \begin{cases} \lambda_j, & (i,j)=(0,j), (i',j')=(1,j), j \geq 0, \\ \theta, & (i,j)=(0,j), (i',j')=(1,j-1), j \geq 1, \\ \lambda_j, & (i,j)=(1,j), (i',j')=(1,j+1), j \geq 0, \\ \alpha, & (i,j)=(1,j), (i',j')=(2,j+1), j \geq 0, \\ \mu, & (i,j)=(1,j), (i',j')=(0,j), j \geq 0, \\ \lambda_j, & (i,j)=(2,j), (i',j')=(2,j+1), j \geq 1, \\ \beta, & (i,j)=(2,j), (i',j')=(0,j), j \geq 1. \end{cases}$$

The graph of transition rates for $Q(t)$ has the form:

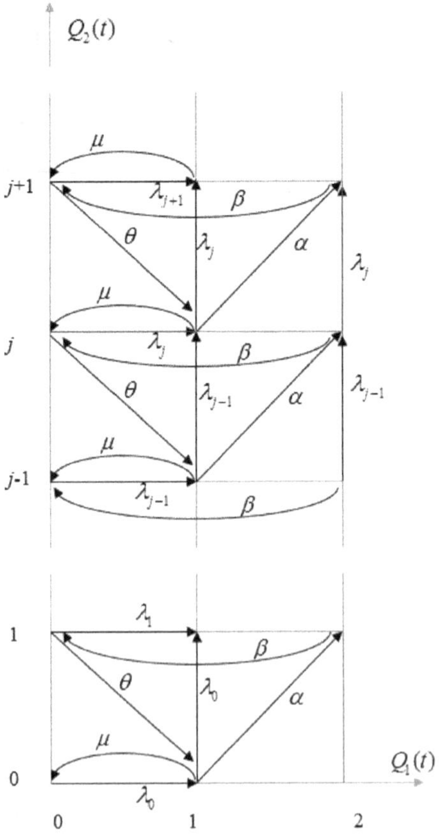

Figure 2. Graph of transition rates for the service process in a system with repeated calls and an unreliable server.

The main goal of the work is to find the conditions of the existence of a stationary regime for $Q(t)$, $t \geq 0$, and to obtain explicit vector-matrix formulas for stationary probabilities.

For the case when a stationary regime for $Q(t)$, $t \geq 0$, exists, we will denote the stationary probabilities by π_{ij}, $(i, j) \in S$. We introduce also the following notations:

$$\pi_j = (\pi_{1j}, \ \ \pi_{2j})^T, j \geq 0, \tilde{A}_{j-1} = \begin{pmatrix} \mu\theta & -\lambda_j(\beta + \lambda_j + \theta) \\ 0 & \lambda_j + \beta \end{pmatrix}, j \geq 1,$$

$$\tilde{A}_{j-1}^{-1} = \frac{1}{\mu\theta(\lambda_j + \beta)} \begin{pmatrix} \lambda_j + \beta & \lambda_j(\beta + \lambda_j + \theta) \\ 0 & \mu\theta \end{pmatrix}, j \geq 1,$$

$$\tilde{B}_{j-1} = \begin{pmatrix} \lambda_{j-1}(\lambda_j + \theta) & 0 \\ \alpha & \lambda_{j-1} \end{pmatrix}, j \geq 1, \quad A_j = \tilde{A}_j^{-1}\tilde{B}_j, j \geq 0, \quad \tilde{\delta}_0 = \left(\frac{\lambda_0}{\mu}, \ \ 0\right)^T, \quad \mathbf{1}^T = (1, \ \ 1).$$

Theorem 1. *Let* $\lambda = \overline{\lim_{k \to \infty}} \lambda_k < \infty$ *and the following condition be satisfied:*

$$\alpha < \frac{\beta(\mu\theta - \lambda(\lambda + \theta))}{\lambda((\lambda + \theta) + \beta)} \tag{1}$$

Then for $Q(t)$, $t \geq 0$, there exists a stationary regime and the stationary probabilities are of the form:

$$\pi_0 = \pi_{00}\delta_0, \quad \pi_j = \pi_{00}A_{j-1} \times \ldots \times A_0\delta_0, \quad j \geq 1 \tag{2}$$

$$\pi_{0\,j+1} = \frac{1}{\theta}(\lambda_j + \alpha, \ \ \lambda_j)\pi_j, \quad j \geq 0 \tag{3}$$

$$\pi_{00} = \left\{1 + \sum_{j=0}^{\infty}\left(\mathbf{1}^T + \frac{1}{\theta}(\lambda_j + \alpha, \ \ \lambda_j)\right)A_{j-1} \times \ldots \times A_0 \cdot \delta_0\right\}^{-1}. \tag{4}$$

Proof. □ When searching for conditions for the existence of a stationary distribution, one can use the Lyapunov functions, on which the analysis of classical models of systems with repeated calls was based (see Falin et al. 1997, Section 2.2).

Let $\breve{x}_k = (0, 0, k)$, $\breve{y}_k = (0, 1, k)$, $\breve{z}_k = (1, 1, k)$. We take $\varphi(i, j, k) = ai + bj + k$ as the Lyapunov function (i.e., $\varphi(\breve{x}_k) = k$, $\varphi(\breve{y}_k) = b + k$, $\varphi(\breve{z}_k) = a + b + k$), where numbers a, b will be chosen later. Under such a choice of the Lyapunov function, the mean shift function $\Delta(s) = \sum_{p \neq s} q_{sp}(\varphi(p) - \varphi(s))$, $s \in S$ is

$$\Delta(\breve{x}_k) = \lambda_k b + \theta(b - 1), \quad \Delta(\breve{y}_k) = -b\mu + \lambda_k + \alpha(a + 1), \quad \Delta(\breve{z}_k) = \lambda_k - \beta(a + b).$$

The function $\Delta(s)$ will be bounded from above according to the condition of the theorem, i.e.,

$$\Delta(s) < \infty, \text{ for all } s \in S.$$

For the existence of $\varepsilon > 0$ such that for all $k \geq k_0 = k_0(\varepsilon)$ holds

$$\begin{cases} \overline{\lim}\lambda_k < \infty, \\ \lambda_k b + \theta(b-1) < -\varepsilon, \; k \geq k_0, \\ -b\mu + \lambda_k + \alpha(a+1) < -\varepsilon, \; k \geq k_0, \\ \lambda_k - \beta(a+b) < -\varepsilon, \; k \geq k_0, \end{cases}$$

it is sufficient that

$$\begin{cases} b < \dfrac{\theta}{\lambda+\theta}, \\ b > a\dfrac{\alpha}{\mu} + \dfrac{\lambda+\alpha}{\mu}, \\ b > \dfrac{\lambda}{\beta} - a. \end{cases}$$

The locus of points (a, b) whose coordinates satisfy the last system of inequalities is shown in the figure below.

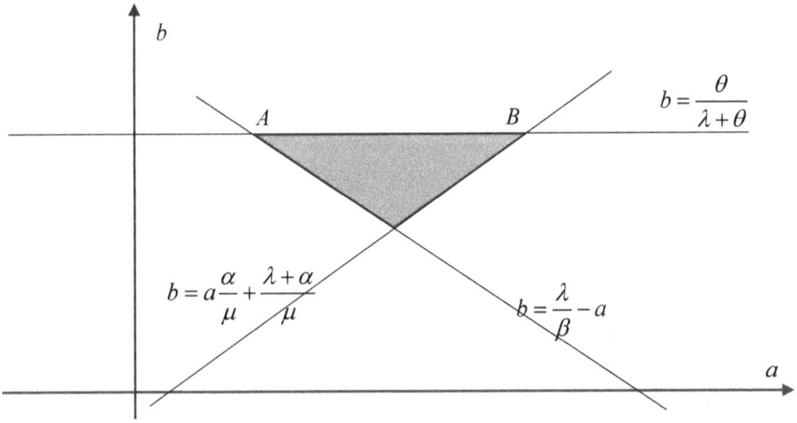

In order for this locus of points to be non-empty, it is necessary and sufficient that point $A\left(\dfrac{\lambda}{\beta} - \dfrac{\theta}{\lambda+\theta}; \dfrac{\theta}{\lambda+\theta}\right)$ be on line $b = \dfrac{\theta}{\lambda+\theta}$ to the left of point $B\left(\frac{\mu\theta}{\alpha(\lambda+\theta)}-\frac{\lambda+\alpha}{\alpha};\ \frac{\theta}{\lambda+\theta}\right)$, i.e., $\frac{\lambda}{\beta}-\frac{\theta}{\lambda+\theta}<\frac{\mu\theta}{\alpha(\lambda+\theta)}-\frac{\lambda+\alpha}{\alpha}$. This inequality is equivalent to (1).

Thus, when (1) is fulfilled, the specified locus of points is not empty. By fixing any point (a, b) from it, we define a Lyapunov function that satisfies the conditions of Statement 8 from Falin et al. 1997, p. 97. This implies the ergodicity of the process $Q(t)$, $t \geq 0$.

Let's deal with construction of stationary probabilities. They satisfy an infinite system of linear equations:

$$\begin{cases}\lambda_0\pi_{00}=\mu\pi_{10},\\ (\mu+\lambda_0+\alpha)\pi_{10}=\lambda_0\pi_{00}+\theta\pi_{01},\\ (\lambda_1+\beta)\pi_{21}=\alpha\pi_{10},\end{cases}\qquad \begin{cases}(\lambda_j+\theta)\pi_{0j}=\mu\pi_{1j}+\beta\pi_{2j},\\ (\lambda_j+\mu+\alpha)\pi_{1j}=\lambda_j\pi_{0j}+\theta\pi_{0j+1}+\lambda_{j-1}\pi_{1j-1},\\ (\lambda_j+\beta)\pi_{2j}=\alpha\pi_{1j-1}+\lambda_{j-1}\pi_{2j-1},\end{cases}$$

$j=1,2,\dots$.

Let $S(j) = \{(p, q) \in S : q \leq j\} \subset S, j = 0, 1, \dots$, be a subset of states of the phase space S. Using the equality of the probability flows in the stationary mode for $S(j)$, we write

$$\theta\pi_{0,j+1}=\left(\lambda_j+\alpha\right)\pi_{1j}+\lambda_j\pi_{2j},\ j\geq 0. \tag{5}$$

Substituting (5) into the original system of linear equations, we find

$$\begin{cases}\lambda_0\pi_{00}=\mu\pi_{10},\\ (\lambda_1+\beta)\pi_{21}=\alpha\pi_{10},\end{cases}\qquad \begin{cases}(\lambda_j+\theta)\pi_{0j}=\mu\pi_{1j}+\beta\pi_{2j},\\ \lambda_j\pi_{0j}=\mu\pi_{1j}-\lambda_j\pi_{2j}-\lambda_{j-1}\pi_{1,j-1}, \quad j=1,2,\dots .\\ (\lambda_j+\beta)\pi_{2j}=\alpha\pi_{1j-1}+\lambda_{j-1}\pi_{2j-1},\end{cases}$$

If we now express $\pi_{0j}=\frac{\mu\pi_{1j}+\beta\pi_{2j}}{\lambda_j+\theta}$, $j \geq 1$, from the equations $\left(\lambda_j+\theta\right)\pi_{0j}=\mu\pi_{1j}+\beta\pi_{2j}$, $j \geq 1$, and substitute them into the rest of the equations of the system, then we obtain a system of linear equations for π_{1j}, π_{2j}, $j \geq 1$:

$$\begin{cases}\mu\theta\pi_{1j}-\lambda_j\left(\beta+\lambda_j+\theta\right)\pi_{2j}=\lambda_{j-1}\left(\lambda_j+\theta\right)\pi_{1j-1},\\ \left(\lambda_j+\beta\right)\pi_{2j}=\alpha\pi_{1j-1}+\lambda_{j-1}\pi_{2j-1}.\end{cases}$$

We write this system of linear equations in vector-matrix form $\tilde{A}_{j-1}\pi_j=\tilde{B}_{j-1}\pi_{j-1}$, $\pi_j=\tilde{A}_{j-1}^{-1}\tilde{B}_{j-1}\pi_{j-1}$ or $\pi_j=A_{j-1}\pi_{j-1}$, whence (2) follows.

Expressions (3) are vector form (5), and (4) follows from (2), (3) and the normalization condition. The theorem is proved. ■

Note, that when constructing the algorithm for calculating stationary probabilities based on formulas (2)–(4), it is necessary to solve the problem of estimating the remainder of the series

$$T(n)=\sum_{j=n}^{\infty}\left(\mathbf{1}^{T}+\frac{1}{\theta}\left(\lambda_j+\alpha,\ \ \lambda_j\right)\right)A_{j-1}\times\ldots\times A_0\cdot\delta_0.$$

Obviously, this will allow controlling the accuracy of the calculation π_{00} and other stationary probabilities.

Let us choose $\lambda' > \lambda$. In order to estimate $T(n)$, we need the Perron root h_1 of the matrix

$$G=\begin{pmatrix}\frac{\lambda'}{\mu}+\frac{(\lambda')^2}{\mu\theta} & \frac{(\lambda')^2}{\mu(\lambda'+\beta)}+\frac{(\lambda')^2}{\mu\theta}\\ \frac{\alpha}{\mu}+\frac{\alpha\lambda'}{\mu\theta} & \frac{\lambda'(\alpha+\mu)}{\mu(\lambda'+\beta)}+\frac{\alpha\lambda'}{\mu\theta}\end{pmatrix}.$$

It can be found that

$$h_1=\frac{\lambda'\left(\theta\left(\mu+\beta+\alpha+\lambda'\right)+\left(\alpha+\lambda'\right)\left(\beta+\lambda'\right)+\sqrt{D}\right)}{2\mu\theta\left(\beta+\lambda'\right)}<1,\ \text{where}$$

$$D=\left(\theta\left(\mu+\beta+\alpha+\lambda'\right)+\left(\alpha+\lambda'\right)\left(\beta+\lambda'\right)\right)^2-4\mu\theta\left(\beta+\lambda'\right)\left(\theta+\lambda'\right)>0.$$

For $T(n)$ such an upper bound is true.

Theorem 2. *Let the conditions of Theorem 1 be satisfied and the choice of $\lambda' > \lambda$ be given. Then*

$$T(n)\le C\cdot h_1^n,\ \text{as}\ n\ge N=N(\lambda'),$$

where the constant $C = C(\lambda')$ is independent of n.

Proof. □ Let's start estimating the remainder of the series

$$T(n)=\sum_{j=n}^{\infty}\left(1+\frac{\lambda_j+\alpha}{\theta},\ \ 1+\frac{\lambda_j}{\theta}\right)\left(\tilde{A}_{j-1}^{-1}\tilde{B}_{j-1}\ldots\tilde{A}_0^{-1}\tilde{B}_0\left(\frac{\lambda_0}{\mu},\ \ 0\right)^T\right)\le$$

$$\le\frac{\lambda_0}{\mu}\left(1+\frac{\lambda'+\alpha}{\theta}\right)\sum_{j=n}^{\infty}(1,\ \ 1)\left(\tilde{A}_{j-1}^{-1}\tilde{B}_{j-1}\ldots\tilde{A}_0^{-1}\tilde{B}_0(1,\ \ 0)^T\right),$$

where $n \ge N$ and $N = N(\lambda')$ such that for all $j \ge N$ the condition $\lambda_j \le \lambda'$ is satisfied.

It is easy to show that $\tilde{A}_{j-1}^{-1}\tilde{B}_{j-1} = H_j \Lambda_j$, where

$$H_j = \begin{pmatrix} \frac{1}{\mu} + \frac{\lambda_j}{\mu\theta} & \frac{\lambda_j}{\mu(\lambda_j + \beta)} + \frac{\lambda_j}{\mu\theta} \\ 0 & \frac{1}{\lambda_j + \beta} \end{pmatrix}$$ and $\Lambda_j = \begin{pmatrix} \lambda_{j-1} & 0 \\ \alpha & \lambda_{j-1} \end{pmatrix}$. Then

$$T(n) \le \frac{\lambda_0}{\mu}\left(1 + \frac{\lambda' + \alpha}{\theta}\right)\sum_{j=n}^{\infty}(1,\ \ 1)\left(\prod_{k=j}^{1} H_k \Lambda_k\right)(1,\ \ 0)^T =$$

$$= \frac{\lambda_0}{\mu}\left(1 + \frac{\lambda' + \alpha}{\theta}\right)\sum_{j=n}^{\infty}(1,\ \ 1)H_j\left(\prod_{k=j}^{2} \Lambda_k H_{k-1}\right)\Lambda_1(1,\ \ 0)^T,$$

where,

$$\Lambda_k H_{k-1} = \begin{pmatrix} \frac{\lambda_{k-1}}{\mu} + \frac{\lambda_{k-1}^2}{\mu\theta} & \frac{\lambda_{k-1}^2}{\mu(\lambda_{k-1} + \beta)} + \frac{\lambda_{k-1}^2}{\mu\theta} \\ \frac{\alpha}{\mu} + \frac{\alpha\lambda_{k-1}}{\mu\theta} & \frac{\lambda_{k-1}(\alpha + \mu)}{\mu(\lambda_{k-1} + \beta)} + \frac{\alpha\lambda_{k-1}}{\mu\theta} \end{pmatrix} < G$$

and comparison of matrices is performed element by element.

Then $T(n) < \frac{\lambda_0}{\mu}\left(1 + \frac{\lambda' + \alpha}{\theta}\right)\sum_{j=n}^{\infty}(1,\ \ 1)H_j G^{j-1}\Lambda_1(1, 0)^T$.

The eigenvalues of matrix G are

$$h_{1,2} = \frac{\lambda'\left(\theta(\mu + \beta + \alpha + \lambda') + (\alpha + \lambda')(\beta + \lambda') \pm \sqrt{D}\right)}{2\mu\theta(\beta + \lambda')}$$ and it is easy to show that $0 < h_2 < h_1 < 1$.

Let's consider the decomposition $G = UJU^{-1}$ where $J = \begin{pmatrix} h_2 & 0 \\ 0 & h_1 \end{pmatrix}$, $U = \begin{pmatrix} \frac{\lambda'(\theta + \lambda') - \mu\theta h_1}{\alpha(\theta + \lambda')} & \frac{\lambda'(\theta + \lambda') - \mu\theta h_2}{\alpha(\theta + \lambda')} \\ 1 & 1 \end{pmatrix}$,

$$U^{-1} = \frac{\alpha(\theta + \lambda')}{\mu\theta(h_2 - h_1)} \cdot \begin{pmatrix} 1 & \frac{(\lambda')^2 + \theta\lambda' - \mu\theta h_2}{\alpha(\theta + \lambda')} \\ -1 & -\frac{(\lambda')^2 + \theta\lambda' - \mu\theta h_2}{\alpha(\theta + \lambda')} \end{pmatrix}.$$ Then

$$T(n) < \frac{\lambda_0}{\mu}\left(1+\frac{\lambda'+\alpha}{\theta}\right)\sum_{j=n}^{\infty}(1,\ \ 1)H_jU\begin{pmatrix} h_2^{j-1} & 0 \\ 0 & h_1^{j-1}\end{pmatrix}U^{-1}\Lambda_1(1,\ \ 0)^T <$$

$$< \frac{2\lambda_0(\theta+\lambda'+\alpha)(\alpha(\mu+\beta)+\lambda'\beta)(\theta(\lambda_0+\mu h_1)+\lambda'(\theta+\lambda'+\lambda_0))}{\mu^3\beta\theta^2(h_1-h_2)}\sum_{j=n}^{\infty}h_1^{j-1}.$$

Condition $h_1 < 1$ allows us to calculate the sum $\sum_{j=n}^{\infty} h_1^{j-1}$ and write down the estimation $T(n) \le C \cdot h_1^n$, where

$$C = C(\lambda') = \frac{2\lambda_0(\theta+\lambda'+\alpha)(\beta+\lambda')(\alpha(\mu+\beta)+\lambda'\beta)}{\mu^2\beta\theta\lambda'(1-h_1)h_1}\times$$

$$\times\frac{\theta(\lambda_0+\mu h_1)+\lambda'(\theta+\lambda'+\lambda_0)}{\sqrt{(\theta(\mu+\beta+\alpha+\lambda')+(\alpha+\lambda')(\beta+\lambda'))^2-4\mu\theta(\beta+\lambda')(\theta+\lambda')}}$$

as required to prove. ■

Thus, the rate of $T(n)$ decreasing to zero has an exponential upper estimation. The proof of Theorem 2 also yields possibility of constructive definition for $C(\lambda')$ and $N(\lambda')$ on λ'.

In the case of a Poisson flow of primary calls ($\lambda_k = \lambda > 0$), the normalizing constant in formulas (2), (3) (that is probability π_{00}) can be written in explicit form.

Corollary 1. *Let* $\lambda_k = \lambda > 0$, $k \ge 0$, *and condition (1) be satisfied. Then for* $Q(t)$, $t \ge 0$, *there exists a stationary regime, and stationary probabilities are defined with (2), (3), where* π_{00} *is of the following form:*

$$\pi_{00} = \frac{(\alpha+\beta)\lambda^2+(\alpha\beta+\theta(\alpha+\beta))\lambda-\mu\beta\theta}{(\alpha+\beta)\lambda^2+(\alpha\beta+(\theta-\lambda)(\alpha+\beta))\lambda-(\mu\beta\theta+(\alpha\beta+\theta(\alpha+\beta))\lambda)}. \quad (6)$$

3. Numerical experiments

Now let us show how one can use formulas (2), (3) for calculating such an important system characteristic as the blocking probability $P_b = \sum_{j=0}^{\infty}\mathbf{1}^T\pi_j$.

Under fixed rates values $\lambda_k = \lambda = 2$, $\mu = 23$, $\theta = 3$, the system blocking probabilities are given in Table 1 for different rates α and β.

Table 1. System blocking probabilities for different rates α and β under defined rates values $\lambda_k = \lambda = 2, \mu = 23, \theta = 3$.

α \ β	0.1	0.2	0.3	0.4	0.5	0.6	0.7	0.8	0.9	1
0.1	0.173	0.13	0.116	0.109	0.104	0.101	0.099	0.098	0.097	0.096
0.2	0.258	0.174	0.145	0.13	0.122	0.116	0.112	0.109	0.106	0.104
0.3	0.342	0.217	0.174	0.152	0.139	0.13	0.124	0.12	0.116	0.113
0.4	0.423	0.261	0.203	0.174	0.157	0.145	0.137	0.13	0.126	0.122
0.5	0.502	0.304	0.232	0.196	0.174	0.159	0.149	0.141	0.135	0.13
0.6	0.577	0.348	0.261	0.217	0.191	0.174	0.161	0.152	0.145	0.139
0.7	0.646	0.391	0.29	0.239	0.209	0.188	0.174	0.163	0.155	0.148
0.8	0.71	0.434	0.319	0.261	0.226	0.203	0.186	0.174	0.164	0.157
0.9	0.766	0.477	0.348	0.283	0.243	0.217	0.199	0.185	0.174	0.165
1	0.815	0.52	0.377	0.304	0.261	0.232	0.211	0.196	0.184	0.174

Tabular data can be presented graphically.

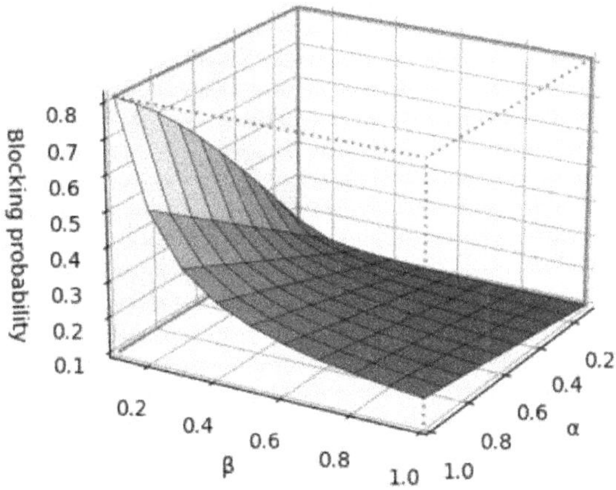

Figure 3. System blocking probabilities for different rates α and β under defined rates values $\lambda_k = \lambda = 2, \mu = 23, \theta = 3$.

Another important characteristic of the system with repeated calls is the occupancy rate $R_o = \sum_{j=0}^{\infty}(1,\ 0).\pi_j = \sum_{j=0}^{\infty}\pi_{1j}$. Under fixed rates values $\lambda_k = 2 + (k+1)^{-1}$, $\mu = 23$, $\theta = 3$, the system occupancy rates are given in Table 2 for different rates α and β.

Table 2. System occupancy rates for different rates α and β under defined rates values $\lambda_k = 2 + (k+1)^{-1}$, $\mu = 23$, $\theta = 3$.

α \ β	0.1	0.2	0.3	0.4	0.5	0.6	0.7	0.8	0.9	1
0.1	0.122	0.125	0.126	0.126	0.126	0.126	0.127	0.127	0.127	0.127
0.2	0.117	0.122	0.124	0.125	0.125	0.125	0.126	0.126	0.126	0.126
0.3	0.113	0.12	0.122	0.123	0.124	0.125	0.125	0.125	0.125	0.126
0.4	0.109	0.117	0.12	0.122	0.123	0.124	0.124	0.125	0.125	0.125
0.5	0.105	0.115	0.119	0.121	0.122	0.123	0.123	0.124	0.124	0.124
0.6	0.102	0.113	0.117	0.12	0.121	0.122	0.123	0.123	0.124	0.124
0.7	0.099	0.111	0.116	0.118	0.12	0.121	0.122	0.122	0.123	0.123
0.8	0.097	0.109	0.114	0.117	0.119	0.12	0.121	0.122	0.122	0.123
0.9	0.095	0.107	0.113	0.116	0.118	0.119	0.12	0.121	0.122	0.122
1	0.093	0.105	0.111	0.115	0.117	0.119	0.12	0.121	0.121	0.122

Tabular data can be presented graphically.

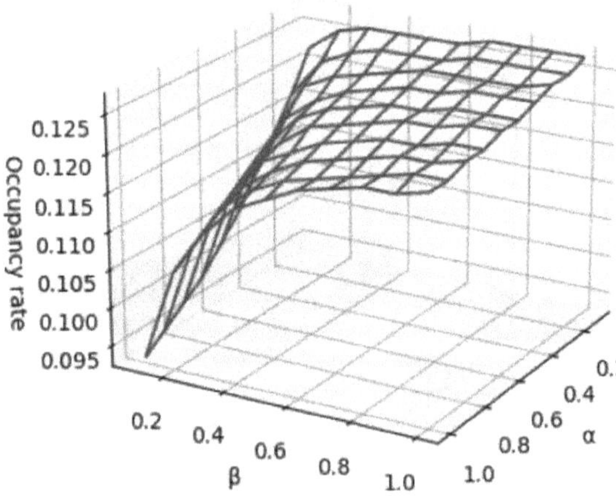

Figure 4. System occupancy rates for different rates α and β under defined rates values $\lambda_k = 2 + (k+1)^{-1}$, $\mu = 23$, $\theta = 3$.

4. Conclusions

For a system with repeated calls and an unreliable server, a condition for the existence of a stationary regime (Condition (1)) is found. It is important that it is convenient to verify Condition 1 in practice.

For calculating the stationary probabilities, vector-matrix formulas (2), (3) are obtained, and the normalizing constant (probability π_{00}) is given by series (4). An estimate of the remainder of this series is obtained in

Theorem 2, which allows us to control the accuracy of the calculations in formulas (2), (3).

For the Poisson flow of primary calls, π_{00} is presented explicitly via the system parameters (formula (6)). The calculation of stationary probabilities and their derivative characteristics is demonstrated by the example of calculating the blocking probability (see Table 1) and occupancy rate (see Table 2).

The obtained results can be used to solve optimization problems in the class of threshold strategies (see, for example, Lebedev et al. 2011).

References

Artalejo, J. and G. Falin. 2002. Standard and retrial queueing systems: A comparative analysis. Revista Matemática Complutense 15(1): 101–129.

Artalejo, J.R. 1994. New results in retrial queueing systems with breakdowns of the servers. Statistica Neerlandica 48: 23–36.

Artalejo, J.R. and A. Gomez-Corral. 2008. Retrial Queueing Systems. A Computational Approach. Springer-Verlag Berlin Heidelberg.

Falin, G.I. and J.G.C. Templeton. 1997. Retrial Queues. London: Chapman & Hall.

Gomez-Corral, A. and M.F. Ramalhoto. 1999. The stationary distribution of a Markovian process arising in the theory of multiserver retrial queueing systems. Mathematical and Computer Modelling 30: 141–158.

Lebedev, E.A. and V.D. Ponomarev. 2011. Retrial queues with variable service rate. Cybernetics and Systems Analysis 47(3): 434–441.

Li, W., D. Shi and X. Chao. 1997. Reliability analysis of M/G/1 queueing systems with server breakdowns and vacations. Journal of Applied Probability 34: 546–555.

Nazarov, A., J. Sztrik and A. Kvach. 2018. A survey of recent results in finite-source retrial queues with collisions. pp. 1–15. *In*: Dudin, A., A. Nazarov and A. Moiseev (eds.). Information Technologies and Mathematical Modelling. Queueing Theory and Applications. ITMM 2018, WRQ 2018. Communications in Computer and Information Science, 912.

Thiruvengadam, K. 1963. Queueing with breakdowns. Operations Research 11: 62–71.

Vinod, B. 1985. Unreliable queueing systems. Computers and Operations Research 12: 322–340.

Walrand, J. 1988. An Introduction to Queueing Networks. Prentice Hall.

Wang, J., J. Cao and Q. Li. 2001. Reliability analysis of the retrial queue with server breakdowns and repairs. Queueing Systems 38: 363–380.

Wartenhorst, P. 1995. N parallel queueing systems with server breakdown and repair. European Journal of Operational Research 82(2): 302–322.

Yang, T. and J.G.C. Templeton. 1987. A survey on retrial queues. Queueing Systems 2: 201–233.

Index

D

E

L

M

N

O

S

For Product Safety Concerns and Information please contact our EU
representative GPSR@taylorandfrancis.com
Taylor & Francis Verlag GmbH, Kaufingerstraße 24, 80331 München, Germany

www.ingramcontent.com/pod-product-compliance
Lightning Source LLC
Chambersburg PA
CBHW060755220326
41598CB00022B/2444
* 9 7 8 1 0 3 2 1 9 6 4 3 5 *